Introduction to Environmental Engineering and Science

Introduction to Environmental Engineering and Science

GILBERT M. MASTERS
Department of Civil Engineering
Stanford University

PRENTICE HALL, Englewood Cliffs, New Jersey 07632

Library of Congress Cataloging-in-Publication Data

MASTERS, GILBERT M.
 Introduction to environmental engineering and science / Gilbert M.
Masters.
  p. cm.
 Includes bibliographical references and index.
 ISBN 0–13–483066–0
 1. Environmental engineering. 2. Environmental protection.
I. Title.
TD145.M33 1991
628—dc20   90–49385
                                                      CIP

Editorial/production supervision: Cece Munson
Interior design: Joan L. Stone
Cover design: Ben Santora
Acquisitions editor: Doug Humphrey
Prepress buyer: Linda Behrens
Manufacturing buyer: Dave Dickey

 © 1991 by Prentice-Hall, Inc.
A Division of Simon & Schuster
Englewood Cliffs, New Jersey 07632

Printed in the United States of America

10 9 8 7 6 5 4 3

ISBN 0-13-483066-0

PRENTICE-HALL INTERNATIONAL (UK) LIMITED, *London*
PRENTICE-HALL OF AUSTRALIA PTY. LIMITED, *Sydney*
PRENTICE-HALL CANADA INC., *Toronto*
PRENTICE-HALL HISPANOAMERICANA, S.A., *Mexico*
PRENTICE-HALL OF INDIA PRIVATE LIMITED, *New Delhi*
PRENTICE-HALL OF JAPAN, INC., *Tokyo*
SIMON & SCHUSTER ASIA PTE. LTD., *Singapore*
EDITORA PRENTICE-HALL DO BRASIL, LTDA., *Rio de Janeiro*

To Billy and Jake

Contents

Preface

The environmental challenges we face today include all of the same ones that we faced more than twenty years ago at the first Earth Day celebration in 1970. In spite of the diligent efforts of environmental engineers, scientists, enlightened legislators, and an aroused public, our environmental problems remain. Many of our large cities continue to be plagued by smog, our beaches are periodically polluted by oil spills, and many of our rivers and lakes still suffer the effects of poorly treated sewage and industrial discharges. We have made progress, however. In fact, when measured in terms of emissions per capita and per unit of economic activity, we have made considerable progress. Most cities have cleaner air and most lakes and rivers are much closer to being "fishable and swimmable" than they were two decades ago.

Unfortunately, as we make progress in some areas, we discover new problems that will prove to be even more intractable than those we have already encountered. These new environmental challenges tend to involve less visible pollutants that are more global in scope and have longer response times. We cannot see the radon in our homes, the carbon dioxide that may lead to climate change, the chlorofluorocarbons (CFCs) that are attacking the earth's ozone layer, and the carcinogens that have contaminated our groundwater. The lake that no longer sustains life looks even "cleaner" than it did before it was sterilized by acid rain. Problems that we cannot see require that we increase our reliance on scientific measurements and computer models. As a result, it will be even more difficult to convince the public, and subsequently the decision makers, that cleaning up or averting these problems is worth the effort and the costs.

The environmental problems that have been added to our list tend to be more global in scope. In the past we worried about the effect of automobile emissions on the air quality of the cities in which we live. Now, in addition, we must consider the impact of those same cars (as well as other sources) on the climate and stratospheric ozone layer of the entire planet. Acid rain does not respect international boundaries, and hazardous wastes that are too expensive to dispose of here all too often wind up halfway across the planet. It is becoming increasingly clear that what we do here in our backyard affects people and ecosystems throughout the world, and, conversely, what they do affects us. The methane released from rice paddies in Indonesia, the massive destruction of rain forests in the Amazon, the exploitation of indigenous coal resources in China, and the inefficient use of energy in the Soviet Union can affect farmers in Nebraska and skiers in California.

The newer environmental problems also seem to have longer response times than those we have become accustomed to. If we were suddenly to stop dumping sewage into a river or to cease driving cars in our cities, in a matter of days the river would cleanse itself and our smog would disappear. On the other hand, cleaning up an aquifer contaminated with toxic chemicals will take years, CFCs emitted today will be around for decades, carbon dioxide from our power plants will be in the atmosphere for centuries, and our radioactive wastes will be problematic for thousands of years.

As our list of environmental concerns grows and the very nature of the problems changes, it has been challenging to find materials suitable for use in the classroom. There has never been a shortage of well-written articles for the layman and there are now many excellent textbooks that provide nontechnical, introductory information for general undergraduate students. There are also numerous scientific journals and specialized environmental engineering texts for advanced students. Most of these technical publications, however, presuppose a working knowledge of fundamental scientific and engineering principles that a beginning student probably would not have.

The purpose of this book is to fill the gap between these general introductory environmental science texts and the more advanced environmental engineering books commonly used in graduate courses. It is not the first book to attempt this bridging, but it does offer some features that I have not seen in other texts. This book does cover the basic, traditional materials in air and water pollution that have been the backbone of many introductory environmental engineering and science courses. And it does provide the necessary fundamental science and engineering principles that are generally assumed to be common knowledge for an advanced undergraduate, but that may be new, or may need to be reviewed, for many students in a mixed class of upper and lower division students. What it adds to the usual engineering introduction, though, is a basis for analyzing and understanding the newer environmental issues that have become the focus of much of the environmental attention in more recent years. New and deserved emphasis is given to such topics as hazardous waste, risk assessment, groundwater contamination, global climate change, stratospheric ozone depletion, and acid deposition.

The book has been organized and presented with three types of courses in mind. First, it could be used, from start to finish, in a standard sophomore- or junior-level environmental engineering course. In a more advanced undergraduate course, the first three chapters, covering fundamentals of mass and energy transfer, chemistry, and mathematics of growth, might be skipped or reviewed only lightly. Emphasis could then be placed on some of the more technical details provided in later chapters, supplemented with current environmental literature. I also envision the book being used as a text for a second course on the environment for less technical students, which would come after the usual general environmental science course that most colleges and universities now offer. Many of the more detailed quantitative aspects of the book can be covered lightly in such courses, if the students are so inclined, while keeping the basic modeling and problem-solving techniques intact.

For most topics covered in this book, pertinent environmental legislation is described, simple engineering models are generated, and qualitative descriptions of treatment technologies are presented. The book has been designed to encourage self-teaching by providing numerous completely worked examples throughout. Virtually every topic that lends itself to quantitative analysis is illustrated with such examples. Each chapter ends with a relatively long list of problems that give added practice for the student and should facilitate the preparation of homework assignments by the professor.

The first three chapters use environmental problems to illustrate the application of certain key principles of engineering and science that are required for any quantitative treatment of environmental problems. The first chapter presents mass and energy transfer, the second reviews some of the essential chemistry, and the third introduces certain mathematical functions of growth that are especially useful in developing future scenarios and projections.

The remaining five chapters are much longer presentations of some of the major environmental problems of the day. These chapters are relatively modular and could be covered in virtually any order. Chapters 4 and 6 cover topics that traditionally have been the essence of undergraduate civil engineering courses on the environment. In Chapter 4, a brief introduction to water resources and water pollutants is followed with a special section on groundwater, a topic that has been somewhat neglected in competing texts. The remaining material on water quality in lakes and rivers is unusual, perhaps, only in its coverage of acidification of lakes. Chapter 5 introduces hazardous substances, toxicology, risk assessment, and the key environmental laws addressing the hazardous waste problem. Chapter 6 introduces the principles of traditional water and wastewater treatment, supplemented with material on hazardous waste treatment. Compared to traditional texts, the coverage of treatment plant design has been minimized here, under the assumption that such material is more suitable for more specialized courses. Chapter 7 presents an introduction to traditional air pollution problems involving criteria pollutants, local meteorology, simple dispersion models, and emission controls. The chapter includes a section on the newly appreciated problem of indoor air quality, especially that associated with radon. Finally, Chapter 8 presents what many would describe as the most crucial environmental problems of the 1990s: global atmospheric change, the greenhouse effect, and stratospheric ozone depletion.

Obviously, for a single author to write a text that covers the range of topics introduced here is risky business. It is a truly unnerving and humbling exercise for a generalist like myself to venture into the domains of real experts who have devoted their lives to studying these subjects in great depth. I hope that what has been gained in terms of a consistent level of presentation and style has not been lost in accuracy. I am greatly in debt to a number of individuals who have reviewed sections of this book and who have given me the benefit of their expertise and advice. In particular, I would like to thank Leonard Ortolano of Stanford, David H. Marks at MIT, Bruce Ritman at the University of Illinois, Earnest F. Gloyna at the University of Texas, Austin, and Mark Benjamin and Eugene B. Welch of the University of Washington for their many helpful suggestions. Special thanks go to Dante Rodriguez for attempting to teach me the difference between "which" and "that," and for his painstaking proofreading of the entire manuscript. Finally, I wish to acknowledge my wife, Mary, whose writing skills and expertise in hazardous waste management have made the book better, and whose encouragement and patience have been essential to its completion.

Gilbert M. Masters

Introduction to Environmental Engineering and Science

CHAPTER 1

Mass and Energy Transfer

*When you can measure what you are speaking about, and
express it in numbers, you know something about it; but
when you cannot measure it, when you cannot express it
in numbers, your knowledge is of a meagre and
unsatisfactory kind; it may be the beginning of knowledge,
but you have scarcely, in your thoughts, advanced to the
stage of science.*
William Thomson, Lord Kelvin (1891)

1.1 INTRODUCTION

While most of the chapters in this book focus on specific environmental problems, such as pollution in surface waters or degradation of air quality, there are a number of important concepts that find application throughout the study of environmental engineering and science.

This chapter begins with a section on units of measurement. Engineers need to be familiar with both the American units of feet, pounds, hours, and degrees Fahrenheit as well as the more recommended International System of units. Both are used in the practice of environmental engineering and both will be used throughout this book.

Next, two fundamental topics, which should be familiar from the study of elementary physics, are presented: the *law of conservation of mass* and the *law of conservation of energy*. These laws tell us that within any environmental system we theoretically should be able to account for the flow of energy and materials

1

into, and out of, that system. The law of conservation of mass, besides giving us an important tool for quantitatively tracking pollutants as they disperse in the environment, reminds us that pollutants have to go somewhere, and that we should be wary of approaches that merely transport them from one medium to another.

In a similar way, the law of conservation of energy is also an essential accounting tool with special environmental implications. When coupled with other thermodynamic principles, it will be useful in a number of applications, including the study of global climate change, thermal pollution, and the dispersion of air pollutants.

1.2 UNITS OF MEASUREMENT

In the United States, environmental quantities are measured and reported in both the *U.S. Customary System* (USCS) and the *International System of Units* (SI) and so it is important to be familiar with both. In this book, preference is given to SI units, though the American system will be used in some circumstances. Table 1.1 gives conversion factors between the SI and USCS systems for some of the most basic units that will be encountered. A more extensive table of conversions is given in the appendix at the end of the book.

In the study of environmental engineering, it is quite common to encounter both extremely large quantities and extremely small ones. The concentration of some toxic substance may be measured in parts per billion, for example, while thermal pollution from a power plant may be measured in hundreds of billions of Btus per day. To describe quantities that may take on such extreme values, it is useful to have a system of prefixes that accompany the units. Some of the most important prefixes are presented in Table 1.2.

Quite often, it is the concentration of some substance in air or water that is of

TABLE 1.1 SOME BASIC UNITS AND CONVERSION FACTORS[a]

Quantity	SI units	SI symbol	×	Conversion factor	=	USCS units
Length	meter	m		3.2808		ft
Mass	kilogram	kg		2.2046		lb
Temperature	Celsius	°C		1.8 (°C) + 32		°F
Area	square meter	m^2		10.7639		ft^2
Volume	cubic meter	m^3		35.3147		ft^3
Energy	kilojoule	kJ		0.9478		Btu
Power	watt	W		3.4121		Btu/hr
Velocity	meter/second	m/s		2.2369		mi/hr
Flow rate	meter³/second	m^3/s		35.3147		ft^3/s
Density	kilogram/meter³	kg/m^3		0.06243		lb/ft^3

[a] See the appendix for a more complete list.

TABLE 1.2 COMMON PREFIXES

Quantity	Prefix	Symbol
10^{-12}	pico	p
10^{-9}	nano	n
10^{-6}	micro	μ
10^{-3}	milli	m
10^{-2}	centi	c
10^{-1}	deci	d
10	deca	da
10^{2}	hecto	h
10^{3}	kilo	k
10^{6}	mega	M
10^{9}	giga	G
10^{12}	tera	T
10^{15}	peta	P
10^{18}	exa	E

interest. In either medium, concentrations may be based on weight, volume, or a combination of the two, which can lead to some confusion.

Liquids

Concentrations of substances dissolved in water are usually expressed in terms of weight of substance per unit volume of mixture. Most often the units are milligrams (mg) or micrograms (μg) of substance per liter (L) of mixture. At times they may be expressed in grams per cubic meter (g/m^3).

Alternatively, concentrations in liquids are expressed as weight of substance per weight of mixture, with the most common units being parts per million (ppm), or parts per billion (ppb). Since most concentrations of pollutants are very small, one liter of mixture weighs essentially 1000 g, so that for all practical purposes we can write

$$1 \text{ mg/L} = 1 \text{ g/m}^3 = 1 \text{ ppm (by weight)} \tag{1.1}$$

$$1 \text{ } \mu\text{g/L} = 1 \text{ mg/m}^3 = 1 \text{ ppb (by weight)} \tag{1.2}$$

In unusual circumstances, the concentration of liquid wastes may be so high that the specific gravity of the mixture is affected, in which case a correction to (1.1) and (1.2) may be required:

$$\text{mg/L} = \text{ppm (by weight)} \times \text{Specific gravity} \tag{1.3}$$

Gases

For most air pollution work, it is customary to express pollutant concentrations in volumetric terms. For example, the concentration of a gaseous pollutant in parts per million (ppm) is the volume of pollutant per million volumes of the air mixture:

$$\frac{1 \text{ volume of gaseous pollutant}}{10^6 \text{ volumes of air}} = 1 \text{ ppm (by volume)} \qquad (1.4)$$

At times, gaseous concentrations are expressed with mixed units of mass per unit volume such as $\mu g/m^3$ or mg/m^3. The relationship between ppm (by volume) and mg/m^3 depends on the density of the pollutant, which, in turn, depends on its pressure and temperature as well as its molecular weight (mol wt). At a temperature of 0 °C and a pressure of 1 atmosphere, one mole of an ideal gas (that is, an amount of gas having weight equal to its molecular weight) occupies a volume of 22.4 L (or 22.4×10^{-3} m^3). Thus we can write

$$mg/m^3 = ppm \times \frac{1 \text{ m}^3 \text{ pollutant}/10^6 \text{ m}^3 \text{ air}}{ppm} \times \frac{\text{mol wt (g/mol)}}{22.4 \times 10^{-3} \text{ m}^3/\text{mol}} \times 10^3 \text{ (mg/g)}$$

or, more simply,

$$mg/m^3 = \frac{ppm \times \text{mol wt}}{22.4} \quad \text{(at 0 °C and 1 atm)} \qquad (1.5)$$

For other temperatures and pressures, corrections to (1.5) need to be applied as follows

$$mg/m^3 = \frac{ppm \times \text{mol wt}}{22.4} \times \frac{273}{T(K)} \times \frac{P(atm)}{1 \text{ atm}} \qquad (1.6)$$

where temperature (T) needs to be expressed in kelvins (K = °C + 273), and pressure (P) is in atmospheres (atm).

Example 1.1 Converting ppm to $\mu g/m^3$

The federal air quality standard for carbon monoxide (based on an 8-hr measurement) is 9.0 ppm. Express this standard as a percentage by volume as well as in mg/m^3 at 1 atm and 25 °C.

Solution Within a million volumes of this air there are 9.0 volumes of CO, no matter what the temperature or pressure (this is the advantage of the ppm units). Hence, the percentage by volume is simply

$$\text{Percent CO} = \frac{9.0}{1 \times 10^6} \times 100 = 0.0009 \text{ percent}$$

To find the concentration in mg/m^3 we need the molecular weight of CO, which is 28 (the atomic weights of C and O are 12 and 16, respectively). There is no pressure correction required, but the temperature correction must be applied. Converting 25 °C to 298 K and substituting into (1.6) yields

$$CO = \frac{9 \times 28}{22.4} \times \frac{273}{298} = 10.3 \text{ mg/m}^3$$

Actually, it is usually given as 10 mg/m^3.

When federal air quality standards are specified in $\mu g/m^3$ or mg/m^3, the assumption is that air temperature is 25 °C and pressure is 1 atm. The conversion under these special circumstances is

$$mg/m^3 = \frac{ppm \times mol\ wt}{24.45} \quad \text{(at 25 °C and 1 atm)} \tag{1.7}$$

1.3 MATERIALS BALANCE

Everything has to go somewhere is a simple way to express one of the most fundamental engineering principles. More precisely, the *law of conservation of mass* says that when chemical reactions take place, matter is neither created nor destroyed (though in nuclear reactions mass can be converted to energy). What this concept allows us to do is track materials, that is, pollutants, from one place to another with mass balance equations.

The first step in a mass balance analysis is to define the particular region in space that is to be analyzed. As examples, the region might include anything from a simple chemical mixing tank, to an entire coal-fired power plant, a lake, a stretch of stream, an air basin above a city, or the globe itself. By picturing an imaginary boundary around the region, as is suggested in Figure 1.1, we can then begin to identify the flow of materials across the boundary as well as the accumulation of materials within the region.

A substance that enters the region has three possible fates: Some of it may leave the region unchanged; some of it may accumulate within the boundary; and some of it may be converted to some other substance (e.g., entering CO may be oxidized to CO_2 within the region). Thus, using Figure 1.1 as a guide, the following materials balance equation can be written for each substance of interest:

$$\begin{pmatrix} \text{Input} \\ \text{rate} \end{pmatrix} = \begin{pmatrix} \text{Output} \\ \text{rate} \end{pmatrix} + \begin{pmatrix} \text{Decay} \\ \text{rate} \end{pmatrix} + \begin{pmatrix} \text{Accumulation} \\ \text{rate} \end{pmatrix} \tag{1.8}$$

Notice the decay term in (1.8) does not imply a violation of the law of conservation of mass. Atoms are conserved, but there is no similar constraint on the chemical reactions, which may change one "substance" into another.

Frequently, (1.8) can be simplified. The most common simplification results when *steady-state* or *equilibrium* conditions can be assumed. Equilibrium simply means that nothing is changing with time; the system has had its inputs held

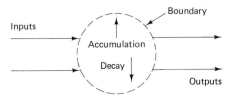

Figure 1.1 A materials balance diagram.

constant for a long enough time that any transients have had a chance to die out. Pollutant concentrations are constant. Hence, the accumulation rate term in (1.8) is set equal to zero and problems can usually be solved using simple algebra.

A second simplification to (1.8) results when a substance is *conserved* within the region in question, meaning there is no radioactive decay, bacterial decomposition, or chemical reaction occurring. For such conservative substances, the decay rate term in (1.8) is zero. Examples of substances that are typically modeled as conservative include total dissolved solids in a body of water or CO_2 in air. Nonconservative substances would include radioactive radon gas in a home or decomposing organic wastes in a lake. Many times problems involving nonconservative substances can be simplified when the reaction rate term is small enough to be ignored.

Steady-State Conservative Systems

The simplest systems to analyze are those in which steady state is assumed and the substance in question is conservative. In these cases, (1.8) simplifies to the following:

$$\text{Input rate} = \text{Output rate} \tag{1.9}$$

Consider the steady-state conservative system shown in Figure 1.2. The system contained within the boundaries might be a lake, a section of a free-flowing stream, or the mass of air above a city. One input to the system is a stream (of water or air, for instance) with a flow rate Q_s (volume/time) and pollutant concentration C_s (mass/volume). The other input is assumed to be a waste stream with flow rate Q_w and pollutant concentration C_w. The output is a mixture with flow rate Q_m and pollutant concentration C_m. If the pollutant is conservative, and if we assume steady-state conditions, then a mass balance based on (1.9) allows us to write the following:

$$C_s Q_s + C_w Q_w = C_m Q_m \tag{1.10}$$

The following example illustrates the use of this equation.

Example 1.2   Two Polluted Streams

A stream flowing at 10.0 m³/s has a tributary feeding into it with a flow 5.0 m³/s. The stream's concentration of chlorides upstream of the junction is 20.0 mg/L and the

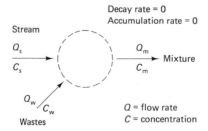

Decay rate = 0
Accumulation rate = 0

Stream
Q_s
C_s

Q_m
Mixture
C_m

Q_w
C_w
Wastes

Q = flow rate
C = concentration

Figure 1.2   A steady-state conservative system. Pollutants enter and leave the region at the same rate.

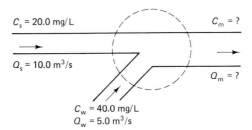

Figure 1.3 Flow rate and chloride concentrations for example stream and tributary.

tributary chloride concentration is 40.0 mg/L. Treating chlorides as a conservative substance, and assuming complete mixing of the two streams, find the downstream chloride concentration.

Solution We begin by sketching the problem and identifying the "region" that we want to analyze, as has been done in Figure 1.3.

Rearranging (1.10) to solve for the chloride concentration downstream gives us

$$C_m = \frac{C_s Q_s + C_w Q_w}{Q_m} = \frac{C_s Q_s + C_w Q_w}{Q_s + Q_w}$$

Note that since the mixture flow is the sum of the two stream flows, $Q_s + Q_w$ has been substituted for Q_m in this expression. All that remains is to substitute the appropriate values for the known quantities into the expression, which brings us to a question of units. The units given for C are mg/L and for Q they are m³/s. Taking the product of concentrations and flow rates yields mixed units of mg/L · m³/s, which we could simplify by applying the conversion factor of 10^3 L = 1 m³. However, if we did so, we should have to reapply that same conversion factor to get the mixture concentration back into the desired units of mg/L. In problems of this sort, it is much easier to simply leave the mixed units in the expression, even though they may look awkward at first, and let them work themselves out in the calculation. The downstream concentration of chlorides is thus

$$C_m = \frac{(20.0 \times 10.0 + 40.0 \times 5.0)\ \text{mg/L} \cdot \text{m}^3/\text{s}}{(10.0 + 5.0)\ \text{m}^3/\text{s}} = 26.7\ \text{mg/L}$$

Steady-State Systems with Nonconservative Pollutants

Many environmental pollutants undergo chemical, biological, or nuclear reactions at a rate sufficient to require us to treat them as nonconservative substances. If we continue to assume that steady-state conditions prevail so that the rate of accumulation is zero, but if the pollutants are nonconservative, then (1.8) becomes

$$\text{Input rate} = \text{Output rate} + \text{Decay rate} \qquad (1.11)$$

The decay of nonconservative substances is frequently modeled as a first-order reaction; that is, it is assumed that the rate of loss of the substance is proportional to the amount of the substance that is present. That is,

$$\frac{dC}{dt} = -KC \tag{1.12}$$

where K is a reaction rate coefficient with dimensions of (1/time), the negative sign implies a loss of substance with time, and C is the pollutant concentration. To solve this differential equation, we can rearrange the terms and integrate

$$\int_{c_0}^{c} \frac{dC}{C} = \int_{0}^{t} (-K) \, dt$$

which yields

$$\ln(C) - \ln(C_0) = \ln \frac{C}{C_0} = -Kt$$

Solving for concentration gives us

$$C = C_0 e^{-Kt} \tag{1.13}$$

where C_0 is the initial concentration. That is, assuming a first-order reaction, the concentration of the substance in question decays exponentially. This exponential function will appear so often in this text that it will be reintroduced and explored more fully later on in Chapter 3.

Equation 1.12 indicates the rate of change of *concentration* of the substance. If we assume the substance is uniformly distributed throughout a volume V, then the total *amount* of substance is CV. The total rate of decay of the amount of a nonconservative substance is thus $d(CV)/dt = V \, dC/dt$, so using (1.12) we can write for a nonconservative substance

$$\text{Decay rate} = KCV \tag{1.14}$$

Substituting Eq. 1.14 into Eq. 1.11 gives us our final, simple yet useful, expression for the mass balance involving a nonconservative pollutant in a steady-state system:

$$\text{Input rate} = \text{Output rate} + KCV \tag{1.15}$$

Implicit in (1.15) is the assumption that the concentration C is uniform throughout the volume V. This complete mixing assumption is common in the analysis of chemical tanks, called *reactors,* and in such cases the idealization is referred to as a *continuously stirred tank reactor* (CSTR) model. In other contexts, such as modeling air pollution, the assumption is referred to as a *complete mix box model.*

Example 1.3   A Polluted Lake

Consider a lake with volume 10.0×10^6 m^3 that is fed by a stream with a flow rate of 5.0 m^3/s and pollution concentration equal to 10.0 mg/L (Figure 1.4). There is also a sewage outfall that discharges 0.5 m^3/s of the same pollutant into the lake. The

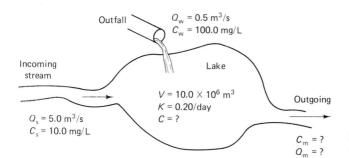

Figure 1.4   A lake with a nonconservative pollutant.

sewage has a concentration of 100.0 mg/L and a reaction rate coefficient of 0.20/day. Assuming the pollution is completely mixed in the lake, and assuming no evaporation or other water losses or gains, find the steady-state concentration.

Solution   Assuming that complete and instantaneous mixing occurs in the lake implies that the concentration in the lake C is the same as the concentration of the mix leaving the lake, C_m. Using (1.15)

$$\text{Input rate} = \text{Output rate} + \text{Decay rate} \tag{1.15}$$

we can find each term as follows:

$$\text{Input rate} = Q_s C_s + Q_w C_w$$

$$= (5.0 \text{ m}^3/\text{s} \times 10.0 \text{ mg/L} + 0.5 \text{ m}^3/\text{s} \times 100.0 \text{ mg/L}) \times 10^3 \text{ L/m}^3$$

$$= 1.0 \times 10^5 \text{ mg/s}$$

$$\text{Output rate} = Q_m C_m = (Q_s + Q_w)C$$

$$= (5.0 + 0.5)\text{m}^3/\text{s} \times C \text{ mg/L} \times 10^3 \text{ L/m}^3$$

$$= 5.5 \times 10^3 C \text{ mg/s}$$

$$\text{Decay rate} = KCV = \frac{0.20/\text{d} \times C \text{ mg/L} \times 10.0 \times 10^6 \text{ m}^3 \times 10^3 \text{ L/m}^3}{24 \text{ hr/d} \times 3600 \text{ s/hr}}$$

$$= 23.1 \times 10^3 C \text{ mg/s}$$

So from (1.15),

$$1.0 \times 10^5 = 5.5 \times 10^3 C + 23.1 \times 10^3 C = 28.6 \times 10^3 C$$

$$C = \frac{1.0 \times 10^5}{28.6 \times 10^3} = 3.5 \text{ mg/L}$$

Idealized models involving nonconservative pollutants in completely mixed, steady-state systems are used to analyze a variety of commonly encountered water pollution problems such as the one shown in the previous example. The same simple models can be applied to certain problems involving air quality, as the following example demonstrates.

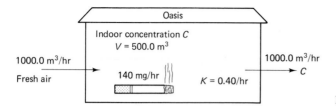

Figure 1.5   Tobacco smoke in a bar.

Example 1.4   A Smoky Bar

A bar with volume 500.0 m³ has 50 smokers in it, each smoking two cigarettes per hour (see Figure 1.5). An individual cigarette emits, among other things, about 1.40 mg of formaldehyde (HCHO). Formaldehyde converts to carbon dioxide with a reaction rate coefficient $K = 0.40$/hr. Fresh air enters the bar at the rate of 1000.0 m³/hr and stale air leaves at the same rate. Estimate the steady-state concentration of formaldehyde in the air, assuming complete mixing. At 25 °C and 1 atm of pressure, how does the result compare with the threshold for eye irritation of about 0.05 ppm?

Solution   The rate at which formaldehyde enters the bar is

$$\text{Input rate} = 50 \text{ smokers} \times 2 \text{ cigs/hr} \times 1.40 \text{ mg} = 140.0 \text{ mg/hr}$$

Since complete mixing is assumed, the concentration of formaldehyde C in the bar is the same as the concentration in the air leaving the bar, so

$$\text{Output rate} = 1000.0 \text{ m}^3/\text{hr} \times C \text{ (mg/m}^3) = 1000.0C \text{ mg/hr}$$

And the decay rate is

$$\text{Decay rate} = KCV = (0.40/\text{hr}) \times (C \text{ mg/m}^3) \times (500.0 \text{ m}^3) = 200.0 \, C \text{ mg/hr}$$

So, from (1.15),

$$\text{Input rate} = \text{Output rate} + \text{Decay rate}$$

$$140.0 = 1000.0C + 200.0C = 1200.0C$$

$$C = 0.117 \text{ mg/m}^3$$

We will use (1.7) to convert mg/m³ to ppm. The molecular weight of formaldehyde is 30, so

$$C(\text{ppm}) = \frac{C(\text{mg/m}^3) \times 24.45}{\text{mol wt}} = \frac{0.117 \times 24.45}{30} = 0.095 \text{ ppm}$$

This is more than enough to cause eye irritation.

Step Function Response

So far we have computed steady-state concentrations in environmental systems that are contaminated with either conservative or nonconservative pollutants. Let us now extend the analysis to include conditions that are not steady state. Quite often, we will be interested in how the concentration will change with time when there is a sudden change in the amount of pollution entering the system. This is known as the *step function response* of the system.

In Figure 1.6 the environmental system that is to be modeled has been drawn as if it were a box of volume V that has equal flows Q into and out of the box. Again, let us assume the contents of the box are at all times completely mixed (a CSTR model) so that the pollutant concentration C in the box is the same as the concentration leaving the box. The total mass of pollutant in the box is therefore VC and the rate of increase of pollutant in the box is $V\, dC/dt$. Let us designate the total rate at which pollution enters the box as S, the source strength, with units of mass per unit time. If the pollutants are nonconservative they will be modeled with a first-order reaction rate coefficient K.

$$\begin{pmatrix} \text{Accumulation} \\ \text{rate} \end{pmatrix} = \begin{pmatrix} \text{Input} \\ \text{rate} \end{pmatrix} - \begin{pmatrix} \text{Output} \\ \text{rate} \end{pmatrix} - \begin{pmatrix} \text{Decay} \\ \text{rate} \end{pmatrix}$$

$$V\frac{dC}{dt} = S - QC - KCV \tag{1.16}$$

where V = box volume (m³)

$\quad C$ = concentration in the box and in the exiting waste stream (g/m³)

$\quad S$ = total rate at which pollutants enter the box (g/hr)

$\quad Q$ = the total flow rate into and out of the box (m³/hr)

$\quad K$ = reaction rate coefficient (hr⁻¹)

The units given above are representative of those that might be encountered; any consistent set will do.

An easy way to find the steady-state solution to (1.16) is simply to set $dC/dt = 0$, which yields

$$C(\infty) = \frac{S}{Q + KV} \tag{1.17}$$

Our concern now though is with the concentration before it reaches steady state, so we must solve (1.16). Rearranging (1.16) gives

$$\frac{dC}{dt} = -\frac{Q + KV}{V}\left(C - \frac{S}{Q + KV}\right) \tag{1.18}$$

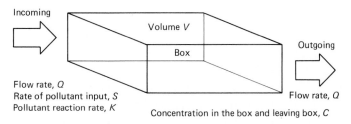

Figure 1.6   A box model for a transient analysis.

One way to solve this differential equation is to make a change of variable. If we let

$$y = C - \frac{S}{Q + KV} \tag{1.19}$$

then

$$\frac{dy}{dt} = \frac{dC}{dt} \tag{1.20}$$

So (1.18) becomes

$$\frac{dy}{dt} = -\left(K + \frac{Q}{V}\right)y \tag{1.21}$$

which is a differential equation just like (1.12) that we solved before. The solution is

$$y = y_0 e^{-(K+Q/V)t} \tag{1.22}$$

where y_0 is the value of y at $t = 0$. If C_0 is the concentration in the box at time $t = 0$, then from (1.19),

$$y_0 = C_0 - \frac{S}{Q + KV} \tag{1.23}$$

and substituting (1.19) and (1.23) into (1.22) yields

$$C - \frac{S}{Q + KV} = \left(C_0 - \frac{S}{Q + KV}\right)e^{-(K+Q/V)t}$$

or

$$C(t) = \left(C_0 - \frac{S}{Q + KV}\right)e^{-(K+Q/V)t} + \frac{S}{Q + KV} \tag{1.24}$$

Equation 1.24 can be made to look slightly simpler by inserting the steady-state solution, $C(\infty)$ from (1.17),

$$C(t) = [C_0 - C(\infty)]e^{-(K+Q/V)t} + C(\infty) \tag{1.25}$$

Equation 3.14 should make some sense. At time $t = 0$, the exponential equals 1 and $C = C_0$. At $t = \infty$, the exponential term equals zero, and $C = C(\infty)$. Equation 1.25 is plotted in Figure 1.7.

Example 1.5   The Smoky Bar Revisited

The bar in Example 1.4 had volume 500.0 m³, with fresh air entering at the rate of 1000.0 m³/hr. Suppose the air in the bar is clean when it opens at 5 pm. If formaldehyde with reaction rate $K = 0.40$/hr is emitted from cigarette smoke at the constant rate of 140.0 mg/hr starting at 5 pm, what would the concentration be at 6 pm?

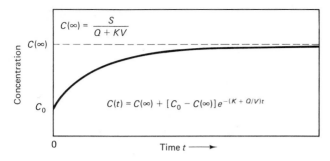

Figure 1.7 Step function response for a complete-mix box model.

Solution   In this case, $Q = 1000.0$ m³/h, $V = 500.0$ m³, $S = 140.0$ mg/hr, and $K = 0.40$/hr. The steady-state concentration, from (1.17), is

$$C(\infty) = \frac{S}{Q + KV} = \frac{140.0 \text{ mg/hr}}{1000.0 \text{ m}^3/\text{hr} + 0.40/\text{hr} \times 500 \text{ m}^3}$$

$$= 0.117 \text{ mg/m}^3$$

which agrees with the result obtained in Example 1.4. To find the concentration at any time after 5 pm, we can apply (1.25) with $C_0 = 0$:

$$C(t) = [C_0 - C(\infty)]e^{-(K+Q/V)t} + C(\infty) = (0 - 0.117)e^{-(0.40+1000.0/500.0)t} + 0.117$$

$$= 0.117(1 - e^{-2.4t})$$

at 6 pm,

$$C(1 \text{ hr}) = 0.117(1 - e^{-2.4 \times 1}) = 0.106 \text{ mg/m}^3$$

To further demonstrate the use of (1.25), let us reconsider the lake analyzed in Example 1.4. This time we will assume there is a sudden drop in the pollutant concentration in the outfall.

Example 1.6   A Sudden Decrease in Pollutants Discharged into the Lake

Consider the 10×10^6-m³ lake analyzed in Example 1.3, which, under the conditions given, was found to have a steady-state pollution concentration of 3.5 mg/L. The pollution is nonconservative with reaction rate constant $K = 0.20$/day. Suppose the condition of the lake is deemed unacceptable and to solve the problem it is decided to completely divert the sewage outfall around the lake, eliminating it as a source of pollution. The incoming stream still has flow $Q_s = 5.0$ m³/s and concentration $C_s = 10.0$ mg/L. Assuming complete-mix conditions, find the concentration of pollution in the lake one week after the diversion and find the new final steady-state concentration.

Solution   For this situation,

$$C_0 = 3.5 \text{ mg/L} \qquad V = 10.0 \times 10^6 \text{ m}^3$$

$$K = 0.20/\text{day}$$

$$Q = 5.0 \text{ m}^3/\text{s} \times 3600 \text{ s/hr} \times 24 \text{ hr/day} = 43.2 \times 10^4 \text{ m}^3/\text{day}$$

The total rate at which pollution is entering the lake from the incoming stream is

$$S = Q_s C_s = 43.2 \times 10^4 \text{ m}^3/\text{day} \times 10.0 \text{ mg/L} \times 10^3 \text{ L/m}^3$$

$$= 43.2 \times 10^8 \text{ mg/day}$$

The steady-state concentration can be obtained from (1.17)

$$C(\infty) = \frac{S}{Q + KV} = \frac{43.2 \times 10^8 \text{ mg/day}}{43.2 \times 10^4 \text{ m}^3/\text{day} + 0.020/\text{day} \times 10 \times 10^6 \text{ m}^3}$$

$$= 1.8 \times 10^3 \text{ mg/m}^3 \times 10^{-3} \text{ m}^3/\text{L} = 1.8 \text{ mg/L}$$

Using (1.25) we can find the concentration in the lake one week after the drop in pollution from the outfall:

$$C(t) = [C_0 - C(\infty)]e^{-(K+Q/V)t} + C(\infty)$$

$$C(7 \text{ days}) = (3.5 - 1.8) \exp\left[-\left(0.20/\text{d} + \frac{43.2 \times 10^4 \text{ m}^3/\text{d}}{10 \times 10^6 \text{ m}^3}\right) \times 7\text{d}\right] + 1.8$$

$$C(7 \text{ days}) = 2.1 \text{ mg/L}$$

1.4 ENERGY FUNDAMENTALS

Just as we are able to use the law of conservation of mass to write mass balance equations that are fundamental to understanding and analyzing the flow of materials, we can in a similar fashion use the *first law of thermodynamics* to write energy balance equations that will help us analyze energy flows. One definition of energy is that it is the capacity for doing work, where work can be described by the product of force and the displacement of an object caused by that force.

A simple interpretation of the *second law of thermodynamics* suggests that when work is done there will always be some inefficiency; that is, some portion of the energy put into the process will end up as waste heat. How that waste heat affects the environment is an important consideration in the study of environmental engineering and science.

Another important term to be familiar with is *power*. Power is the *rate* of doing work so it has units of energy per unit of time. In SI units power is given in J/s or watts (1 J/s = 1 W = 3.412 Btu/hr).

The First Law of Thermodynamics

The first law of thermodynamics says, simply, that energy can neither be created nor destroyed. Energy may change forms in any given process, as when chemical energy in a fuel is converted to heat and electricity in a power plant, or when potential energy of water behind a dam is converted to mechanical energy that spins a turbine in a hydroelectric plant. No matter what is happening, the first law says we should be able to account for every bit of energy as it takes part in the process under study, so that in the end we have just as much as we had in the beginning. With proper accounting, even nuclear reactions involving conversion of mass to energy can be treated.

To apply the first law it is necessary to define the system being studied, much as was done in the analysis of mass flows. Realize that the system can be anything that we want to draw an imaginary boundary around—It can be an automobile engine, or a nuclear power plant, or a volume of gas emitted from a smokestack. Later when we explore the topic of global temperature equilibrium the system will be the earth itself. Once a boundary has been defined, the rest of the universe becomes the *surroundings.* Just because a boundary has been defined, however, does not mean that energy and/or materials cannot flow across that boundary. Systems in which both energy and matter can flow across the boundary are referred to as *open systems,* while those in which energy is allowed to flow across the boundary, but matter is not, are called *closed systems.*

Since energy is conserved, we can write the following for whatever system we have defined

$$\text{Energy in} = \text{Energy out} + \text{Change in internal energy} \qquad (1.26)$$

where *internal energy* is the energy stored in the system. The most common way that changes in internal energy show themselves in the environmental systems considered in this book is by a change in temperature. The change in internal energy that occurs when a substance with mass m undergoes a change in temperature of ΔT is given by

$$\text{Change in internal energy} = mc\Delta T \qquad (1.27)$$

where c is called the *specific heat* of that substance. Specific heat is defined to be the amount of energy required to raise the temperature of a unit of mass of a substance by one degree. The specific heat of water is the basis for two important units of energy, namely the *British thermal unit,* or Btu, which is defined to be the energy required to raise one pound of water by 1 °F, and the *calorie,* which is the energy required to raise one gram of water by 1 °C. In the definitions just given, the assumed temperature of the water is 15 °C (59 °F). Since calories are no longer a preferred energy unit, values of specific heat in the SI system are given in J/kg °C, where 1 cal/g °C = 1 Btu/lb °F = 4184 J/kg °C.

Example 1.7   A Water Heater

How long would it take to heat the water in a 40-gallon electric water heater from 50 to 140 °F if the heating element delivers 5 kW? Assume all of the electrical energy is converted to heat in the water, neglect the energy required to raise the temperature of the tank itself, and neglect any heat losses from the tank to the environment.

Solution   The first thing to note is that the electric input is expressed in kilowatts, which is a measure of the *rate* of energy input (i.e., power). To get total energy input to the water we must multiply rate × time. Letting Δt be the number of hours that the heating element is on gives

$$\text{Energy input} = 5 \text{ kW} \times \Delta t \text{ hr} = 5 \ \Delta t \text{ kWhr}$$

Assuming no losses from the tank and no water withdrawn from the tank during the heating period, there is no energy output:

$$\text{Energy output} = 0$$

The change in energy stored corresponds to the water going from 50 to 140 °F. Using (1.27) gives

$$\text{Change in internal energy} = mc\Delta T$$

$$= 40 \text{ gal} \times 8.34 \text{ lb/gal} \times 1 \text{ Btu/lb °F} \times (140 - 50) \text{ °F}$$

$$= 30 \times 10^3 \text{ Btu}$$

So, setting the energy input equal to the change in internal energy, and converting units using Table 1.1, yields

$$5 \,\Delta t \text{ kWhr} \times 3412 \text{ Btu/kWhr} = 30 \times 10^3 \text{ Btu}$$

$$\Delta t = 1.76 \text{ hr}$$

There are two key assumptions implicit in (1.27). First, the specific heat is assumed to be constant over the temperature range in question, though in actuality it does vary slightly. And second, (1.27) assumes that there is no change of *phase* in the substance as would occur if the substance were to freeze (liquid-to-solid phase change) or boil (liquid-to-vapor phase change).

When a substance changes phase, there is a change in internal energy without a change in temperature. The energy required to cause a phase change from solid to liquid is called the *latent heat of fusion*. Similarly, the energy required to change phase from liquid to vapor is called the *latent heat of vaporization*. For example, 333 kJ will melt 1 kg of ice, while 2258 kJ are required to convert 1 kg of water at 100 °C to steam at the same temperature. When steam condenses or when water freezes, those same amounts of energy are released. To account for phase changes, we can write

$$\text{Change in internal energy (due to a phase change)} = mH_L \qquad (1.28)$$

where m is the mass and H_L is the latent heat.

Figure 1.8 illustrates the concept of latent heat for water, the most important substance that we will encounter in this book.

Values of specific heat, latent heat, and density for water are given in Table 1.3 for both SI and USCS units. An additional entry has been included in the table which adds the latent heat of vaporization to the amount of energy required to raise water from 15 to 100 °C. This is a useful number that can be used to estimate the amount of energy required to cause surface water on the earth to evaporate. The value of 15 °C has been picked as the starting temperature since that is approximately the current average surface temperature of the globe.

One way to demonstrate the concept of latent heat while at the same time introducing an important component of the global energy balance that will be

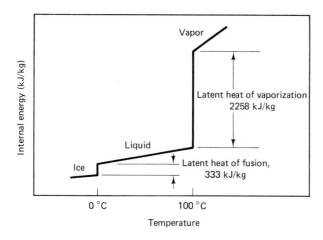

Figure 1.8 The latent heats of fusion and vaporization for water. The slope of the liquid phase is the specific heat of water, about 4.18 kJ/kg°C, whereas for ice it is about 2.0 kJ/kg°C.

encountered in Chapter 7 is to estimate the energy required to power the global hydrologic cycle.

Example 1.8   Global Precipitation

Global precipitation has been estimated to average about 1 m of water per year across the entire globe. If the surface area of the earth is 5.10×10^{14} m^2, find the energy required to cause that much water to evaporate each year. Compare your answer to the estimated 1987 world energy consumption of 3.3×10^{17} kJ and to the average rate at which sunlight strikes the surface of the earth, which is 167 W/m^2.

Solution   In Table 1.3 the energy required to vaporize 1 kg of 15 °C water (which is about the average global temperature) is given as 2613 kJ. The increase in internal energy associated with vaporizing that water is

$$\text{Increase in internal energy} = 1 \text{ m/yr} \times 5.10 \times 10^{14} \text{ m}^2 \times 10^3 \text{ kg/m}^3 \times 2613 \text{ kJ/kg}$$

$$= 1.33 \times 10^{21} \text{ kJ/yr}$$

This is about 4000 times the world energy consumption figure of 3.3×10^{17} kJ/year. Averaged over the globe, this energy rate is

$$\frac{1.33 \times 10^{24} \text{ J/yr} \times 1 \text{ W/(J/s)}}{365 \text{ d/yr} \times 24 \text{ hr/d} \times 3600 \text{ s/yr} \times 5.10 \times 10^{14} \text{ m}^2}$$

TABLE 1.3 IMPORTANT PHYSICAL PROPERTIES OF WATER

Property	SI units	USCS units
Specific heat (at 15 °C)	4.184 kJ/kg°C	1.00 Btu/lb-°F
Latent heat of vaporization	2258 kJ/kg	972 Btu/lb
Heat to vaporize 15 °C water	2613 kJ/kg	1123 Btu/lb
Latent heat of fusion	333 kJ/kg	144 Btu/lb
Density (at 4 °C)	1000.00 kg/m³	62.4 lb/ft³ (8.34 lb/gal)

or

$$82.7 \text{ W/m}^2$$

which is equivalent to almost 50% of the 167 W/m^2 of incoming sunlight striking the earth's surface. It might also be noted that the energy required to raise the water vapor high into the atmosphere, once it has evaporated, is negligible compared to the heat of vaporization. (See Problem 1.14 at the end of the chapter.)

Many practical environmental engineering problems involve the flow of both matter and energy across the system boundaries (open systems). For example, it is quite common for a hot liquid, usually water, to be used to deliver heat to a pollution control process, or the opposite, for water to be used as a coolant to remove heat from a process. In such cases, there are energy flow rates and fluid flow rates and Eq. 1.27 needs to be modified as follows:

$$\text{Rate of change in internal energy} = \dot{m}c \, \Delta T \qquad (1.29)$$

where $\dot{m}$ is the mass flow rate across the system boundary, given by the product of fluid flow rate and density, c is specific heat, and ΔT is the change in temperature of the fluid that is carrying the heat to, or away from, the process. For example, if water is being used to cool a steam power plant, then $\dot{m}$ would be the mass flow rate of coolant and ΔT would be the increase in temperature of the cooling water as it passes through the steam plant's condenser. Typical units for energy rates include watts, Btu/hr, and J/s, while mass flow rates might typically be in kg/s or lb/hr.

The use of a local river for power plant cooling is common and the following example illustrates the approach that can be taken to compute the increase in river temperature that results. In Chapter 4, some of the environmental impacts of this thermal pollution will be explored.

Example 1.9   Thermal Pollution of a River

A coal-fired power plant converts fuel energy into electrical energy with an efficiency of 33.3 percent. The electrical power output of the plant is 1000 MW. The other two-thirds of the energy content of the fuel is rejected to the environment as waste heat. About 15 percent of the waste heat goes up the smokestack and the other 85 percent is taken away by cooling water that is drawn from a nearby river. The river has an upstream flow of 100.0 m^3/s and a temperature of 20.0 °C.

  a. If the cooling water is allowed to rise in temperature by only 10.0 °C, what flow rate from the stream would be required?
  b. What would be the river temperature just after it receives the heated cooling water?

Solution   Since 1000 MW represents 33.3 percent of the power delivered to the plant by fuel, the total rate at which energy enters the power plant P_i is

$$P_i = \frac{1000 \text{ MW}}{0.333} = 3000 \text{ MW}$$

Total losses to the cooling water and stack are therefore 3000 MW − 1000 MW = 2000 MW. Of that 2000 MW,

$$\text{Stack losses} = 0.15 \times 2000 \text{ MW} = 300 \text{ MW}$$

and

$$\text{Coolant losses} = 0.85 \times 2000 \text{ MW} = 1700 \text{ MW}$$

a. Finding the cooling water needed to remove 1700 MW with a temperature increase ΔT of 10.0 °C, will require the use of (1.29) along with the specific heat of water, 4184 J/kg°C, given in Table 1.3:

$$\text{Rate of change in internal energy} = \dot{m}c\Delta T$$

$$1700 \text{ MW} = \dot{m} \text{ kg/s} \times 4184 \text{ J/kg°C} \times 10.0 \text{ °C} \times 1 \text{ MW}/(10^6 \text{ J/s})$$

$$\dot{m} = \frac{1700}{4184 \times 10.0 \times 10^{-6}} = 40.6 \times 10^3 \text{ kg/s}$$

Using the conversion 1000 kg/m³ gives a flow rate of 40.6 m³/s.

b. To find the new temperature of the river we can use (1.29) with 1700 MW being released into a full river flow rate of 100.0 m³/s.

$$\text{Rate of change in internal energy} = \dot{m}c\Delta T$$

$$\Delta T = \frac{1700 \text{ MW} \times [(1 \times 10^6 \text{ J/s})/\text{MW}]}{100.0 \text{ m}^3/\text{s} \times 10^3 \text{ kg/m}^3 \times 4184 \text{ J/kg°C}} = 4.1 \text{ °C}$$

So the temperature of the river will be elevated by 4.1 °C, making it 24.1 °C. The results of the calculations just performed are shown in Figure 1.9. Note the use of the subscript t for energy quantities that are thermal and e for

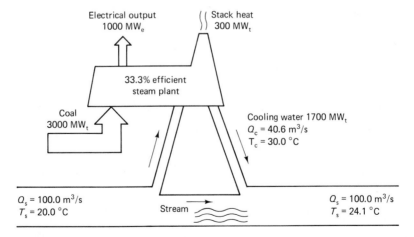

Figure 1.9 Cooling water energy balance for the 33.3-percent efficient, 1000-MW$_e$ power plant of Example 1.9.

electrical quantities. This is a convenient way to distinguish between the two, and, though it is not really necessary, it is sometimes used.

The Second Law of Thermodynamics

In Example 1.9, you will notice that a relatively modest fraction of the fuel energy contained in the coal actually was converted to the desired output, namely electrical power, and a rather large amount of the fuel energy ended up as waste heat rejected to the environment. The second law of thermodynamics says that there will always be some waste heat; that is, it is impossible to devise a machine that can convert heat to work with 100 percent efficiency. There will always be "losses" (though, by the first law, the energy is not lost, it is merely converted into the less useful form of lower temperature heat).

The steam-electric plant just described is an example of a *heat engine,* a device studied at some length in thermodynamics. One way to view the steam plant is that it is a machine that takes heat from a high-temperature source (the burning fuel), converts some of it into work (the electrical output), and rejects the remainder into a low-temperature reservoir (the river and the atmosphere). It turns out that the maximum efficiency that our steam plant can possibly have is critically dependent on both the source temperature and the temperature of the reservoir accepting the rejected heat.

Figure 1.10 shows a theoretical heat engine operating between two heat reservoirs, one at temperature T_h and one at T_c. An amount of heat energy Q_h is transferred from the hot reservoir to the heat engine. The engine does work W and rejects an amount of waste heat Q_c to the cold reservoir.

The efficiency of this engine is the ratio of the work delivered by the engine to the amount of heat energy taken from the hot reservoir:

$$\eta = \frac{W}{Q_h} \qquad\qquad (1.30)$$

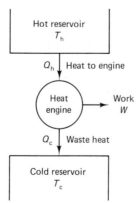

Figure 1.10 Definition of terms for a Carnot engine.

The most efficient heat engine that could possibly operate between the two heat reservoirs is called a *Carnot* engine after the French engineer Sadi Carnot who first developed the explanation in the 1820s. Analysis of Carnot engines (e.g., Thorndike, 1976) shows that the most efficient engine possible, operating between two temperatures, T_h and T_c, has an efficiency of

$$\eta_{max} = 1 - \frac{T_c}{T_h} \qquad (1.31)$$

where these are absolute temperatures measured using either the Kelvin or the Rankine scale. Conversions from Celsius to Kelvin, and Fahrenheit to Rankine are

$$K = {}^{\circ}C + 273.15 \qquad (1.32)$$

$$R = {}^{\circ}F + 459.67 \qquad (1.33)$$

One immediate observation that can be made from (1.31) is that the maximum possible heat engine efficiency increases as the temperature of the hot reservoir increases or the temperature of the cold reservoir decreases. In fact, since neither infinitely hot temperatures nor absolute zero temperatures are possible, we must conclude that no real engine has 100 percent efficiency, which is just a restatement of the second law.

Equation 1.31 can help us understand the seemingly low efficiency of thermal power plants such as the one diagrammed in Figure 1.11. In this plant, fuel is burned in a firing chamber surrounded by metal tubing. Water circulating through this boiler tubing is converted to high-pressure, high-temperature steam. During this conversion of chemical to thermal energy, losses on the order of 10 percent occur due to incomplete combustion and loss of heat up the smokestack. Later,

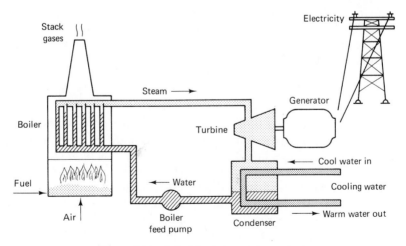

Figure 1.11   A fuel-fired steam power plant.

we shall consider local and regional air pollution effects caused by these emissions as well as their possible role in global warming.

The steam produced in the boiler then enters a steam turbine, which is in some ways similar to a child's pinwheel. The high-pressure steam expands as it passes through the turbine blades, causing a shaft that is connected to the generator to spin. While the turbine in Figure 1.11 is shown as a single unit, in actuality turbines have many stages, with steam exiting one stage and entering another, gradually expanding and cooling as it goes. The generator converts the rotational energy of a spinning shaft into electrical power which goes out onto transmission lines for distribution. A well-designed turbine may have an efficiency that approaches 90 percent while the generator may have a conversion efficiency even higher than that.

The spent steam from the turbine undergoes a phase change back to the liquid state as it is cooled in the condenser. This phase change creates a partial vacuum which helps pull steam through the turbine, thereby increasing the turbine efficiency. The condensed steam is then pumped back to the boiler to be reheated.

The heat released during the steam condensation is transferred to cooling water which circulates through the condenser. Usually cooling water is drawn from a lake or river, heated in the condenser, and returned to that body of water, in which case the process is called *once-through cooling*. A more expensive approach that requires less water involves use of cooling towers which transfer the heat directly into the atmosphere rather than into a receiving body of water. In either case, the rejected heat is released into the environment. In terms of the heat engine concept shown in Figure 1.11, our cold reservoir temperature is thus determined by the temperature of the environment.

Let us estimate the maximum possible efficiency that a thermal power plant such as that diagrammed in Figure 1.11 can have. A reasonable estimate of T_h might be the temperature of the steam from the boiler, which is typically around 600 °C. For T_c, we might use an ambient temperature of about 20 °C. Using these values in (1.31), and remembering to convert temperatures to the absolute scale, gives

$$\eta_{max} = 1 - \frac{20 + 273}{600 + 273} = 0.66 = 66 \text{ percent}$$

New fossil-fuel fired power plants have efficiencies around 40 percent. Nuclear plants have materials constraints that force them to operate at somewhat lower temperatures than fossil plants, which results in efficiencies of around 33 percent. The average efficiency of all thermal plants actually in use in the United States, including new and old (less efficient) plants, fossil and nuclear, is close to 33 percent. That suggests the following convenient rule-of-thumb: *For every 3 units of energy entering the average thermal power plant, approximately 1 unit is converted to useful electricity and 2 units are rejected to the environment as waste heat.*

The following example uses this rule of thumb for power plant efficiency combined with other emission factors to develop a mass and energy balance for a typical coal-fired power plant.

Example 1.10   Mass and Energy Balance for a Coal-Fired Power Plant

Typical coal burned in power plants in the United States has an energy content of approximately 24 kJ/g and an average carbon content of about 62 percent. For almost all new coal plants, Clean Air Act emission standards limit sulfur emissions to 260 g of sulfur dioxide (SO_2) per million kilojoules of heat input to the plant (130 g of elemental sulfur per 10^6 kJ). They also restrict particulate emissions to 13 g/10^6 kJ. Suppose the average plant burns fuel with 2 percent sulfur content and 10 percent unburnable minerals called *ash*. About 70 percent of the ash is released as *fly ash* and about 30 percent settles out of the firing chamber and is collected as *bottom ash*. Assume this is a typical coal plant with 3 units of heat energy required to deliver 1 unit of electrical energy.

a. Per kilowatt-hour of electrical energy produced, find the emissions of SO_2, particulates, and carbon (assume all of the carbon in the coal is released to the atmosphere).

b. How efficient must the sulfur emission control system be to meet the sulfur emission limitations?

c. If we assume all of the particulate matter is made up of fly ash, how efficient must the particulate control system be to meet the particulate emission limits?

Solution

a. We first need the heat input to the plant. Since 3 kWhr of heat is required for each 1 kWhr of electricity delivered,

$$\frac{\text{Heat input}}{\text{kWhr electricity}} = 3 \text{ kWhr heat} \times \frac{1 \text{ kJ/s}}{\text{kW}} \times 3600 \text{ s/hr} = 10\ 800 \text{ kJ}$$

The sulfur emissions are thus restricted to

$$\text{S emissions} = \frac{130 \text{ g S}}{10^6 \text{ kJ}} \times 10\ 800 \text{ kJ/kWhr} = 1.40 \text{ g S/kWh}$$

(1 g of S corresponds to 2 g of SO_2, so 2.8 g SO_2/kWhr would be emitted.) The particulate emissions would be

$$\text{Particulate emissions} = \frac{13 \text{ g}}{10^6 \text{ kJ}} \times 10\ 800 \text{ kJ/kWhr} = 0.14 \text{ g/kWhr}$$

To find carbon emissions, let us first find the amount of coal burned per kilowatt-hour:

$$\text{Coal input} = \frac{10\ 800 \text{ kJ/kWhr}}{24 \text{ kJ/g coal}} = 450 \text{ g coal/kWhr}$$

So, since the coal is 62 percent carbon,

$$\text{Carbon emissions} = \frac{0.62 \text{ g C}}{\text{g coal}} \times \frac{450 \text{ g coal}}{\text{kWhr}} = 280 \text{ gC/kWhr}$$

b. Burning 450 g coal containing 2 percent sulfur will release $0.02 \times 450 = 9.0$ g of S. Since the allowable emissions are 1.4 g, the removal efficiency must be

$$S \text{ removal efficiency} = 1 - \frac{1.4}{9.0} = 0.85 = 85 \text{ percent}$$

c. Since 10 percent of the coal is ash, and 70 percent of that is fly ash, the total fly ash generated will be

$$\text{fly ash generated} = 0.70 \times 0.10 \times 450 \text{ g coal/kWhr} = 31.5 \text{ g fly ash/kWhr}$$

The allowable particulate matter is restricted to 0.14 g/kWhr, so

$$\text{Particulate removal efficiency} = 1 - \frac{0.14}{31.5} = 0.995 = 99.5 \text{ percent}$$

The complete mass and energy balance for this coal plant is diagrammed in Figure 1.12. In this diagram it has been assumed that 85 percent of the waste heat is removed by cooling water and the remaining 15 percent is lost in stack gases (corresponding to the conditions given in Example 1.9).

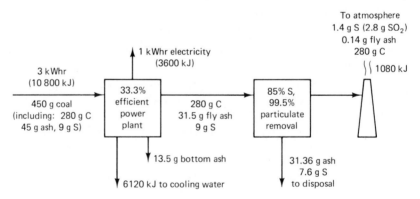

Figure 1.12   Energy and mass balance for a coal-fired power plant generating 1 kWhr of electricity (see Example 1.10).

The Carnot efficiency limitation provides insight into the likely performance of other types of thermal power plants in addition to the steam plants just described. For example, there have been many proposals to build power plants that would take advantage of the temperature difference between the relatively warm surface waters of the ocean and the rather frigid waters found below. In some locations, the sun heats the ocean's top layer to as much as 30 °C while several hundred meters down, the temperature is a constant 4 or 5 °C. Power plants, called *ocean thermal energy conversion* (OTEC) systems, could be designed to operate on these small temperature differences in the ocean, but as the following example shows, they would be quite inefficient.

Example 1.11   OTEC System Efficiency

Consider an OTEC system operating between 30 and 5 °C. What would be the maximum possible efficiency for an electric generating station operating with these temperatures?

Solution   Using (1.31), we find

$$\eta_{max} = 1 - \frac{(5 + 273)}{(30 + 273)} = 0.08 = 8 \text{ percent}$$

An even lower efficiency, estimated at 2 to 3 percent for a real plant, would be expected.

Conductive and Convective Heat Transfer

When two objects are at different temperatures, heat will be transferred from the hotter object to the colder one. That heat transfer can be by *conduction* when there is direct physical contact between the objects; by *convection* when there is a liquid or gas between them; or by *radiation,* which can take place even in the absence of any physical medium between the objects.

Conductive heat transfer is usually associated with solids, as one molecule vibrates the next in the lattice. The rate of heat transfer in a solid is proportional to the thermal conductivity of the material. Metals tend to be good thermal conductors, which makes them very useful when high heat transfer rates are desired. Other materials are much less so. Some are particularly poor thermal conductors, which makes them potentially useful as thermal insulation.

Convective heat transfer occurs when a fluid at one temperature comes in contact with a substance at another temperature. For example, warm air in a house in the winter that comes in contact with a cool wall surface will transfer heat to the wall. As that warm air loses some of its heat, it becomes cooler and denser, and it will sink, to be replaced by more warm air from the interior of the room. Thus, there is a continuous movement of air around the room and with it a transferrence of heat from the warm room air to the cool wall. The cool wall, in turn, conducts heat to the cold exterior surface of the house where outside air removes the heat by convection.

While we could separately consider the convection and conduction processes through our hypothetical wall, it is more convenient to combine them (along with radiation) into a single, overall heat transfer process that can be characterized by the following simple equation:

$$q = \frac{A(T_i - T_o)}{R} \tag{1.34}$$

where q is the rate of heat transfer through a barrier (e.g., the wall of a building), given a surface area A perpendicular to the heat flow, air temperature T_i on one side of the barrier and T_o on the other, and an overall thermal resistance to that heat flow given by R. In SI units, R is expressed as m²-K/W, while in the more common American system, R has units of (ft²-°F-hr/Btu). If you buy insulation at the hardware store, it will be designated as having an R value in the American unit system; for example, $3\frac{1}{2}$-in.-thick fiberglass insulation is marked *R-11*, while 6 in. of the same material is *R-19*.

As the following example illustrates, improving the efficiency with which we use energy saves not only money but also the pollutants that would have been emitted while generating that saved energy. This important connection between energy efficiency and pollution control is unfortunately all too often overlooked and unappreciated.

Example 1.12 Reducing Pollution by Adding Ceiling Insulation

A home with 1500 ft² of poorly insulated ceiling is located in an area with an 8-month heating season during which time the outdoor temperature averages 40 °F while the inside temperature is kept at 70 °F (this could be Chicago, for example). It has been proposed to the owner that $1000 be spent to add more insulation to the ceiling raising its total R value from 11 to 40 (ft²-°F-hr/Btu). The house is heated with electricity that costs 8 cents/kWhr.

    a. How much money would the owner expect to save each year and how long would it take for the energy savings to pay for the cost of insulation?

    b. Suppose 1 million homes served by coal plants like the one analyzed in Example 1.10 could achieve similar energy savings. Estimate the annual reduction in SO_2, particulate, and carbon emissions that would be realized.

Solution

    a. Using (1.34) to find the heat loss rate with the existing insulation gives

$$q = \frac{A(T_i - T_o)}{R} = \frac{1500 \text{ ft}^2 \times (70 - 40) \text{ °F}}{11 \text{ (ft}^2 \text{ °F hr/Btu)}} = 4090 \text{ Btu/hr}$$

After adding the insulation, the new heat loss rate will be

$$q = \frac{A(T_i - T_o)}{R} = \frac{1500 \text{ ft}^2 \times (70 - 40) \text{ °F}}{40 \text{ (ft}^2 \text{ °F hr/Btu)}} = 1125 \text{ Btu/hr}$$

The annual energy savings can be found by multiplying the rate at which energy is being saved by the number of hours in the heating season:

$$\text{Energy saved} = \frac{(4090 - 1125) \text{ Btu/hr}}{3412 \text{ Btu/kWhr}} \times 24 \text{ hr/day} \times 30 \text{ day/mo} \times 8 \text{ mo/yr}$$

$$= 5005 \text{ kWhr/yr}$$

The annual savings in dollars would be

$$\text{Dollar savings} = 5005 \text{ kWhr/yr} \times \$0.08/\text{kWhr} = \$400/\text{yr}$$

Since the estimated cost of adding extra insulation is $1000, the reduction in electricity bills would pay for this investment in about $2\frac{1}{2}$ heating seasons.

b. One million such houses would save a total of 5 billion kWhr/year (nearly the entire annual output of a typical 1000 MW$_e$ power plant). Using the emission factors derived in Example 1.10, the reduction in air emissions would be

$$\text{Carbon reduction} = 280 \text{ g C/kWhr} \times 5 \times 10^9 \text{ kWhr/yr} \times 10^{-3} \text{ kg/g}$$

$$= 1400 \times 10^6 \text{ kg/yr}$$

$$SO_2 \text{ reduction} = 2.8 \text{ g } SO_2/\text{kWhr} \times 5 \times 10^9 \text{ kWhr/yr} \times 10^{-3} \text{ kg/g}$$

$$= 14 \times 10^6 \text{ kg/yr}$$

$$\text{Particulate reduction} = 0.14 \text{ g/kWhr} \times 5 \times 10^9 \text{ kWhr/yr} \times 10^{-3} \text{ kg/g}$$

$$= 0.7 \times 10^6 \text{ kg/yr}$$

Radiation Heat Transfer

Heat transfer by thermal radiation is the third way that one object can warm another. Unlike conduction and convection, radiant energy is transported by electromagnetic waves and does not require a medium to carry the energy. As is the case for other forms of electromagnetic phenomena, such as radio waves, x rays, and gamma rays, thermal radiation can be described either in terms of wavelengths or, using the particle nature of electromagnetic radiation, in terms of discrete photons of energy. In this book, the most important application of radiant heat transfer will come in Chapter 8 when the effects of various greenhouse gases on global climate will be discussed.

Every object emits thermal radiation. The usual way to describe how much radiation a real object emits, as well as other characteristics of the wavelengths emitted, is to compare it to a theoretical abstraction called a *blackbody*. A blackbody is defined to be a perfect emitter as well as a perfect absorber. As a perfect emitter, it radiates more energy per unit of surface area than any real object at the same temperature. As a perfect absorber, it absorbs all radiation that impinges upon it; that is, none is reflected and none is transmitted through it.

For a blackbody with surface area A and absolute temperature T, the total rate at which radiant energy is emitted is given by the *Stefan–Boltzmann law of radiation:*

$$E = \sigma A T^4 \qquad (1.35)$$

where E = total blackbody emission rate (W)
  σ = the Stefan–Boltzmann constant = 5.67×10^{-8} W/m²-K⁴
  T = absolute temperature (K)
  A = surface area of the object (m²)

Actual objects do not emit as much radiation as this hypothetical blackbody. The ratio of the amount of radiation an actual object would emit to the amount that a blackbody would emit at the same temperature is known as the emittance, ε. The emittance of most natural materials is quite high and is not particularly related to color. The emittance of desert sand, dry ground, and most woodlands is estimated to be approximately 0.90 while water, wet sand, and ice all have estimated emittances of 0.95 (Kreith and Kreider, 1978).

While Stefan–Boltzmann's law gives the total rate at which energy is radiated from a blackbody, it does not tell us anything about the wavelengths emitted. A blackbody emits radiation with a range of wavelengths that can be described with a spectral distribution such as the one shown in Figure 1.13. The wavelength at which the spectrum reaches its maximum is given by the following, known as *Wien's displacement rule:*

$$\lambda_{max}(\mu m) = \frac{2898}{T(K)} \tag{1.36}$$

where wavelength is specified in micrometers and temperature is in kelvins.

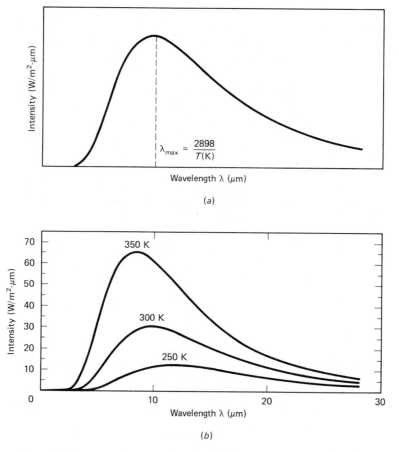

(a)

(b)

Figure 1.13   Spectral emissive power of a blackbody (*a*) showing Wien's rule for the wavelength at which power is a maximum, and (*b*) showing the effect of temperature.

The way to interpret a spectral diagram is to realize that the area under the curve between any two wavelengths is equal to the power radiated by the object within that band of wavelengths. Hence, the total area under the curve is equal to the total power radiated, given by the Stefan–Boltzmann law. Objects at higher temperatures have higher curves (greater area) and, in addition, Wien's rule indicates that objects with higher temperatures reach their maximum emissive intensity at shorter wavelengths, so their spectral curves are also shifted toward the left, as is shown in Figure 1.13*b*.

Example 1.13   The Earth's Spectrum

Consider the earth to be a blackbody with average temperature 15.0 °C and surface area equal to 5.1×10^{14} m^2. Find the rate at which energy is radiated by the earth and the wavelength at which maximum power is radiated.

Solution   Using (1.35), the earth radiates

$$E = \sigma A T^4$$

$$= 5.67 \times 10^{-8} \text{ W/m}^2\text{-K}^4 \times 5.1 \times 10^{14} \text{ m}^2 \times (15.0 + 273.15 \text{ K})^4$$

$$= 2.0 \times 10^{17} \text{ W}$$

The wavelength at which the maximum point is reached in the earth's spectrum is, by (1.35),

$$\lambda_{max}(\mu m) = \frac{2898}{T(K)} = \frac{2898}{288.15} = 10.1 \ \mu m$$

This tremendous rate of energy emission by the earth is balanced by the rate at which the earth absorbs energy from the sun. The sun, however, being much higher in temperature, radiates energy to the earth with much shorter wavelengths than the earth radiates back again. This wavelength shift plays a crucial role in the greenhouse effect. As described in Chapter 8, carbon dioxide and other greenhouse gases are relatively transparent to the incoming short wavelengths from the sun, but they tend to absorb the outgoing, longer wavelengths radiated by the earth. As those greenhouse gases accumulate in our atmosphere, they act like a blanket that envelops the planet, upsets the radiation balance, and raises the earth's temperature.

Example 1.14   The Temperature of the Sun

Measurements made outside the earth's atmosphere indicate that the solar spectrum peaks at 0.48 μm. If the sun is considered to be a blackbody, what would its temperature be?

Solution   From Wien's rule,

$$T(K) = \frac{2898}{\lambda_{max}(\mu m)} = \frac{2898}{0.48} = 6040 \text{ K}$$

It should be noted that the temperature deep within the sun is many millions of degrees and the calculation based on Wien's rule merely gives us an effective temperature of the surface.

PROBLEMS

1.1. The air quality standard for ozone (O_3) is 0.08 ppm. Express that standard in $\mu g/m^3$ at 1 atm pressure and 20 °C.

1.2. The exhaust gas from an automobile contains 1.0 percent by volume of carbon monoxide. Express this concentration in mg/m^3 at 25 °C and 1 atm.

1.3. Suppose the average concentration of SO_2 is measured to be 400 $\mu g/m^3$ at 25 °C and 1 atm. Does this exceed the (24-hr) air quality standard of 0.14 ppm? (See Table 2.1 for atomic weights.)

1.4. Five million gallons per day (MGD) of a conservative substance, with concentration 10.0 mg/L, is released into a stream having an upstream flow of 10.0 MGD and substance concentration of 3.0 mg/L. Assume complete mixing.
   a. What is the concentration in ppm just downstream?
   b. What is the mass rate of the substance in lb/day passing just downstream of the junction? *Note:* 1.0 m^3 = 264 gal.

1.5. A river with 400.0 ppm of salts (a conservative substance) and an upstream flow of 25.0 m^3/s receives an agricultural discharge of 5.0 m^3/s carrying 2000.0 mg/L of salts (see Figure P1.5). The salts quickly become uniformly distributed in the river. A municipality just downstream withdraws water and mixes it with enough pure water (no salt) from another source to deliver water having no more than 500 ppm salts to its customers.

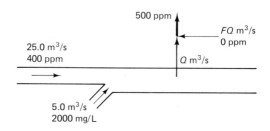

Figure P1.5

What should be the mixture ratio F of pure water to river water?

1.6. A lake with constant volume 10.0×10^6 m^3 is fed by a pollution-free stream with flow rate 50.0 m^3/s. A factory dumps 5.0 m^3/s of a nonconservative waste with concentration 100.0 mg/L into the lake. The pollution has a reaction rate coefficient K of 0.25/ day. Assume the pollution is well mixed in the lake. Find the steady-state concentration of pollution in the lake.

1.7. The two-pond system shown in Figure P1.7 is fed by a stream with flow rate 1.0 MGD (million gallons per day) and a BOD (a nonconservative pollutant) concentration of 20.0 mg/L. The rate of decay of BOD is 0.30/day. The volume of the first lake is 5.0 million gallons and the second is 3.0 million. Assuming complete mixing within each lake, find the BOD concentration leaving each lake.

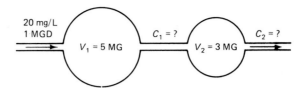

Figure P1.7

1.8. A lagoon is to be designed to accommodate an input flow of 0.10 m³/s of nonconservative pollutant with concentration 30.0 mg/L and reaction rate 0.20/day. The effluent from the lagoon must have pollutant concentration of less than 10.0 mg/L. Assuming complete mixing, how large must the lagoon be?

1.9. A simple way to model air pollution over a city is with a box model that assumes complete mixing and limited ability for the pollution to disperse horizontally or vertically except in the direction of the prevailing winds (as, for example, a town located in a valley with an inversion layer above it). Consider a town having an inversion at 250.0 m, a 20.0-km horizontal distance perpendicular to the wind, a wind speed of 2.0 m/s, and a carbon monoxide (CO) emission rate of 60.0 kg/s (see Figure P1.9). Assume the CO is conservative and completely mixed in the box.

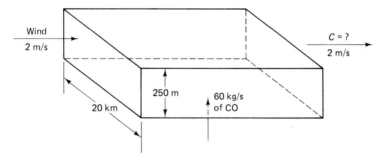

Figure P1.9

What would be the CO concentration in the box?

1.10. A 4 × 8-ft solar collector has water circulating through it at the rate of 1.0 gallon per minute (gpm) while exposed to sunlight with an intensity of 300.0 Btu/ft²-hr (see

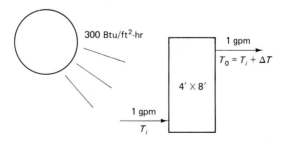

Figure P1.10

Figure P1.10). Fifty percent of that sunlight is captured by the collector and heats the water flowing through it. What would be the temperature rise of the water as it leaves the collector?

1.11. An uncovered swimming pool at 80 °F loses 1.0 in. of water off of its 1000.0-ft² surface each week due to evaporation. The cost of energy to heat the pool is $10 per million Btu. A salesman claims a $500 pool cover that reduces evaporative losses by two-thirds will pay for itself in one 15-week swimming season? Can it be true?

1.12. A 33.3-percent efficient 1000 MW$_e$ nuclear power plant rejects the other two-thirds of the fuel energy into cooling water that is withdrawn from a local river (there are no stack losses as is the case for a fossil-fuel-fired plant). The river has an upstream flow of 100.0 m³/s and a temperature of 20.0 °C.

  a. If the cooling water is allowed to rise in temperature only by 10 °C, what flow rate from the river would be required? Compare it to the coal plant in Example 1.9.

  b. How much would the river temperature rise as it receives the heated cooling water? Again compare it to Example 1.9.

1.13. Mars radiates energy with a peak wavelength of 13.2 μm. Treating it as a blackbody, what would its temperature be?

1.14. Compare the energy required to evaporate a pound of water at 15 °C to that required to raise that pound 10 000 feet into the air. (1.0 Btu = 778 ft-lb)

1.15. A 600-MW$_e$ power plant has an efficiency of 36 percent, with 15 percent of the waste heat being released to the atmosphere as stack heat and the other 85 percent taken away in the cooling water (see Figure P1.15). Instead of drawing water from a river, heating it, and returning it to the river, this plant uses an evaporative cooling tower wherein heat is released to the atmosphere as cooling water is vaporized.

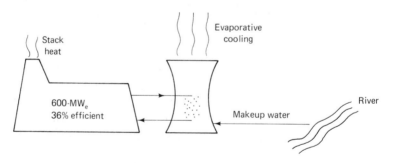

Figure P1.15

At what rate must 15 °C makeup water be provided from the river to offset the water lost in the cooling tower?

1.16. An electric water heater held at 140.0 °F is kept in a 70.0 °F room. When purchased its insulation is equivalent to R-5. An owner puts a 25.0-ft² blanket on the water heater, raising its total R value to 15.0. If electricity costs 8.0 cents/kWhr, how much money will be saved in energy each year?

1.17. A 15-W compact fluorescent light bulb produces the same amount of light as a 60-W incandescent while using only one-fourth the power. Over the 9000-hr lifetime of one of these bulbs, compute the carbon, SO$_2$, and particulate emissions that would be saved if one of these bulbs replaces an incandescent and the electricity comes from the coal-fired power plant described in Example 1.10.

1.18. No. 6 fuel oil has a carbon content 20.0 kg carbon per 10^9 J. If it is burned in a 40.0-percent efficient power plant, find the carbon emissions per kilowatt-hour of electricity produced, assuming all of the carbon in the fuel is released into the atmosphere. By law, in new oil-fired power plants SO_2 emissions are limited to 86 g/10^6 kJ (of heat input) and emissions of nitrogen oxides (NO_x) are limited to 130 g NO_x/10^6 kJ. Estimate the maximum allowable SO_2 and NO_x emissions per kilowatt-hour.

1.19. Consider the air over a city to be a box 100 km on a side that reaches up to an altitude of 1.0 km. Clean air is blowing into the box along one of its sides with a speed of 4 m/s. Suppose an air pollutant with reaction rate $K = 0.20$/hr is emitted into the box at a total rate of 10.0 kg/s. Find the steady-state concentration if the air is assumed to be completely mixed.

1.20. If the wind in Problem 1.19 suddenly dropped to 1 m/s, estimate the concentration of pollutants 2 hr later.

1.21. A lagoon with volume 1200 m^3 has been receiving a steady flow of a conservative waste at the rate of 100 m^3/day for a long enough time to assume that steady-state conditions apply. The waste entering the lagoon has a concentration of 10 mg/L. Assuming complete mix conditions, what would be the concentration of pollutant in the effluent leaving the lagoon? If the input waste concentration suddenly increases to 100 mg/L, what would the concentration in the effluent be 7 days later?

1.22. Repeat Problem 1.21 for a nonconservative pollutant with reaction rate $K = 0.20$/day.

REFERENCES

HARTE, J., 1985, *Consider a Spherical Cow, A Course in Environmental Problem Solving,* Kaufman, Los Altos, CA.

KREITH, F., and J. F. KREIDER, 1978, *Principles of Solar Engineering,* McGraw-Hill, New York.

THORNDIKE, E. H., 1976, *Energy and Environment, A Primer for Scientists and Engineers,* Addison-Wesley, Reading, MA.

WANIELISTA, M. P., Y. A. YOUSEF, J. S. TAYLOR, and C. D. COOPER, 1984, *Engineering and the Environment,* Brooks/Cole Engineering Division, Monterey, CA.

CHAPTER 2

Environmental Chemistry

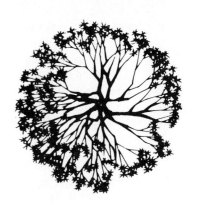

*It often matters much how given atoms combine, in what
arrangement, with what others, what impulse they receive,
and what impart. The same ones make up earth, sky, sea,
and stream; the same the sun, the animals, grain, and
trees, but mingling and moving in ever different ways.*
Lucretius (95–52 B.C.) in *The Nature of Things*

2.1 INTRODUCTION

Almost every pollution problem that we face has a chemical basis. Even the most
qualitative descriptions of such problems as the greenhouse effect, ozone depletion, toxic wastes, groundwater contamination, air pollution, and acid rain, to
mention a few, require at least a rudimentary understanding of some basic chemical concepts. And, of course, an environmental engineer who must design an
emission control system or a waste treatment plant must be well grounded in
chemical principles and the techniques of chemical engineering. In this brief
chapter, the topics have been selected with the goal of providing only the essential
chemical principles required to understand the nature of the pollution problems
that we face and the engineering approaches to their solutions.

2.2 STOICHIOMETRY

When a chemical reaction is written down, it provides both qualitative and quantitative information. Qualitatively, we can see what chemicals are interacting to produce what end products. Quantitatively, the principle of conservation of mass can be applied to give information about how much of each compound is involved to produce the results shown. The balancing of equations so that the same number of each kind of atom appears on each side of the equation and the subsequent calculations which can be used to determine amounts of each compound involved is known as *stoichiometry*.

The first step is to balance the equation. For example, suppose we want to investigate the combustion of methane (CH_4), the principal component of natural gas. Methane combines with oxygen to produce carbon dioxide and water, as the following reaction suggests:

$$CH_4 + O_2 \longrightarrow CO_2 + H_2O$$

The equation is not balanced. One atom of carbon appears on each side, which is fine, but there are four atoms of hydrogen on the left and only two on the right, and there are only two atoms of oxygen on the left while there are three on the right. We might try to double the water molecules on the right to balance the hydrogen on each side, but then there would be an imbalance of oxygen with two on the left and four on the right. So try doubling the oxygen on the left. This sort of trial-and-error approach to balancing simple reactions usually converges pretty quickly. In this instance the following is a balanced equation with the same number of C, H, and O atoms on each side of the arrow:

$$CH_4 + 2O_2 \longrightarrow CO_2 + 2H_2O \qquad (2.1)$$

This balanced chemical equation can be read as follows: One molecule of methane reacts with two molecules of oxygen to produce one molecule of carbon dioxide and two molecules of water. It is of more use, however, to be able to describe this reaction in terms of the mass of each substance, that is, how many grams of oxygen are required to react with how many grams of methane, and so on. To do so requires that we know something about the mass of individual atoms and molecules.

The *atomic weight* of an atom is the mass of the atom measured in *atomic mass units* (amu), where one amu is defined to be exactly one-twelfth the mass of a carbon atom having six protons and six neutrons in its nucleus. While this might suggest that if we look up the atomic weight of carbon we would expect to find it to be exactly 12 amu, that is not the case. All carbon atoms do have six protons, but they do not all have six neutrons, so they do not all have the same atomic weight. Atoms having the same number of protons but differing numbers of neutrons are called *isotopes*. What is reported in tables of atomic weights, such as Table 2.1, is the average based on the relative abundance of different isotopes found in nature. Also shown in Table 2.1 is the *atomic number*, which is the number of

TABLE 2.1 ATOMIC NUMBERS AND ATOMIC WEIGHTS

Element	Symbol	Atomic number	Atomic weight	Element	Symbol	Atomic number	Atomic weight
Actinium	Ac	89	227.03	Mercury	Hg	80	200.59
Aluminum	Al	13	26.98	Molybdenum	Mo	42	95.94
Americium	Am	95	243	Neodymium	Nd	60	144.24
Antimony	Sb	51	121.75	Neon	Ne	10	20.18
Argon	Ar	18	39.95	Neptunium	Np	93	237.05
Arsenic	As	33	74.92	Nickel	Ni	28	58.70
Astatine	At	85	210	Niobium	Nb	41	92.91
Barium	Ba	56	137.33	Nitrogen	N	7	14.01
Berkelium	Bk	97	247	Nobelium	No	102	259
Berylium	Be	4	9.01	Osmium	Os	76	190.2
Bismuth	Bi	83	208.98	Oxygen	O	8	16.00
Boron	B	5	10.81	Palladium	Pd	46	106.4
Bromine	Br	35	79.90	Phosphorus	P	15	30.97
Cadmium	Cd	48	112.41	Platinum	Pt	78	195.09
Calcium	Ca	20	40.08	Plutonium	Pu	94	244
Californium	Cf	98	251	Polonium	Po	84	209
Carbon	C	6	12.01	Potassium	K	19	39.09
Cerium	Ce	58	140.12	Praeseodymium	Pr	59	140.91
Cesium	Cs	55	132.90	Promethium	Pm	61	145
Chlorine	Cl	17	35.45	Protactinium	Pa	91	231.04
Chromium	Cr	24	51.99	Radium	Ra	88	226.03
Cobalt	Co	27	58.93	Radon	Rn	86	222
Copper	Cu	29	63.55	Rhenium	Re	75	186.2
Curium	Cm	96	247	Rhodium	Rh	45	102.91
Dysprosium	Dy	66	162.50	Rubidium	Rb	37	85.45
Einsteinium	Es	99	254	Ruthenium	Ru	44	101.07
Erbium	Er	68	167.26	Samarium	Sm	62	150.4
Europium	Eu	63	151.96	Scandium	Sc	21	44.96
Fermium	Fm	100	257	Selenium	Se	34	78.96
Fluorine	F	9	19.00	Silicon	Si	14	28.09
Francium	Fr	87	223	Silver	Ag	47	107.89
Gadolinium	Gd	64	157.25	Sodium	Na	11	22.99
Gallium	Ga	31	69.72	Strontium	Sr	38	87.62
Germanium	Ge	32	72.59	Sulfur	S	16	32.06
Gold	Au	79	196.97	Tantalum	Ta	73	180.95
Hafnium	Hf	72	178.49	Technetium	Tc	43	97
Helium	He	2	4.00	Tellurium	Te	52	127.60
Holmium	Ho	67	164.93	Terbium	Tb	65	158.93
Hydrogen	H	1	1.01	Thallium	Tl	81	204.37
Indium	In	49	114.82	Thorium	Th	90	232.04
Iodine	I	53	126.90	Thulium	Tm	69	168.93
Iridium	Ir	77	192.22	Tin	Sn	50	118.69
Iron	Fe	26	55.85	Titanium	Ti	22	47.90
Krypton	Kr	36	83.80	Tungsten	W	74	183.85
Lanthanum	La	57	138.91	Uranium	U	92	238.03
Lawrencium	Lr	103	260	Vanadium	V	23	50.94
Lead	Pb	82	207.2	Xenon	Xe	54	131.30
Lithium	Li	3	6.94	Ytterbium	Yb	70	173.04
Lutetium	Lu	71	174.97	Yttrium	Y	39	88.91
Magnesium	Mg	12	24.31	Zinc	Zn	30	65.38
Manganese	Mn	25	54.94	Zirconium	Zr	40	91.22
Mendelevium	Md	101	258				

protons in the nucleus. All isotopes of a given element have the same atomic number.

The *molecular weight* of a molecule is simply the sum of the atomic weights of all of the constituent atoms. If we divide the mass of a substance by its molecular weight, the result is the mass expressed in *moles* (mol). Usually the mass is expressed in grams, in which case the moles are *g-moles;* in like fashion, if the mass is expressed in pounds, the result would be *lb-moles*. One g-mole contains 6.022×10^{23} molecules (Avogadro's number) and one lb-mole about 2.7×10^{26} molecules.

$$\text{Moles} = \frac{\text{Mass}}{\text{Molecular weight}} \tag{2.2}$$

The special advantage of expressing amounts in moles is that one mole of any substance contains exactly the same number of molecules.

Thus, we now have two ways to express the reaction given in Eq. 2.1, repeated here:

$$CH_4 + 2O_2 \longrightarrow CO_2 + 2H_2O$$

On a molecular level we can say that one molecule of methane reacts with two molecules of oxygen to produce one molecule of carbon dioxide and two molecules of water. On a larger scale, we can say that one mole of methane reacts with two moles of oxygen to produce one mole of carbon dioxide and two moles of water. Since we know how many grams are contained in each mole, we can express our mass balance in those terms as well.

To express the above methane reaction in grams, we need first to find the number of grams per mole for each substance. Using Table 2.1, we find that the atomic weight for C is 12, for H is 1, and for O is 16. Notice that these values have been rounded slightly, which is common engineering practice. Thus, the molecular weights and hence the number of grams per mole are

$$CH_4 = 12 + 4 \times 1 \ = 16 \text{ g/mol}$$

$$O_2 = \ 2 \times 16 \quad\ = 32 \text{ g/mol}$$

$$CO_2 = 12 + 2 \times 16 = 44 \text{ g/mol}$$

$$H_2O = \ 2 \times 1 + 16 = 18 \text{ g/mol}$$

Summarizing these various ways to express the oxidation of methane, we obtain

CH_4	$+$	$2O_2$	$\longrightarrow$	CO_2	$+$	$2H_2O$
1 molecule of methane	$+$	2 molecules of oxygen	$\longrightarrow$	1 molecule of carbon dioxide	$+$	2 molecules of water

or

1 mol of methane	$+$	2 mol of oxygen	$\longrightarrow$	1 mol of carbon dioxide	$+$	2 mol of water

or

$$\underset{\text{of methane}}{16 \text{ g}} + \underset{\text{of oxygen}}{64 \text{ g}} \longrightarrow \underset{\text{carbon dioxide}}{44 \text{ g of}} + \underset{\text{of water}}{36 \text{ g}}$$

Notice that mass is conserved in the last expression; that is, there are 80 g on the left and 80 g on the right.

Example 2.1   Combustion of Butane

What mass of carbon dioxide would be given off if 100 g of butane (C_4H_{10}) is completely oxidized to carbon dioxide and water?

Solution   First write down the reaction:

$$C_4H_{10} + O_2 \longrightarrow CO_2 + H_2O$$

then balance it:

$$2C_4H_{10} + 13O_2 \longrightarrow 8CO_2 + 10H_2O$$

Find the grams per mole for butane:

$$C_4H_{10} = 4 \times 12 + 10 \times 1 = 58 \text{ g/mol}$$

We already know that there are 44 g/mol of CO_2, so we do not need to recalculate that. Two moles of butane ($2 \times 58 = 116$ g) yields 8 mol of carbon dioxide (8 mol $\times$ 44 g/mol = 352 g CO_2). So, we can set up the following proportion:

$$\frac{116 \text{ g } C_4H_{10}}{352 \text{ g } Co_2} = \frac{100 \text{ g } C_4H_{10}}{X \text{ g } CO_2}$$

Thus, X = $100 \times 352/116 = 303$ g of CO_2 produced.

Many environmental problems involve concentrations of substances dissolved in water. In Chapter 1 we introduced two common sets of units, mg/L and ppm. However, it is also useful to express concentrations in terms of *molarity,* which is simply the number of moles of substance per liter of solution. A one molar (1 M) solution has 1 mol of substance dissolved into enough water to make the mixture have a volume of 1 L. Molarity is related to mg/L concentrations by the following:

$$\text{mg/L} = \text{Molarity (mol/L)} \times \text{Molecular weight (g/mol)} \times 10^3 \text{ (mg/g)} \qquad (2.3)$$

The following example illustrates the use of molarity and at the same time introduces another important concept having to do with the amount of oxygen required to break down (oxidize) a given substance.

Example 2.2   Theoretical Oxygen Demand

Consider a 1.67×10^{-3} M glucose solution ($C_6H_{12}O_6$) that is completely oxidized to CO_2 and H_2O. Find the amount of oxygen required to complete the reaction.

Solution   To find the oxygen required to completely oxidize this glucose, we first write a balanced equation with the accompanying weights:

$$C_6H_{12}O_6 \quad + \quad 6O_2 \quad \longrightarrow \quad 6CO_2 \quad + \quad 6H_2O$$

$$180 \qquad 6 \times 32 = 192 \qquad 6 \times 44 = 264 \quad 6 \times 18 = 108$$

Thus, it takes 192 g of oxygen to oxidize 180 g of glucose. From (2.3), the concentration of glucose is

$$mg/L = 1.67 \times 10^{-3} \text{ mol/L} \times 180 \text{ g/mol} \times 10^3 \text{ mg/g} = 300 \text{ mg/L}$$

so the oxygen requirement would be

$$300 \text{ mg/L glucose} \times \frac{192 \text{ g O}_2}{180 \text{ g glucose}} = 320 \text{ mg/L O}_2$$

If the chemical composition of a substance is known, then the amount of oxygen required to oxidize it to carbon dioxide and water can be calculated using stoichiometry, as was done in the above example. That oxygen requirement is known as the *theoretical oxygen demand*. If that oxidation is carried out by bacteria using the substance for food, then the amount of oxygen required is known as the *biochemical oxygen demand* or *BOD*. The BOD will be somewhat less than the theoretical oxygen since some of the original carbon is incorporated into bacterial cell tissue rather than being oxidized to carbon dioxide. Oxygen demand is an important measure of the likely impact that wastes will have on a receiving body of water and much more will be said about it in Chapter 4.

The convenience of using moles to describe amounts of substances also helps when calculating atmospheric concentrations of pollutants. It was Avogadro's hypothesis, made back in 1811, that equal volumes of all gases, at a specified temperature and pressure, contain equal number of molecules. In fact, since one mole of any substance has Avogadro's number of molecules, it follows that one mole of gas, at a specified temperature and volume, will occupy a predictable volume. At standard temperature and pressure (STP), corresponding to 0 °C and 1 atm (760 mm of mercury, 101.3 kPa), one g-mole of an ideal gas occupies 22.4 L, or 0.0224 m^3, and contains 6.022×10^{23} molecules. This fact was used in Chapter 1 to derive relationships between concentrations expressed in $\mu g/m^3$ and ppm (by volume).

Let us demonstrate the usefulness of Avogadro's hypothesis for gases by applying it to a very modern concern, that is, the rate at which we are pouring carbon dioxide into the atmosphere as we burn up our fossil fuels.

Example 2.3   Worldwide CO_2 Emissions

Worldwide fossil fuel energy consumption in 1987 was about 3.0×10^{20} J/year. If all of that energy would have been supplied by methane gas having an energy content of 3.9×10^7 J/m^3 (at STP), at what rate would CO_2 have been emitted to the atmosphere? Also, express that emission rate as (metric) tons of carbon (not CO_2) per year.

Solution We first need to express that consumption rate in moles. Converting joules of energy into moles of methane is straightforward:

$$\text{moles } CH_4 = \frac{3.0 \times 10^{20} \text{ J/yr}}{3.9 \times 10^7 \text{ J/m}^3} \times \frac{1}{22.4 \times 10^{-3} \text{ m}^3/\text{mol}} = 3.4 \times 10^{14} \text{ mol}$$

We know from the balanced chemical reaction given in (2.1) that each mole of CH_4 yields one mole of CO_2, so there will be 3.4×10^{14} mol of CO_2 emitted. And, since the molecular weight of CO_2 is 44, the mass of CO_2 emitted is

$$\text{mass } CO_2 = 3.4 \times 10^{14} \text{ mol/yr} \times 44 \text{ g/mol} = 1.5 \times 10^{16} \text{ g/yr}$$

To express the emissions as tons of C per year, we must convert grams to tons and then sort out the fraction of CO_2 that is carbon. The fraction of CO_2 that is C is simply the ratio of the atomic weight of carbon (12) to the molecular weight of carbon dioxide (44):

$$\text{mass C} = 1.5 \times 10^{16} \text{ g/yr} \times \frac{1 \text{ kg}}{1000 \text{ g}} \times \frac{1 \text{ ton}}{1000 \text{ kg}} \times \frac{12}{44}$$

$$= 4.1 \times 10^9 \text{ ton/yr}$$

The 4.1 gigatons of carbon found in the above example is somewhat less than the actual rate of about 5.5 gigatons. In the example it was assumed, for simplicity, that all of our fossil fuel energy came from the combustion of methane. In actuality, most of our fossil fuel energy is derived from petroleum and coal which emit considerably more carbon per unit of energy. Furthermore, the natural gas we burn is a mix of hydrocarbons which liberates more carbon per unit of energy than pure methane. These important influences on carbon emission rates will be seen again in Chapter 8 when we discuss the problem of global warming.

2.3 CHEMICAL EQUILIBRIA

In the reactions considered so far, the assumption has been that they proceed in one direction only. Most chemical reactions are, to some extent, reversible, proceeding in both directions at once. When the rates of reaction are the same, that is, products are being formed on the right at the same rate as they are being formed on the left, the reaction is said to have reached *equilibrium*.

The following represents a generalized reversible reaction

$$aA + bB \rightleftharpoons cC + dD \tag{2.4}$$

in which the lowercase letters a, b, c, and d are coefficients corresponding to the number of molecules or ions of the respective substances that result in a balanced equation. The capital letters A, B, C, and D are the chemical species themselves. The double arrow designation indicates that the reaction proceeds in both directions at the same time.

In equilibrium, we can write that

$$\frac{[C]^c[D]^d}{[A]^a[B]^b} = K \tag{2.5}$$

where the [ ] designation represents concentrations of the substances in equilibrium, expressed in moles per liter. Do not use concentrations in mg/L! K is called the *equilibrium constant*. It should also be emphasized that Eq. 2.5 is valid only when chemical equilibrium is established, if ever. Natural systems are often subject to constantly changing inputs, and since some reactions occur very slowly, equilibrium may never be established. A practicing environmental engineer must therefore use this important equation with a certain degree of caution.

Many molecules, when dissolved in water, separate into positively charged ions, called *cations,* and negatively charged ions, called *anions.* Equation 2.5 can be applied to the dissociation of such molecules, in which case K is referred to as a *dissociation constant* or an *ionization constant.* In other circumstances, the quantity on the left is a solid and the problem is to determine the degree to which that solid enters solution. In such cases the equilibrium constant is called the *solubility product.*

Finally, often when dealing with very large and very small numbers, it is helpful to introduce the following logarithmic measure:

$$K = 10^{-pK} \tag{2.6}$$

or

$$pK = -\log K \tag{2.7}$$

Acid–Base Reactions

Water dissociates slightly into hydrogen ions (protons, H^+) and hydroxide ions (OH^-) as the following reaction suggests:

$$H_2O \;\rightleftharpoons\; H^+ + OH^- \tag{2.8}$$

The corresponding equilibrium expression for this reaction is

$$\frac{[H^+][OH^-]}{[H_2O]} = K \tag{2.9}$$

The molar concentration of water in its pure state is 1000 g/L divided by 18 g/mol, or 55.56 mol/L. Since water dissociates only slightly, the molar concentration after ionization is not changed enough to be of significance, so $[H_2O]$ is essentially a constant that can be included in the equilibrium constant. The result is the following:

$$[H^+][OH^-] = K_w = 10^{-14} \quad \text{at 25 °C} \tag{2.10}$$

where K_w is the dissociation constant for water. For dilute aqueous solutions in general, $[H_2O]$ is considered constant and is included in the equilibrium contant. K_w is temperature dependent but unless otherwise stated, the value given in (2.10) at 25 °C will be the assumed value.

It is important to point out that (2.10) holds no matter what the source of hydrogen ions or hydroxide ions. That is, it is valid even if other substances dissolved in the water make their own contributions to the hydrogen and hydroxide supplies. It is always one of the equations that must be satisfied when chemical equilibria problems are analyzed.

It is customary to express $[H^+]$ and $[OH^-]$ concentrations using the logarithmic measure introduced in (2.6) and (2.7). To express hydrogen ion concentrations, the pH scale is used, where

$$pH = -\log[H^+] \tag{2.11}$$

or

$$[H^+] = 10^{-pH} \tag{2.12}$$

With the pH scale, it is easy to specify whether a solution is acidic, basic, or neutral. A *neutral* solution corresponds to the case where the concentration of hydrogen ions $[H^+]$ equals the concentration of hydroxide ions $[OH^-]$. From (2.10), for a neutral solution

$$[H^+][OH^-] = [H^+][H^+] = [H^+]^2 = 10^{-14}$$

so

$$[H^+] = 10^{-7}$$

that is, a neutral solution has a pH of 7 (written pH 7).

An *acidic* solution is one for which $[H^+]$ is greater than $[OH^-]$; that is, the hydrogen ion concentration is greater than 10^{-7} mol/L, and its pH is less than 7. A *basic* solution is the other way around, with more hydroxide ions than hydrogen ions and pH greater than 7. Notice that for every unit change in pH, the concentration of hydrogen ions changes by a factor of 10.

Figure 2.1 illustrates the pH scale, showing example values of pH for several common solutions. Notice in the figure that a distinction is made between distilled water and "pure" rainfall. As will be seen in the next section on carbonates, as rainwater falls it absorbs carbon dioxide from the air and carbonic acid is formed. Even without any pollution in the air, the pH of rainfall is thus lowered to around 5.65. By the usual definition, acid rain, caused by industrial pollutants, has pH lower than 5.6. Actual acid rain and acid fog has been recorded with pH below 2.0.

Example 2.4   pH of Tomato Juice

Find the hydrogen ion concentration and the hydroxide ion concentration in tomato juice having a pH of 4.1.

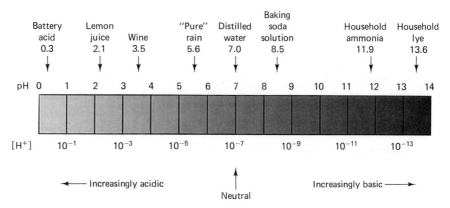

Figure 2.1 The pH scale.

Solution   From (2.12), the hydrogen ion concentration is

$$[H^+] = 10^{-pH} = 10^{-4.1} = 7.94 \times 10^{-5} \text{ mol/L}$$

and from (2.10) the hydroxide ion concentration is

$$[OH^-] = \frac{10^{-14}}{[H^+]} = \frac{10^{-14}}{7.94 \times 10^{-5}} = 1.25 \times 10^{-10} \text{ mol/L}$$

Acid–base reactions are among the most important in environmental engineering. Often, to protect the local ecosystem, wastes will require neutralization before being released into the environment. Most aquatic forms of life, for example, are very sensitive to the pH of their habitat. In other circumstances, by forcing the pH toward one end of the spectrum or the other, chemical equilibrium equations can be shifted toward the left or the right, possibly resulting in unwanted substances being driven out of solution as precipitates or gases.

As an example of the value of being able to control pH, consider the problem of removing nitrogen from municipal wastewater. One reason we might want to remove nitrogen is to keep it from stimulating the growth of algae in the receiving body of water. Another reason might be to prevent excessive nitrate $[NO_3^-]$ levels in drinking water from causing a potentially lethal condition in babies known as *methemoglobinemia.*

One way to remove nitrogen during wastewater treatment is with a process known as ammonia stripping. When organic matter decomposes, nitrogen is first released in the form of ammonia (NH_3) or ammonium ion (NH_4^+). Ammonium ions are highly soluble in water, while NH_3 is not. By driving the equilibrium reaction

$$NH_3 + H_2O \; \rightleftharpoons \; NH_4^+ + OH^- \tag{2.13}$$

toward the left, less soluble ammonia gas is formed, which can then be encouraged to leave solution and enter the air in a *gas stripping tower.* In such a tower, water is allowed to trickle downward over slats or corrugated surfaces while clean air is blown in from the bottom aerating the dripping water. Gas stripping previously was used only in wastewater treatment, to remove such gases as ammonia and hydrogen sulfide, but now it is beginning to be used to remove *volatile organic chemicals* (VOCs) from contaminated groundwater (as described in Chapters 5 and 6). So, how can (2.13) be driven toward the formation of ammonia? The reaction can be driven to the left by increasing $[OH^-]$, that is, by raising the pH, as the following example demonstrates.

Example 2.5 Ammonia Stripping

Nitrogen in a wastewater treatment plant is in the form of ammonia and ammonium ion. Find the fraction of the nitrogen that is in the ammonia form (and hence strippable) as a function of pH (at 25 °C) and draw a graph. The equilibrium constant is 1.82×10^{-5}.

Solution   The equilibrium equation for the reaction given in (2.13) is

$$\frac{[NH_4^+][OH^-]}{[NH_3]} = K_{NH_3} = 1.82 \times 10^{-5} \tag{2.14}$$

Notice K_{NH_3} is like K_w in that it treats the concentration of H_2O as a constant that is already accounted for in the equilibrium constant itself. What we want to find is the fraction of nitrogen in the form of ammonia, or

$$NH_3 \text{ fraction} = \frac{[NH_3]}{[NH_3] + [NH_4^+]} = \frac{1}{1 + [NH_4^+]/[NH_3]} \tag{2.15}$$

Equation 2.10 must also be satisfied:

$$[H^+][OH^-] = K_w = 10^{-14} \tag{2.10}$$

Rearranging (2.14) and substituting (2.10) into it gives

$$\frac{[NH_4^+]}{[NH_3]} = \frac{K_{NH_3}}{[OH^-]} = \frac{K_{NH_3}}{K_w/[H^+]} \tag{2.16}$$

and putting this into (2.15) gives

$$NH_3 \text{ fraction} = \frac{1}{1 + K_{NH_3}[H^+]/K_w}$$

$$= \frac{1}{1 + 1.82 \times 10^{-5} \times \dfrac{10^{-pH}}{10^{-14}}} \tag{2.17}$$

$$= \frac{1}{1 + 1.82 \times 10^{(9-pH)}}$$

A table of some values can easily be generated from (2.17) and the results are plotted in Figure 2.2.

As Example 2.5 suggests, to drive the reaction significantly toward the ammonia side requires a pH in excess of about 10. Since typical wastewaters seldom have a pH that high, it is necessary to add chemicals, the usual being lime (CaO),

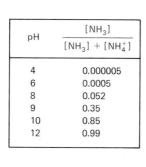

pH	$\dfrac{[NH_3]}{[NH_3] + [NH_4^+]}$
4	0.000005
6	0.0005
8	0.052
9	0.35
10	0.85
12	0.99

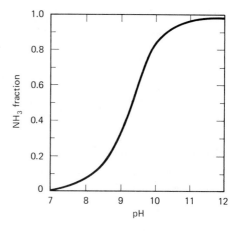

Figure 2.2   Dependence of the ammonia fraction on pH (Example 2.5).

to sufficiently raise the pH. The lime, unfortunately, reacts with CO_2 in the air to form a calcium carbonate scale that can accumulate on stripping surfaces and which must be removed periodically, creating a sludge disposal problem of its own.

The above example was but one illustration of the value of being able to control pH in treatment processes. In other circumstances, pollution itself affects pH, driving reactions in directions that may be undesirable. A notable example of this phenomenon is the mobilization of trace metals, especially aluminum, when pH drops during acidification of a watershed. Aluminum is very toxic to fish but its normal form is a quite insoluble solid that is relatively safe. However, when acid deposition acidifies a lake, the reduced pH drives equilibria reactions toward liberation of mobile Al^{3+} ions which are very toxic. Mobilization of trace metals will be discussed further in Chapter 4.

Solubility Product

The above comments about solids and gases in water suggests that we consider such phenomena a bit more carefully. We will deal with gases in the next section, after this brief introduction to solids.

All solids are to some degree soluble, some much more so than others. A generalized equation describing the equilibrium condition in which a solid is dissociating into its ionic components (*dissolution*) at the same rate that ionic components are recombining into the solid form (*precipitation*) is given as follows:

$$\text{Solid} \quad \rightleftharpoons \quad a\text{A} + b\text{B} \tag{2.18}$$

where A and B are the ionic components that make up the solid. Applying (2.5) yields

$$\frac{[\text{A}]^a[\text{B}]^b}{[\text{solid}]} = K \tag{2.19}$$

As long as there is still solid present in equilibrium, its affect can be incorporated into the equilibrium constant:

$$[A]^a[B]^b = K_{sp} \tag{2.20}$$

where K_{sp} is called the *solubility product*. Table 2.2 gives a short list of solubility products of particular importance in environmental engineering.

As an example of the use of (2.20), consider the fluoridation of water with calcium fluoride, CaF_2.

Example 2.6 Fluoride Solubility

Find the equilibrium concentration of fluoride ions in pure water caused by the dissociation of CaF_2. Express the answer both in units of mol/L and mg/L.

Solution   From Table 2.2, the reaction and solubility product are

$$CaF_2 \rightleftharpoons Ca^{2+} + 2F^- \qquad K_{sp} = 3.9 \times 10^{-11}$$

Remembering to square the term for the fluoride ion, the mass action equation becomes

$$[Ca^{2+}][F^-]^2 = 3.9 \times 10^{-11} \tag{2.21}$$

If we let the solubility of Ca^{2+} in mol/L be s, then the concentration of F^- will be $2s$ since 2 mol of F^- is formed for every mole of Ca^{2+}. Thus,

$$[Ca^{2+}] = s$$

$$[F^-] = 2s$$

and from (2.21)

$$K_{sp} = s \times (2s)^2 = 4s^3 = 3.9 \times 10^{-11}$$

$$s = [Ca^{2+}] = 2.1 \times 10^{-4} \text{ mol/L}$$

$$2s = [F^-] = 4.2 \times 10^{-4} \text{ mol/L}$$

To find the concentration of fluoride ions in mg/L, we need the atomic weight of fluorine, which from Table 2.1 is 19 g/mol.

$$[F^-] = 4.2 \times 10^{-4} \text{ mol/L} \times 19 \text{ g/mol} \times 10^3 \text{ mg/g} = 8.0 \text{ mg/L}$$

TABLE 2.2   TYPICAL SOLUBILITY-PRODUCT CONSTANTS AT 25 °C

Equilibrium equation	K_{sp}	Significance in environmental engineering
$CaCO_3 \rightleftharpoons Ca^{2+} + CO_3^{2-}$	4.7×10^{-9}	Hardness removal, scaling
$CaSO_4 \rightleftharpoons Ca^{2+} + SO_4^{2-}$	2.4×10^{-5}	Flue gas desulfurization
$Cu(OH)_2 \rightleftharpoons Cu^{2+} + 2OH^-$	1.6×10^{-19}	Heavy metal removal
$Al(OH)_3 \rightleftharpoons Al^{3+} + 3OH^-$	5×10^{-33}	Coagulation, acidification
$Ca_3(PO_4)_2 \rightleftharpoons 3Ca^{2+} + 2PO_4^{3-}$	1.3×10^{-32}	Phosphate removal
$CaF_2 \rightleftharpoons Ca^{2+} + 2F^-$	3.9×10^{-11}	Fluoridation

Source: Dickerson (1979)

The fluoride concentration obtained in Example 2.6 is far above recommended drinking water levels of 1.8 mg/L. Fluoride concentrations of approximately 1 mg/L in drinking water help prevent dental cavities in children, but discoloration of teeth, called *mottling*, is relatively common when concentrations exceed 2.0 mg/L.

Solubility of Gases in Water

When air comes in contact with water, some of it dissolves into the water. Different constituents of air dissolve to different degrees and in amounts that vary with temperature and water purity. The behavior of gases in contact with water was reported by W. Henry in England in 1903, and the resulting relationship is known as *Henry's law*:

$$X = K_H P_g \qquad (2.22)$$

where X = mole fraction of the gas dissolved in liquid
  K_H = Henry's law constant (atm^{-1})
  P_g = the partial pressure of the gas in air (atm)

The quantity P_g is the partial pressure of the gas in air which is simply its volumetric concentration times the air pressure. For example, molecular oxygen makes up about 21 percent of the atmosphere, so P_g would be 0.21 times the atmospheric pressure. The units suggested for P_g are atmospheres (atm) where 1 atm corresponds to a barometric pressure of 760 mm of mercury or 101 325 pascals (Pa), and 1 Pa equals one newton per square meter.

When a number of gases are dissolved in pure water, the mole fraction of any one of them can be written as

$$X_1 = \frac{[gas_1]}{[H_2O] + [gas_1] + [gas_2] + \cdots} \cong \frac{[gas_1]}{[H_2O]} \qquad (2.23)$$

where $[gas_i]$ = the concentration of dissolved gas i (mol/L)

Since $[H_2O]$ = (1000 g/L)/(18 g/mol) = 55.56 mol/L, we can rewrite (2.22) in the following more convenient form:

$$[gas] = 55.56 K_H P_g \qquad (2.24)$$

Each gas–liquid system has its own value for Henry's coefficient. The coefficient itself varies both with temperature (solubility decreases as temperature increases) and with concentration of other dissolved gases and solids (the solubility decreases as other dissolved material in the liquid increases). It should be noted that Henry's law is expressed in various ways with different units for the coefficient depending on the method of expression. The user must be careful, then, to check the units given for Henry's constant before applying the law. Table 2.3 gives some values of K_H for CO_2 and O_2, two of the gases that we will be most concerned with.

TABLE 2.3   HENRY'S LAW
COEFFICIENTS FOR CO_2 AND O_2 (atm^{-1})

T (°C)	$K_{H_{CO_2}}$	$K_{H_{O_2}}$
0	0.001374	0.0000391
5	0.001137	0.0000330
10	0.000967	0.0000303
15	0.000823	0.0000271
20	0.000701	0.0000244
25	0.000611	0.0000222

Source: Tchobanoglous and Schroeder
(1985).

Another factor that must sometimes be accounted for when computing P_g is the decrease in air pressure that occurs as altitude increases. One estimate for atmospheric pressure as a function of altitude is the following (Thomann and Mueller, 1987):

$$P = P_0 - 1.15 \times 10^{-4} H \tag{2.25}$$

where P = atmospheric pressure at altitude H (atm)

H = altitude (m)

P_0 = atmospheric pressure at sea level (atm)

Example 2.7   Solubility of Oxygen in Water

By volume, the concentration of oxygen in air is about 21 percent. Find the equilibrium concentration of O_2 in water (in mol/L and mg/L) at 25 °C and 1 atm of pressure. Recalculate it for Denver at an altitude of 1525 m.

Solution   Air is 21 percent oxygen, so its partial pressure at 1 atm is

$$P_g = 0.21 \times 1 \text{ atm} = 0.21 \text{ atm}$$

From Table 2.3, at 25 °C, $K_{H_{O_2}} = 2.22 \times 10^{-5}$/atm; so, from (2.24),

$$[O_2] = 55.56 \, K_H P_g = 55.56 \text{ mol/L} \times 2.22 \times 10^{-5}/\text{atm} \times 0.21 \text{ atm}$$

$$= 2.59 \times 10^{-4} \text{ mol/L}$$

$$= 2.59 \times 10^{-4} \text{ mol/L} \times 32 \text{ g/mol} \times 10^3 \text{ mg/g} = 8.28 \text{ mg/L}$$

In Denver, at 1525 m, atmospheric pressure can be estimated using (2.25):

$$P = P_0 - 1.15 \times 10^{-4} H = 1 - 1.15 \times 10^{-4} \times 1525 = 0.825 \text{ atm}$$

so the partial pressure of O_2 at that altitude would be

$$P_g = 0.825 \text{ atm} \times 0.21 = 0.173 \text{ atm}$$

and, using (2.24),

$$[O_2] = 55.56 \text{ mol/L} \times 2.22 \times 10^{-5}/\text{atm} \times 0.173 \text{ atm} = 2.13 \times 10^{-4} \text{ mol/L}$$

$$= 2.13 \times 10^{-4} \text{ mol/L} \times 32 \text{ g/mol} \times 10^3 \text{ mg/g} = 6.84 \text{ mg/L}$$

It should be emphasized that calculations based on Henry's law provide equilibrium concentrations of dissolved gases, or, as they are frequently called, *saturation* values. It is often the case that actual values differ considerably from those at equilibrium. There may be more than the saturation value of a dissolved gas, as when photosynthesis by plants pumps air into the water at a faster rate than it can leave from the air/water interface. It is more common for dissolved gases to be less than the saturation value, as occurs when bacteria decompose large quantities of waste, drawing oxygen from the water (possibly leading to an oxygen deficiency that can kill fish). In either case, when an excess or a deficiency of a dissolved gas occurs pressures act to try to bring the amount dissolved to the saturation level.

Later, in Chapter 4, the importance of saturated values of dissolved oxygen will be discussed and a convenient table of values (Table 4.8) will be presented.

The Carbonate System

The carbonate system about to be described is the most important acid–base system in natural waters because it controls pH. It is composed of the following chemical species:

Aqueous carbon dioxide	$CO_{2\,(aq)}$
Carbonic acid	H_2CO_3
Bicarbonate ion	HCO_3^-
Carbonate ion	CO_3^{2-}

Aqueous CO_2 is formed when atmospheric CO_2 dissolves in water; we can find its concentration in fresh water using Henry's law (2.22):

$$[CO_{2\,(aq)}] = 55.56\,K_H P_{CO_2} \qquad (2.26)$$

where the concentration is in mol/L and P_{CO_2} is the partial pressure of gaseous CO_2 in the atmosphere (about 350 ppm or 350×10^{-6} atm). Aqueous CO_2 then forms carbonic acid (H_2CO_3) which, in turn, ionizes to form hydrogen ions (H^+) and bicarbonate (HCO_3^-):

$$CO_{2\,(aq)} + H_2O \rightleftharpoons H_2CO_3 \rightleftharpoons H^+ + HCO_3^- \qquad (2.27)$$

The bicarbonate (HCO_3^-) ionizes to form more hydrogen ion (H^+) and carbonate (CO_3^{2-}):

$$HCO_3^- \rightleftharpoons H^+ + CO_3^{2-} \qquad (2.28)$$

In addition, if a calcium carbonate solid, such as limestone ($CaCO_3$), is either present initially or forms by reaction between the CO_3^{2-} and dissolved Ca^{2+}, then the solubility reaction for $CaCO_3$ (s) applies in the system:

$$CaCO_3 \rightleftharpoons Ca^{2+} + CO_3^{2-} \qquad (2.29)$$

If sufficient time is allowed for the system to reach equilibrium, then the equilibrium constants for reactions (2.27–2.29) can be used to analyze the system. Reaction (2.27) results in

$$\frac{[H^+][HCO_3^-]}{[CO_{2\,(aq)}]} = K_1 = 4.47 \times 10^{-7} \text{ mol/L} \qquad (2.30)$$

and (2.28) yields

$$\frac{[H^+][CO_3^{2-}]}{[HCO_3^-]} = K_2 = 4.68 \times 10^{-11} \qquad (2.31)$$

while (2.29) is governed by

$$[Ca^{2+}][CO_3^{2-}] = K_{sp} = 4.57 \times 10^{-9} \qquad (2.32)$$

The values of K_1, K_2, and K_{sp} are temperature dependent. The values given in (2.30)–(2.32) are at 25 °C.

There are four possible conditions that determine the behavior of the carbonate system. There may or may not be a solid source of carbonate as indicated in (2.29) and the solution may or may not be exposed to the atmosphere. If it is exposed to the atmosphere the system is said to be *open*; if it is not, the system is *closed*. We will analyze the common situation of an open system not in contact with a carbonate solid, such as occurs when rainwater passes through the atmosphere absorbing CO_2 and forming carbonic acid.

Before proceeding to that calculation, it is useful to compare the relative concentrations of carbonate and bicarbonate as a function of pH. Dividing (2.31) by $[H^+]$ gives

$$\frac{[CO_3^{2-}]}{[HCO_3^-]} = \frac{K_2}{[H^+]} = \frac{K_2}{10^{-pH}}$$

and then incorporating the value of K_2 given in (2.31) yields

$$\frac{[CO_3^{2-}]}{[HCO_3^-]} = 4.68 \times 10^{(pH-11)} \qquad (2.33)$$

Equation 2.33 indicates that unless pH is extremely high, the carbonate concentration is usually negligible compared to the bicarbonate concentration, a fact that we will find useful in the following example where we calculate the pH of pristine rainwater.

Example 2.8   The pH of Natural Rainwater

Estimate the pH of natural rainwater assuming that the only substance affecting it is CO_2 that is absorbed from the atmosphere. Assume that the concentration of CO_2 is 350 ppm, and the temperature and pressure are 25 °C and 1 atm.

Solution   We have a number of equations to work with that are based on equilibrium constants, but there is another relationship that needs to be introduced, based on charge neutrality. The rainwater is assumed to have started without any electrical charge, and no net charge is added to it by the absorption of carbon dioxide. To have this neutrality maintained, the total positive charge contributed by the hydrogen ions (H^+) must equal the total negative charge of the bicarbonate (HCO_3^-), carbonate (CO_3^{2-}), and hydroxyl (OH^-) ions:

$$[H^+] = [HCO_3^-] + 2[CO_3^{2-}] + [OH^-] \qquad (2.34)$$

Notice that since each carbonate ion (CO_3^{2-}) has two negative charges, its charge contribution is twice as much per mole; hence, its coefficient is 2.

Knowing that rainfall is likely to be slightly acidic will let us simplify (2.34). Equation 2.33 indicates that for an acidic solution, $[CO_3^{2-}] \ll [HCO_3^-]$; so (2.34) becomes

$$[H^+] \approx [HCO_3^-] + [OH^-] \qquad (2.35)$$

Another equation that must be satisfied is (2.10):

$$[H^+][OH^-] = K_w = 1 \times 10^{-14} \qquad (2.10)$$

Substituting (2.10) into (2.35) gives

$$[H^+] = [HCO_3^-] + \frac{10^{-14}}{[H^+]} \qquad (2.36)$$

Putting $[HCO_3^-]$ from (2.30) into (2.36) yields

$$[H^+] = \frac{K_1[CO_{2(aq)}] + 10^{-14}}{[H^+]} \qquad (2.37)$$

So,

$$[H^+]^2 = K_1[CO_{2(aq)}] + 10^{-14} \qquad (2.38)$$

We can find the saturation value of dissolved CO_2 from (2.26):

$$
\begin{aligned}
[CO_{2(aq)}] &= 55.56 K_H P_{CO_2} \\
&= 55.56 \text{ mol/L} \times 0.000611/\text{atm} \times 350 \times 10^{-6} \text{ atm} \qquad (2.26) \\
&= 1.18 \times 10^{-5} \text{ mol/L}
\end{aligned}
$$

where K_H was taken from Table 2.3 and P_{CO_2} is given as 350 ppm or 350×10^{-6} atm. Equation 2.38 now gives us

$$[H^+]^2 = (4.47 \times 10^{-7} \times 1.18 \times 10^{-5} + 10^{-14})(\text{mol/L})^2$$

or

$$[H^+] = 2.30 \times 10^{-6} \text{ mol/L}$$

$$pH = -\log [H^+] = 5.64$$

Example 2.8 indicates that the pH of pristine rainwater is not the same as the pH of pure water due to the presence of carbon dioxide in the atmosphere. There are other naturally occurring substances in the atmosphere that also affect pH. Sulfur dioxide (SO_2) lowers pH, while ammonia and alkaline dust raise it. When all such natural substances are accounted for, rainfall is likely to have a pH value somewhere between 5 and 6 before it is influenced by human activities. As a result, some define acid rain caused by human activities to be rainfall with a pH of 5 or below while others prefer to define it in terms of the carbonate calculation given above, that is, as precipitation with a pH less than 5.6. Acid deposition is discussed more fully in Chapters 4 and 7.

2.4 ORGANIC CHEMISTRY

The term *organic chemistry* has come to mean the chemistry of the compounds of carbon. This term is broader than the term *biochemistry* which can be described as the chemistry of life. The need for a distinction between the two began with a discovery by Fredrich Wöhler in 1828. He accidentally converted an inorganic compound, ammonium cyanate, into urea, which until that time had been thought to be strictly associated with living things. That discovery demolished the *vital-force theory* which held that organic and inorganic compounds were distinguished by some sort of "life force." By this definition DDT is just as organic as yogurt.

The science of organic chemistry is incredibly complex and varied. There are literally millions of different organic compounds known today and 100 000 or so of these are products of synthetic chemistry, unknown in nature. About all that can be done here in a few pages is to provide the barest introduction to the origins of some of these names, so that they will appear a bit less alien when we encounter them in the future.

One way to visualize the bonding of atoms is with electron-dot formulas which represent valence electrons in the outermost orbitals. These diagrams were developed by G. N. Lewis, in 1916, and are referred to as *Lewis structures*. According to Lewis' theory of covalence, atoms form bonds by losing, gaining, or sharing enough electrons to achieve the outer electronic configuration of a noble gas. For hydrogen that means bonding with two shared electrons; for many other elements, including carbon, it means achieving a pattern with eight outermost electrons. For example, the following Lewis structure for the compound butane (C_4H_{10}) clearly shows each hydrogen atom sharing a pair of electrons and each carbon atom sharing four pairs of electrons:

n-Butane

A common way to simplify Lewis structures is to use a straight-line bond to represent pairs of shared electrons. Further simplification can sometimes be achieved by lumping commonly occurring subunits, such as $-CH_2-$ and CH_3-, as shown below.

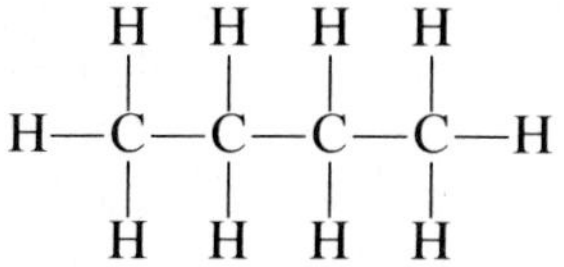

$$
\begin{array}{cccc}
H & H & H & H \\
| & | & | & | \\
H-C- & C- & C- & C-H \\
| & | & | & | \\
H & H & H & H
\end{array}
\qquad CH_3-CH_2-CH_2-CH_3
$$

n-Butane

Butane occurs in another form known as isobutane. Isobutane has the same molecular formula, C_4H_{10}, but has a different structural formula:

$$
\begin{array}{c}
H \\
| \\
H-C-H \\
\quad \\
H \quad\quad H \\
| \quad\quad | \\
H-C-C-C-H \\
| \quad | \quad | \\
H \quad H \quad H
\end{array}
$$

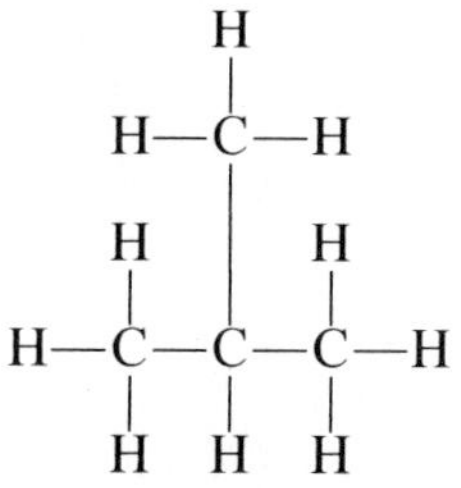

Isobutane

Compounds having the same molecular formula but different structural formulas are known as structural *isomers*. Isomers may have very different chemical and physical properties. As the number of carbon atoms per molecule increases, the number of possible isomers increases dramatically. Decane ($C_{10}H_{22}$), for example, has 75 possible isomers.

Butane and decane are examples of an important class of compounds called *hydrocarbons*, compounds containing only H and C atoms. Hydrocarbons may have linear structures (e.g., *n*-butane), branch structures (e.g., isobutane), or ring structures. The compound benzene (C_6H_6) is one such ring structure:

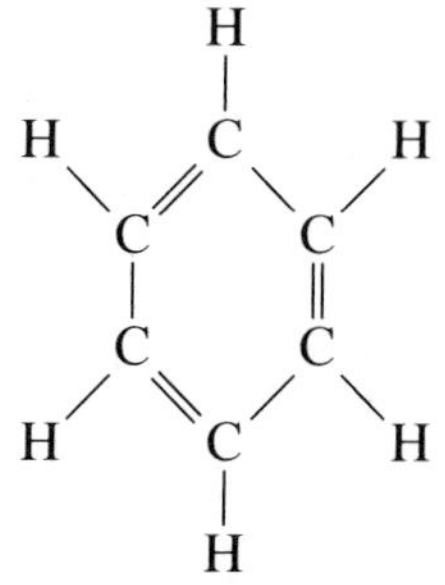

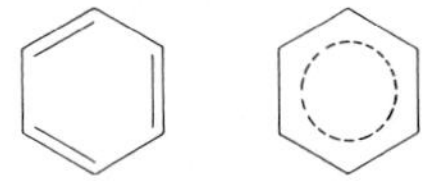

Simplified diagrams

Benzene

In the figure for benzene, notice each carbon atom has one double bond with a neighboring carbon. A double bond means four electrons are shared in the bond. The benzene ring is usually simplified by leaving off the hydrogen atoms, as shown.

Hydrocarbons in which each carbon atom forms four single bonds to other atoms are called *saturated hydrocarbons*, *paraffins*, or *alkanes*. Alkanes form a series beginning with methane (CH_4), ethane (C_2H_6), propane (C_3H_8), and butane (C_4H_{10}), which are all gases. The series continues with pentane (C_5H_{12}) and on up through ($C_{20}H_{42}$), which are liquids and include gasoline and diesel fuel. Beyond that the alkanes are waxy solids. Notice that each of these compounds ends in *-ane*. When one of the hydrogens is removed from an alkane, the result is called a *radical*, and the suffix becomes *-yl*. Thus, for example, removing one hydrogen from methane (CH_4) produces the methyl radical (CH_3—). Table 2.4 lists some of the methane-series radicals.

Hydrocarbons appear as building blocks in a great number of chemicals of environmental importance. By substituting other atoms or groups of atoms for some of the hydrogens, new compounds are formed. To get a feel for the naming of these compounds consider the replacement of some of the hydrogen atoms with chlorine in an ethane molecule. Each carbon atom is given a number and the corresponding hydrogen replacements are identified by that number. For example, 1,1,1-trichloroethane, better known as TCA, has three chlorines all attached to the same carbon, while 1,2-dichloroethane (1,2-DCA) has one chlorine attached to each of the two carbon atoms. Similarly, 1,1,2-trichloro-1,2,2-trifluoroethane has chlorine and fluorine atoms attached to each carbon:

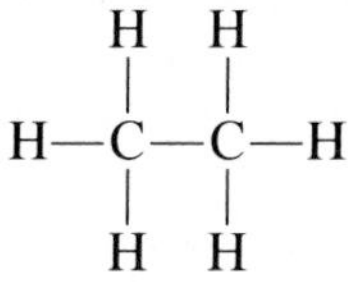

Ethane

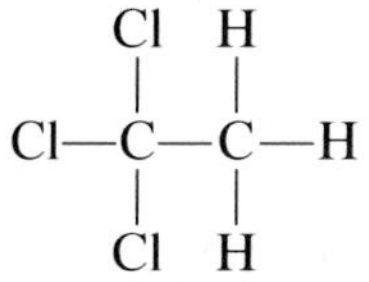

1,1,1-Trichloroethane (TCA)

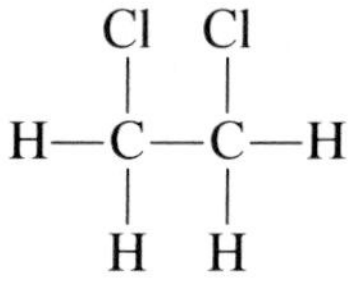

1,2-Dichloroethane (1,2-DCA)

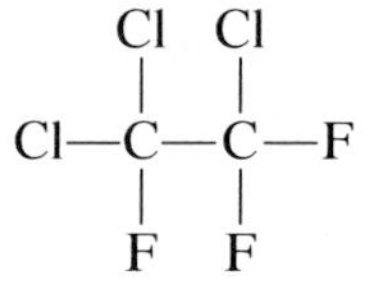

1,1,2-Trichloro-1,2,2-trifluoroethane (Freon 113)

TABLE 2.4 NAMES OF METHANE-SERIES RADICALS

Parent compound	Formula	Radical	Formula
Methane	CH_4	Methyl	CH_3—
Ethane	C_2H_6	Ethyl	C_2H_5—
Propane	C_3H_8	*n*-Propyl	C_3H_7—
Propane	C_3H_8	Isopropyl	$(CH_3)_2CH$—
n-Butane	C_4H_{10}	*n*-Butyl	C_4H_9—

Source: Sawyer and McCarty (1978).

Other names for TCA include methyl chloroform and methyltrichloromethane, names that remind us that it contains the methyl radical (CH_3—). TCA is commonly used in the electronics industry for flux cleaning and degreasing operations.

Example 2.9   Isomers of Trichloropropane

How many different isomers are there of trichloropropane, and how would they be numbered?

Solution   Propane has three carbons. The three chlorine atoms can be (1) all attached to an end carbon atom, (2) two attached to an end and one in the middle, (3) two attached to an end and one on the other end, (4) one on the end and two in the middle, or (5) one on each carbon atom. They would be numbered 1,1,1; 1,1,2; 1,1,3; 1,2,2; 1,2,3, respectively.

Most of the examples so far have been of saturated hydrocarbons. Unsaturated compounds contain at least two carbon atoms joined by more than one covalent bond. When there are double bonds the chemical name ends in *-ene*. An example compound is trichloroethylene, which is more commonly known as TCE, with the following structure:

$$\begin{array}{cc} Cl & Cl \\ | & | \\ Cl—C{=}C—H \end{array}$$

Trichloroethylene (TCE)

TCE is a solvent that was widely used until it was shown to cause cancer in animals. It has been replaced in many applications by TCA. Both TCE and TCA have been used extensively in the electronics industry and they are now quite commonly found in the groundwater beneath such facilities.

Molecules in which the hexagonal benzene ring occur are called *aromatic compounds*. This category includes DDT, PCBs, and the defoliants used in Vietnam, 2,4,5-T and 2,4-D. When a benzene ring attaches to another molecule, the C_6H_5— that remains of the benzene ring is called a *phenyl* group and so the name *phenyl* frequently appears in the compound name. Another physical characteristic of these compounds that is described by the nomenclature is the location of the attachments to the benzene ring. The terms *ortho, meta,* or *para,* or a numerical designation, specify the relative locations of the attachments, as suggested in Figure 2.3.

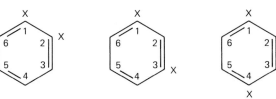

1, 2- or *ortho*   1, 3- or *meta*   1, 4- or *para*

Figure 2.3   Designating the attachment points for phenyl.

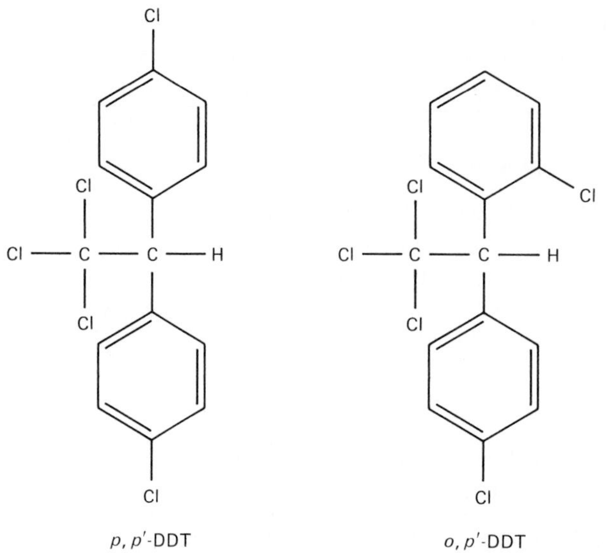

p,p'-DDT   o,p'-DDT

Figure 2.4 Two isomers of dichloro-diphenyltrichloroethane (DDT).

Figure 2.4 shows examples of the *o, m, p* nomenclature for two isomers of dichlorodiphenyltrichloroethane (DDT). Hopefully, this 30-letter name is somewhat less intimidating than it was before you read this section. Reading the name from right to left, the ethane on the end tells us it starts as a simple 2-carbon ethane molecule; the trichloro refers to the three chlorines around one of the carbons; the diphenyl tells us there are two phenyl groups; and the dichloro refers to the chlorines on the phenyls. *p,p'*-DDT is an isomer with the chlorine and carbon attachments to each benzene located in the para position; the *o,p'*-DDT has one chlorine in the ortho position and one in the para position. Actually, dichlorodiphenyltrichloroethane as a name is insufficient to unambiguously describe DDT. An even more informative name is 1,1,1-trichloro-2-2-bis(*p*-chlorophenyl)ethane, where *-bis* means "taken twice."

In Figure 2.3, the corners of a phenyl group are numbered and that too provides a way to help describe the structure of a molecule. The herbicides 2,4-dichlorophenoxyacetic acid (2,4-D), and 2,4,5-trichlorophenoxyacetic acid (2,4,5-T) shown in Figure 2.5 provide examples of this numerical designation. Though

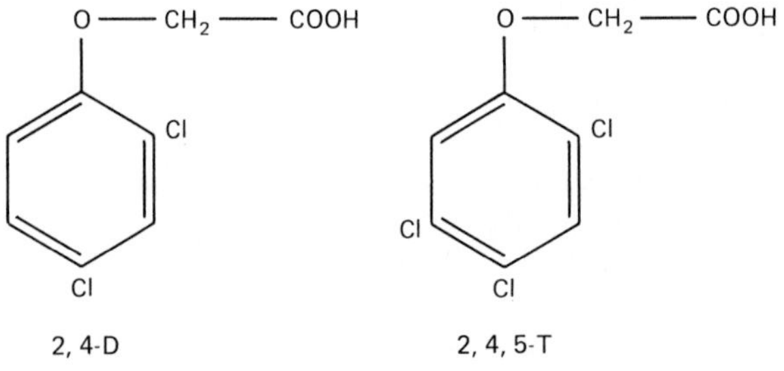

2, 4-D   2, 4, 5-T

Figure 2.5 2,4-Dichlorophenoxyacetic acid (2,4-D) and 2,4,5-trichlorophenoxyacetic acid (2,4,5-T).

we have not gone into it here, the —COOH group is characteristic of organic acids and the single —CH_2— between it and the oxygen is an indication that this is acetic acid.

Providing unambiguous names for complex organic molecules is extremely difficult and what has been provided here is an almost trivial introduction. Even these structural formulas, which convey much more information than a name like DDT, are still highly simplified representations, and there are yet higher levels of descriptions based on quantum mechanics.

2.5 NUCLEAR CHEMISTRY

As a simple but adequate model, we may consider an atom to consist of a dense nucleus containing a number of uncharged neutrons and positively charged protons, surrounded by a cloud of negatively charged electrons. The number of protons is the *atomic number,* and that is what defines a particular element. The sum of the number of protons and neutrons is the *mass number.* Elements with the same atomic number but differing mass numbers are called *isotopes.* The usual way to identify a given isotope is by giving its chemical symbol with the mass number written at the upper left and the atomic number at the lower left. For example, the two most important isotopes of uranium (which has 92 protons) are

$$^{235}_{92}U \qquad \text{and} \qquad ^{238}_{92}U$$

$$\text{Uranium-235} \qquad\qquad \text{Uranium-238}$$

When referring to a particular element, it is common to drop the atomic number subscript since it adds little information to the chemical symbol. Thus, U-238 is a common way to describe that isotope. When discussing changes to an element as it decays, the full notation is helpful.

Some atomic nuclei are unstable, that is, *radioactive,* and during the spontaneous changes that take place within the nucleus, various forms of *radiation* are emitted. While all elements having more than 83 protons are naturally radioactive, it is possible to artificially produce unstable isotopes, or *radionuclides,* of virtually every element in the periodic table. Radioactivity is of interest in the study of environmental engineering both because it poses a hazard when organisms are exposed to it, and because it is useful as a tracer to help measure the flow of materials in the environment.

There are three kinds of radiation: *alpha, beta,* and *gamma* radiation. *Alpha* (α) radiation consists of the emission of two protons and two neutrons from the nucleus. When an unstable nucleus emits an α particle, its atomic number decreases by 2 units and its mass number decreases by 4 units. The following example shows the decay of plutonium into uranium:

$$^{239}_{94}Pu \longrightarrow \ ^{235}_{92}U + ^{4}_{2}\alpha + \gamma$$

In this reaction, not only is an α particle ejected from the nucleus but electromagnetic radiation is emitted as well, indicated by the γ.

As an α particle passes through an object, its energy is gradually dissipated through interaction with other atoms. Its positive charge attracts electrons in its path, raising their energy levels and possibly removing them completely from their nuclei (*ionization*). Alpha particles are relatively massive and easy to stop. Our skin is sufficient protection for sources that are external to the body, but taken internally, such as by inhalation, α particles can be extremely dangerous.

Beta (β) particles are electrons that are emitted from an unstable nucleus as a result of the spontaneous transformation of a neutron into a proton plus electron. As a result the mass number remains unchanged while the atomic number increases by one. A γ ray may or may not accompany the transformation. The following example shows the decay of strontium-90 into yttrium:

$$\ce{^{90}_{38}Sr} \longrightarrow \ce{^{90}_{39}Y} + \beta$$

As negatively charged β particles pass orbital electrons, the Coulomb repulsive force between the two can raise the energy level of orbiting electrons, possibly kicking them free of their corresponding atoms. While such ionizations occur less frequently than with α particles, β penetration is much deeper. Alpha particles travel less than 100 μm into tissue, while β particles may travel several centimeters. They can be stopped with a modest amount of shielding, however. For example, to stop a β particle about one centimeter thickness of aluminum is sufficient.

Gamma (γ) rays have no charge or mass, being simply a form of electromagnetic radiation that travels at the speed of light. As such, they can be thought of either as waves with characteristic wavelengths or as photon particles. Gamma rays have very short wavelengths in the range of 10^{-3}–10^{-7} μm. Having such short wavelengths means they are capable of causing ionizations. As such they are biologically damaging. These rays are very difficult to contain and may require several centimeters of lead to provide adequate shielding.

All of these forms of radiation are dangerous to living things. The electron excitations and ionizations that are caused by such radiation cause molecules to become unstable, resulting in the breakage of chemical bonds and other molecular damage. The chain of chemical reactions that follows will result in the formation of new molecules that did not exist before the irradiation. Then, on a much slower time scale, the organism responds to the damage; so slow, in fact, that it may be years before the final effects become visible. Low-level exposures can cause *somatic* and/or *genetic* damage. Somatic effects may be observed in the organism that has been exposed and include higher risk of cancer, leukemia, sterility, cataracts, and a reduction in lifespan. Genetic damage, by increasing the mutation rate in chromosomes and genes, affects future generations.

As will be discussed in Chapter 7, one naturally occurring radiation exposure that is thought to be a major cause of lung cancer is caused by the inhalation of

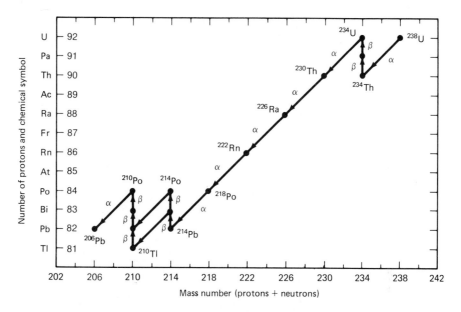

Figure 2.6   The uranium-238 decay chain.

radon (Rn-222) and radon decay products (various isotopes of polonium). Radon is a chemically inert gas and an alpha emitter that seeps out of the soil and can accumulate in houses. It is an intermediate product in a naturally occurring decay chain that starts with uranium-238 and ends with a stable isotope of lead. Figure 2.6 diagrams this radioactive series. There are three other similar series that can be drawn: One begins with uranium-235 and ends with lead-207, another begins with thorium-232 and ends with lead-208, and the other begins with an artificial isotope plutonium-241 and ends with bismuth-209.

An important parameter that characterizes a given radioactive isotope is its half-life, which is the time required for half of the atoms to spontaneously transform, or decay, into other elements. For example, if we start with 100 g of an isotope which has a half-life of 1 year, we would find 50 g remaining after 1 year, 25 g after 2 years, 12.5 g after 3 years, and so on. While the half-life for a given isotope is constant, half-lives of radionuclides in general vary from fractions of a second to billions of years. Half-lives for the radon portion of the ^{238}U decay chain are given in Table 2.5.

There are a number of commonly used radiation units, which unfortunately can easily be confused. The *curie* (Ci) is the basic unit of decay rate; 1 Ci corresponds to the disintegration of 3.7×10^{10} atoms per second, which is approximately the decay rate of 1 g of radium. A similar unit is the *becquerel* (Bq), where 1 Ci = 3.7×10^{10} Bq. While the curie is a measure of the rate at which radiation is emitted by a source, it tells us nothing about the radiation dose that is actually absorbed by an object.

TABLE 2.5 HALF-LIFE OF RADON DECAY CHAIN

Isotope	Emission	Half-life
^{222}Rn	α	3.8 day
^{218}Po	α	3.05 min
^{214}Pb	β	26.8 min
^{214}Bi	β	19.7 min
^{214}Po	α	160 μs
^{210}Pb	β	19.4 yr
^{210}Tl	β	1.32 min
^{210}Bi	β	4.85 day
^{210}Po	α	1.38 day
^{206}Pb		Stable

The *roentgen* (R) is defined in terms of the number of ionizations produced in a given amount of air by x or γ rays. Of more interest is the amount of energy actually absorbed by tissue, be it bone, fat, muscle, or whatever. The *rad* (radiation *a*bsorbed *d*ose) corresponds to an absorption of 100 ergs of energy per gram of any substance. The rad has the further advantage that it may be used for any form of radiation, α, β, γ, or x. For water and soft tissue, the rad and roentgen are approximately equal.

Another unit, the *rem* (roentgen *e*quivalent *m*an) has been introduced to take into account the different biological effects that various forms of radiation have on organisms. Thus, for example, if a 10-rad dose of β particles produces the same biological effect as a 1-rad dose of α particles, both doses would have the same value when expressed in rems. This unit is rather loosely defined, making it difficult to convert from rads to rems. However, in many situations involving x rays, γ rays, and β radiation, rads and rems are approximately the same. The rem and millirem are units most often used when dealing with human radiation exposure.

Nuclear Fission

Nuclear reactors obtain their energy from the heat that is produced when uranium atoms *fission*. In a fission reaction, uranium-235 captures a neutron and becomes unstable uranium-236, which splits apart, discharging two *fission fragments,* two or three neutrons, and γ rays (Figure 2.7). Most of the energy released is in the form of kinetic energy in the fission fragments. That energy is used to heat water, producing steam which powers a turbine and generator in much the same way as was diagrammed in Figure 1.11.

The fission fragments produced are always radioactive and concerns for their proper disposal have created much of the controversy surrounding nuclear reactors. Typical fission fragments include cesium-137, which concentrates in

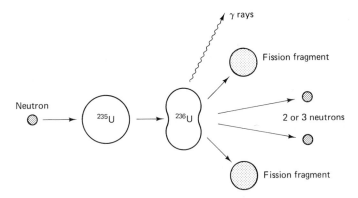

Figure 2.7   The fissioning of uranium-235 creates two radioactive fission fragments, neutrons, and γ rays (Masters, 1974).

muscles and has a half-life of 30 years; strontium-90, which concentrates in bone and has a half-life of 28 years; and iodine-131, which concentrates in the thyroid gland and has a half-life of 8.1 days. The half-lives of fission fragments tend to be no longer than a few tens of years so that after a period of several hundred years their radioactivity will decline to relatively insignificant levels.

Reactor wastes also include some radionuclides with very long half lives. Of major concern is plutonium, which has a half-life of 24 390 years. Only a few percent of the uranium atoms in reactor fuel is the fissile isotope ^{235}U, while essentially all of the rest is ^{238}U, which does not fission. Uranium-238 can, however, capture a neutron and be transformed into plutonium as the following reactions suggest:

$$^{238}_{92}\text{U} + \text{n} \longrightarrow {}^{239}_{92}\text{U} \xrightarrow{\beta} {}^{239}_{93}\text{Np} \xrightarrow{\beta} {}^{239}_{94}\text{Pu} \qquad (2.39)$$

This plutonium, along with several other long-lived radionucides, makes nuclear wastes dangerously radioactive for tens of thousands of years, which greatly increases the difficulty of providing safe disposal. Removing the plutonium from nuclear wastes before disposal has been proposed as a way to shorten the decay period but that introduces another problem. Plutonium not only is radioactive and highly toxic, it is also the critical ingredient in the production of atomic bombs. A single nuclear reactor produces enough plutonium each year to make dozens of small atomic bombs and some have argued that if the plutonium is separated from nuclear wastes the risk of illegal diversions for such weapons would be increased.

As Figure 2.8 suggests, the power reactor itself is only one of several sources in the complete nuclear fuel cycle that creates waste disposal problems. Uranium ores must first be mined and the uranium extracted. Mine tailings typically contain toxic metals such as arsenic, cadmium, and mercury, as well as the radionuclides associated with the decay of ^{238}U shown in Figure 2.6. Only about

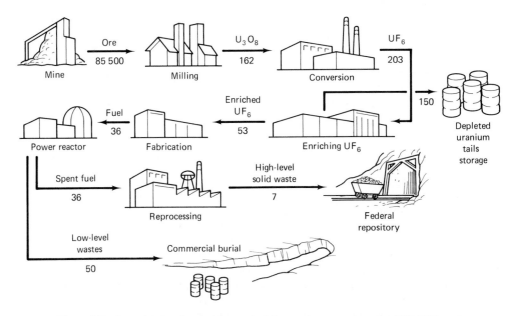

Figure 2.8 Annual tons of materials required for routine operation of a 1000-MW nuclear reactor. (*Source:* Henry/Heinke, *Environmental Science and Engineering,* © 1989. Reprinted by permission of Prentice Hall, Inc., Englewood Cliffs, New Jersey.)

0.72 percent of naturally occurring uranium is the desired isotope ^{235}U and an enrichment facility is needed to increase that concentration to 2 or 3 percent for use in the reactor. The enriched uranium is then fabricated into fuel pellets and shipped to a reactor. Highly radioactive wastes from the reactor are temporarily stored on site until they can be reprocessed and readied for ultimate disposal in a federal repository. Such a repository is not expected to become operational until the early part of the twenty-first century. Low level wastes can be disposed of in specially designed landfills to be described in Chapter 6.

Providing proper disposal of radioactive wastes at every stage in the nuclear fuel cycle is a challenging engineering task but there are many who argue it is well within our capabilities.

PROBLEMS

2.1. Consider the following reaction representing the combustion of propane.

$$C_3H_8 + O_2 \longrightarrow CO_2 + H_2O$$

   a. Balance the equation.
   b. How many moles of oxygen are required to burn one mole of propane?

c. How many grams of oxygen are required to burn 100 g of propane?

d. At standard temperature and pressure, what volume of oxygen would be required to burn 100 g of propane? If oxygen is 21 percent of air, what volume of air at STP would be required?

2.2. Trinitrotoluene (TNT), $C_7H_5N_3O_6$, combines explosively with oxygen to produce CO_2, water, and N_2. Write a balanced chemical equation for the reaction and calculate the grams of oxygen required for each 100 g of TNT.

2.3. What is the molarity of 10 g of salt (NaCl) dissolved in 1 L of water?

2.4. 86-proof whiskey is 43 percent ethyl alcohol (CH_3CH_2OH) by volume. If the density of ethyl alcohol is 0.79 kg/L, what is its molarity in whiskey?

2.5. The earth's ozone layer is under attack in part by chlorine released when ultraviolet radiation breaks apart certain chlorofluorocarbons (CFCs). Consider three CFCs known as F-11 (CCl_3F), F-12 (CCl_2F_2), and F-22 (CHF_2Cl).

a. Calculate the percent chlorine in each of these CFCs.

b. If F-22 could replace F-11, molecule for molecule, by what percentage would the mass of chlorine emissions be reduced?

2.6. Suppose total world energy consumption of fossil fuels, equal to 3×10^{20} J/year, were to be obtained entirely by combustion of petroleum with the approximate chemical formula C_2H_3. Petroleum has an energy content of about 43×10^6 J/kg.

a. Estimate the emissions of CO_2 per year.

b. What is the ratio of grams of C emitted per unit of energy for petroleum versus methane (use the results of Example 2.3)?

2.7. Hydrochloric acid (HCl) completely ionizes when dissolved in water. Calculate the pH of a solution containing 25 mg/L of HCl.

2.8. What is the pH of a solution containing 3×10^{-4} mg/L of OH^- (25 °C)?

2.9. Find the theoretical oxygen demand of a solution containing 200 mg/L of acetic acid, CH_3COOH.

2.10. Water is frequently disinfected with chlorine gas, forming hypochlorous acid (HOCl), which partially ionizes to hypoclorite and hydrogen ions as follows:

$$HOCl \rightleftarrows H^+ + OCl^- \quad \text{with equilibrium constant } K = 2.9 \times 10^{-8}$$

The amount of [HOCl], which is the desired disinfectant, depends on the pH. Find the fraction that is hypochlorous acid (that is, $[HOCl]/\{[HOCl] + [OCl^-]\}$) as a function of pH. What would the hypochlorous fraction be for pH 6, 8, and 10?

2.11. Hydrogen sulfide (H_2S) is an odorous gas that can be stripped from solution in a process similar to that described in Example 2.5 for ammonia. The reaction is

$$H_2S \rightleftarrows H^+ + HS^- \quad \text{with equilibrium constant } K = 0.86 \times 10^{-7}$$

Find the fraction of hydrogen sulfide in H_2S form at pH 6 and 8.

2.12. Solid aluminum phosphate ($AlPO_4$) is in equilibrium with its ions in solution:

$$AlPO_4 \rightleftarrows Al^{3+} + PO_4^{3-} \quad \text{with } K_{sp} = 10^{-22}$$

Find the equilibrium concentration of phosphate ions (in mg/L).

2.13. Calculate the equilibrium concentration of dissolved oxygen in water at 15 °C and 1 atm and again at 2000 m elevation.

2.14. Suppose the gas above the soda in a bottle of pop is pure CO_2 at a pressure of 2 atm. Calculate $[CO_2]$ at 25 °C.

2.15. Calculate the pH of the soda pop in Problem 2.14.

2.16. It has been estimated that the concentration of CO_2 in the atmosphere before the industrial revolution was about 275 ppm. If the accumulation of CO_2 in the atmosphere continues as many predict it will, then by the middle of the next century it will probably be around 600 ppm. Calculate the pH of rainwater at 25 °C (neglecting the effect of any other gases) in each of these times.

2.17. One strategy for dealing with the acidification of lakes is to periodically add lime ($CaCO_3$) to them. Calculate the pH of a lake at 15 °C that has more than enough lime in it to saturate the water with its ions Ca^{2+} and CO_3^{2-}. *Suggestions:* Begin with the carbonate system equations (2.26) and (2.30) through (2.33), then add a charge balance equation:

$$[H^+] + 2[Ca^{2+}] = [HCO_3^-] + 2[CO_3^{2-}] + [OH^-] \approx [HCO_3^-] + [OH^-]$$

(The above holds for pH values that are less than about 10. Is that a valid assumption?) Assume 350 ppm of CO_2 in the air.

By the way, this calculation is the same as would be made to estimate the pH of the oceans, which are saturated with $CaCO_3$.

2.18. Draw the Lewis structure and the more common line structure for ethylene C_2H_4.

2.19. From the names alone, write the structures for the following organics:
  a. Dichloromethane
  b. Trichloromethane (chloroform)
  c. 1,1-Dichloroethylene
  d. Trichlorofluoromethane (F-11)
  e. 1,1,2,2-Tetrachlorethane
  f. o-Dichlorobenzene
  g. Tetrachloroethene (PCE)
  h. Dichlorofluoromethane (F-21)

2.20. What values of a and b would complete each of the following (X and Y are not meant to be any particular elements)?

$$^{266}_{88}X \longrightarrow \alpha + {}^{a}_{b}Y$$
$$^{a}_{15}X \longrightarrow \beta + {}^{32}_{b}Y$$

2.21. The half-life of iodine-125 is about 60 days. If we were to start with 64 g of it, about how much would remain after 1 year?

REFERENCES

DICKERSON, R. E., H. B. GRAY, and G. P. HAIGHT, 1979, *Chemical Principles,* 3d ed., Benjamin/Cummings, Menlo Park, CA.

HENRY, J. G., and G. W. HEINKE, 1989, *Environmental Science and Engineering,* Prentice Hall, Englewood Cliffs, NJ.

MASTERS, G. M., 1974, *Introduction to Environmental Science and Technology,* Wiley, New York.

PEAVY, H. S., D. R. ROWE, and G. TCHOBANOGLOUS, 1985, *Environmental Engineering,* McGraw-Hill, New York.

SAWYER, C. N., and P. L. MCCARTY, 1978, *Chemistry for Environmental Engineering,* 3d ed., McGraw-Hill, New York, p. 32.

SNOEYINK, V. L., and D. JENKINS, 1980, *Water Chemistry,* Wiley, New York.

STUMM, W., and J. J. MORGAN, 1981, *Aquatic Chemistry, An Introduction Emphasizing Chemical Equilibria in Natural Waters,* 2d ed., Wiley, New York.

TCHOBANOGLOUS, G., and E. D. SCHROEDER, 1985, *Water Quality,* Addison Wesley, Reading, MA, p. 122.

THOMANN, R. V., and J. A. MUELLER, 1987, *Principles of Surface Water Quality and Control,* Harper & Row, New York, p. 270.

CHAPTER 3

Mathematics of Growth

It is very difficult to make an accurate prediction,
especially about the future.
Niels Bohr

3.1 INTRODUCTION

A sense of the future is an essential component in the mix of elements that should be influencing the environmental decisions we make today. In some circumstances, only a few factors may need to be considered and the time horizon may be relatively short. For example, a wastewater treatment plant operator may want to predict the growth rate of bacteria in a digester over a period of hours or days. The designer of the plant, on the other hand, probably needed to estimate the local population growth rate over the next decade or two to size the facility. At the other extreme, to make even the crudest estimate of future carbon emissions and their effect on global warming, scientists need to forecast population and economic growth rates, anticipated improvements in energy efficiency, the global fossil fuel resource base and the consumption rate of individual fuels, the rate of deforestation or reforestation that could be anticipated, and so on. And these estimates need to be made for periods of time that are often measured in hundreds of years.

As the above quote of Niels Bohr suggests, we cannot expect to make accurate predictions of the future, especially when the required time horizon may be extremely long. We can, however, often make simple estimates that are robust

66

enough that the insight they provide is most certainly valid. We can say, for example, with considerable certainty that world population growth at today's rates cannot continue for another hundred years. We can also use simple mathematical models to develop very useful "what if" scenarios: *If* population growth continues at a certain rate, and *if* energy demand is proportional to economic activity, and so forth, *then* the following would occur.

The purpose of this chapter is to develop some simple but powerful mathematical tools that can shed considerable light on the future of a number of environmental problems. We begin with what is probably the most useful and powerful mathematical function encountered in environmental studies, the *exponential function*. Other growth functions encountered often are also explored, including the *logistic* and the *Gaussian* functions. Applications that will be demonstrated using these functions include population growth, resource consumption, pollution accumulation, and radioactive decay.

3.2 EXPONENTIAL GROWTH

Exponential growth occurs in any situation where the increase in some quantity is proportional to the amount currently present. This type of growth is quite common, and the mathematics required to represent it is relatively simple, yet extremely important. We will approach this sort of growth first as discrete, year-by-year increases, and then in the more usual way as a continuous growth function.

Suppose something grows by a fixed percentage each year. For example, if we imagine our savings at a bank earning 5 percent each year, compounded once a year, then the amount of increase in savings over any given year is 5 percent of the amount available at the beginning of that year. If we start now with $1000, then at the end of one year we would have $1050 ($1000 + 0.05 \times 1000$); at the end of two years we would have $1102.50 ($1050 + 0.05 \times 1050$); and so on. We can represent this mathematically as follows:

$$N_0 = \text{initial amount}$$

$$N_t = \text{amount after } t \text{ years}$$

$$r = \text{growth rate (fraction per year)}$$

then

$$N_{t+1} = N_t + rN_t = N_t(1 + r)$$

For example, $N_1 = N_0(1 + r)$; $N_2 = N_1(1 + r) = N_0(1 + r)^2$; and, in general,

$$N_t = N_0(1 + r)^t \tag{3.1}$$

Example 3.1   U.S. Electricity Growth (I)

In 1986, the United States produced 2.4×10^{12} kWhr of electricity. The average annual growth rate of U.S. electricity demand in the 15 years after the 1973 oil embargo was about 2.2 percent. (For comparison, it had been about 7 percent per year for many decades before the embargo.) Estimate the electricity consumption in 2086 if the 2.2 percent per year growth rate were to remain constant.

Solution   From (3.1), we have

$$N_{100} = N_0(1 + r)^{100}$$

$$= 2.4 \times 10^{12} \times (1 + 0.022)^{100}$$

$$= 21.1 \times 10^{12} \text{ kWhr/yr}$$

Continuous Compounding

For most events of interest in the environment, it is usually assumed that the growth curve is a smooth, continuous function without the annual jumps that (3.1) is based on. In financial calculations this is referred to as continuous compounding. With the use of a little bit of calculus, the growth curve becomes the true exponential function that we will most often want to use.

One way to state the condition that leads to exponential growth is that the quantity grows in proportion to itself; that is, the rate of change of the quantity N is proportional to N. The proportionality constant r is called the rate of growth and has units of (time^{-1}).

$$\frac{dN}{dt} = rN \qquad (3.2)$$

This is essentially the same as (1.12) solved in Chapter 1. The solution is

$$N = N_0 e^{rt} \qquad (3.3)$$

which is plotted in Figure 3.1.

Example 3.2   U.S. Electricity Growth (II)

Suppose we repeat Example 3.1, but now consider the 2.2-percent growth rate to be continuously compounded. Starting with the 1986 electricity consumption of 2.4×10^{12} kWhr/year, what would it be in 100 years if the growth rate remains constant?

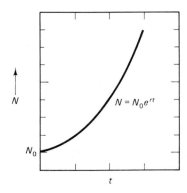

Figure 3.1   The exponential function.

Solution   Using (3.3)

$$N = N_0 e^{rt}$$

$$= 2.4 \times 10^{12} \times e^{0.022 \times 100}$$

$$= 21.6 \times 10^{12} \text{ kWhr/yr}$$

Example 3.1, in which increments were computed once each year, and Example 3.2, in which growth was continuously compounded, have produced nearly identical results. As either the period of time in question or the growth rate increases, the two approaches begin to diverge. At 12-percent growth, for example, the answers would differ by nearly a factor of two. In general, it is better to express growth rates as if they are continuously compounded so that (3.3) becomes the appropriate expression.

Doubling Time

Calculations involving exponential growth can sometimes be made without a calculator by taking advantage of the following unusual characteristic of the exponential function. A quantity that is growing exponentially requires a fixed amount of time to double in size, regardless of the starting point. That is, it takes the same amount of time to grow from N_0 to $2N_0$ as it does to grow from $2N_0$ to $4N_0$, and so on, as shown in Figure 3.2.

The *doubling time* (T_d) of a quantity that grows at a fixed exponential rate r is easily derived. From

$$N = N_0 e^{rt} \tag{3.3}$$

the doubling time can be found by setting $N = 2N_0$ at $t = T_d$:

$$2N_0 = N_0 e^{rT_d}$$

Notice how N_0 appears on both sides of the equation and can be canceled out, which is another way of saying the length of time required to double the quantity

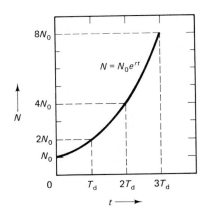

Figure 3.2   Illustrating the concept of doubling time.

does not depend on how much you start with. So, canceling N_0 and taking the natural log of both sides gives

$$\ln 2 = rT_d$$

or

$$T_d = \frac{\ln 2}{r} \cong \frac{0.693}{r} \tag{3.4}$$

If the growth rate r is expressed as a percentage instead of as a fraction, we get the following important result:

$$T_d \cong \frac{69.3}{r(\%)} \cong \frac{70}{r(\%)} \tag{3.5}$$

Equation 3.5 is well worth memorizing; that is, *the length of time required to double a quantity growing at r percent is about equal to 70 divided by r percent.* If your savings at the bank earns 7 percent interest, it will take about 10 years to double what you have in the account. If the population of a country grows continuously at 2 percent, then it will double in size in 35 years, and so on.

Example 3.3   Historical World Population Growth Rate

It took the world about 300 years to increase in population from 0.5 billion to 4.0 billion. If we assume exponential growth at a constant rate over that period of time, what would that growth rate be? Do it first with a calculator and (3.3), then with the rule of thumb suggested in (3.5).

Solution   Using (3.3), we have

$$N = N_0 e^{rt}$$

$$4.0 \times 10^9 = 0.5 \times 10^9 \, e^{r300}$$

Dividing through and then taking the natural log of both sides gives

$$r = \frac{\ln(4.0/0.5)}{300} = 0.00693 = 0.693 \text{ percent}$$

Using the rule-of-thumb approach given by Eq. 3.5, we can say that three doublings would be required to have the population grow from 0.5 billion to 4.0 billion. Three doublings in 300 years means each doubling took 100 years.

$$T_d = 100 \text{ yr} \cong \frac{70}{r(\%)}$$

so

$$r(\%) \cong \frac{70}{100} = 0.7 \text{ percent}$$

Our answers would have been exactly the same if the rule of thumb in (3.5) had not been rounded off.

TABLE 3.1 EXPONENTIAL GROWTH
FACTORS FOR VARIOUS NUMBERS
OF DOUBLING TIMES

Number of doublings (n)	Growth factor (2^n)
1	2
2	4
3	8
4	16
5	32
10	1024
20	$\approx 1.05 \times 10^6$
30	$\approx 1.07 \times 10^9$

Exponential growth is deceptively fast. As Table 3.1 suggests, quantities growing at only a few percent per year increase in size with incredible speed after just a few doubling times. For example, at the 1989 rate of world population growth ($r = 1.8$ percent), the doubling time was about 39 years. *If that rate were to continue* for just 5 doubling times, or about 200 years, world population would increase by a factor of 32, from 5.2 billion to 166 billion. In 20 doubling times, there would be over 5 million billion (quadrillion) people, or more than one person for each square foot of surface area of the earth. The absurdity of these figures simply points out the impossibility of the underlying assumption that exponential growth, even at this relatively low-sounding rate, could continue for such periods of time.

Half-Life

When the rate of *decrease* of a quantity is proportional to the amount present, exponential growth becomes exponential decay. We have already dealt with quantities that decay exponentially in Chapter 1, when nonconservative pollutants were discussed, and in Chapter 2, when radioactivity was introduced. Exponential decay can be described using either a *reaction rate coefficient* (K) or a *half-life* ($T_{1/2}$) and the two are easily related to each other. A reaction rate coefficient for an exponential decay plays the same role that r played for exponential growth.

Exponential decay can be expressed as

$$N = N_0 e^{-Kt} \tag{3.6}$$

where K is a reaction rate coefficient (time^{-1}), N_0 is an initial amount, and N is an amount at time t.

A plot of (3.6) is shown in Figure 3.3 along with an indication of *half-life*. To relate half-life to reaction rate, we set $N = N_0/2$ at $t = T_{1/2}$, giving

$$\frac{N_0}{2} = N_0 e^{-KT_{1/2}}$$

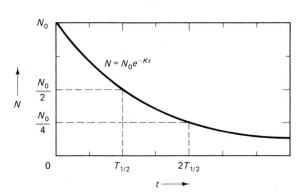

Figure 3.3   Exponential decay and half-life.

Canceling the N_0 term and taking the natural log of both sides gives us the desired expression:

$$T_{1/2} = \frac{\ln 2}{K} \cong \frac{0.693}{K} \tag{3.7}$$

Example 3.4   Radon Half-life

If we start with a 1.0-Ci radon-222 source, what would its activity be after 5 days?

Solution   Table 2.5 indicates that the half-life for radon is 3.8 days. Rearranging (3.7) to give a reaction rate coefficient results in

$$K = \frac{\ln 2}{T_{1/2}} = \frac{0.693}{3.8 \text{ day}} = 0.182/\text{day}$$

Using (3.6) gives

$$N = N_0 e^{-Kt} = 1 \text{ Ci} \times e^{-0.182/d \times 5d} = 0.40 \text{ Ci}$$

Disaggregated Growth Rates

Quite often the quantity that we want to model can be considered to be the product of a number of individual factors. For example, to estimate future automobile gasoline consumption we might *disaggregate* demand into the following four factors:

(gallons of gasoline) = (gallons/mile) × (miles/car) × (cars/person) × (population)

By considering demand this way, we would hope to increase the accuracy of our forecast by estimating separately the rates of change of automobile fuel economy, miles driven, cars per person, and population.

If we can express the quantity of interest as the product of individual factors

$$P = P_1 P_2 \cdots P_n \tag{3.8}$$

and if each factor is itself growing exponentially

$$P_i = p_i e^{r_i t}$$

then

$$P = (p_1 e^{r_1 t})(p_2 e^{r_2 t}) \cdots (p_n e^{r_n t})$$
$$= (p_1 p_2 \cdots p_n) e^{(r_1 + r_2 + \cdots + r_n)t}$$

which can be written as

$$P = P_p e^{rt}$$

where

$$P_p = (p_1 p_2 \cdots p_n) \tag{3.9}$$

and

$$r = r_1 + r_2 + \cdots + r_n \tag{3.10}$$

Equation 3.10 is a very simple, useful result. That is, if a quantity can be expressed as a product of factors, each growing exponentially, then *the total rate of growth is just the sum of the individual rates of growth*.

Example 3.5   Future U.S. Energy Demand

One way to disaggregate U.S. energy consumption is with the following product:

Energy consumption = (energy/GNP) × (GNP/person) × (population)

where GNP is the gross national product. Suppose we project per capita GNP to grow at 2.0 percent and population at 0.7 percent. Suppose we assume that through conservation efforts we expect energy required per dollar of GNP to *decrease* exponentially at the rate that it did between 1973 (the year of the oil embargo) and 1986. In 1973, (energy/GNP) was 27.1 kBtu/$ and in 1986 it was 20.1 kBtu/$ (in constant 1982 dollars). Total U.S. energy demand in 1986 was 74.0 quads (quadrillion Btu). If the above rates of change continue, what would energy demand be in the year 2000?

Solution   First, let us find the exponential rate of decrease of (energy/GNP) over the 13 years between 1973 and 1986. Rearranging (3.3) and solving for the growth rate gives

$$r = \frac{1}{t} \ln \frac{N}{N_0} = \frac{1}{13} \ln \frac{20.1}{27.1} = -0.022 = -2.2 \text{ percent}$$

Notice that the algebra automatically produces the negative sign. The overall energy growth rate is projected to be the sum of the three rates:

$$r = -2.2 \text{ percent} + 2.0 \text{ percent} + 0.7 \text{ percent} = 0.5 \text{ percent}$$

(For comparison, energy growth rates in the United States before the 1973 embargo were typically 3–4 percent per year, but during the ensuing 13 years it was 0 percent.) If we project out 14 years to the year 2000, (3.3) gives

$$\text{Energy}_{2000} = 74.0e^{0.005 \times 14} = 79.4 \text{ quads}$$

3.3 RESOURCE CONSUMPTION

To maintain human life on earth, we depend on a steady flow of energy and materials to meet our needs. Some of the energy we use is *renewable* (e.g., hydroelectric power, windpower, solar heating systems, and even firewood if the trees are replanted), but most comes from *nonrenewable* fossil fuels. The materials such as copper, steel, and concrete with which we build our infrastructure are nonrenewable as well. The future of our way of life is to a large extent dependent on the availability of abundant supplies of inexpensive, nonrenewable energy and materials.

How long will those resources last? The answer, of course, depends on how much there is and how quickly we use it. We will explore two different ways to model the rate of consumption of a resource: One is based on the simple exponential function that we have been working with, and the other is based on a more realistic bell-shaped curve.

Exponential Resource Production Rates

When a mineral is extracted from the earth, geologists traditionally say the resource is being *produced* (rather than *consumed*). If we plot the rate of production of a resource versus time, as has been illustrated in Figure 3.4, the area under the curve between any two times will represent the total resource that has been produced during that time interval. That is, if P is production rate (e.g., barrels of oil per day, tons of aluminum per year), and Q is the resource consumed (total barrels, total tons) between times t_1 and t_2, we can write

$$Q = \int_{t_1}^{t_2} P \, dt \qquad (3.11)$$

If we assume that the production rate of a resource grows exponentially, we can easily determine the total amount produced during any time interval. Con-

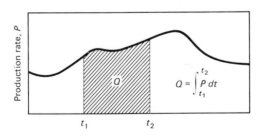

Figure 3.4 The total amount of a resource produced between times t_1 and t_2 is the cross-hatched area under the curve.

versely, if we know the total amount to be produced, we can estimate the length of time that it will take to produce it. The basic assumption of exponential growth is probably not a good one if the time frame is very long, but, nonetheless, it will give us some very useful insights. In the next section we will work with a more reasonable production rate curve.

Figure 3.5 shows an exponential rate of growth in production. If the time interval of interest begins with $t = 0$, we can write

$$Q = \int_0^t P_0 e^{rt}\, dt = \frac{P_0}{r}\, e^{rt}\Big|_0^t$$

which has as a solution

$$Q = \frac{P_0}{r}\, (e^{rt} - 1) \tag{3.12}$$

where Q = the total resource produced by time t
 P_0 = the initial production rate
 r = the exponential rate of growth in production

Equation 3.12 tells us how much of the resource is used up in a given period of time if the production rate grows exponentially. If we want to know how long it will take to use up a given amount of the resource, we can rearrange (3.12) to give

$$T = \frac{1}{r} \ln \left(\frac{rQ}{P_0} + 1 \right) \tag{3.13}$$

where T = the length of time required to use an amount Q.

Applications of (3.13) often lead to startling results, as should be expected anytime we assume that exponential rates of growth will continue for a prolonged period of time.

Example 3.6   World Coal Production

World coal production in 1986 was equal to 5.0 billion (short) tons and the estimated total recoverable reserves of coal were estimated at 1.00 trillion tons. Growth in

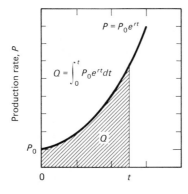

Figure 3.5   Total consumption of a resource experiencing exponential growth.

world coal production in the previous decade averaged 2.7 percent per year. How long would it take to use up those reserves if that growth rate continues unchanged?

Solution   Equation 3.13 gives

$$T = \frac{1}{r} \ln \left(\frac{rQ}{P_0} + 1 \right)$$

$$= \frac{1}{0.027} \ln \left(\frac{0.027 \times 1.0 \times 10^{12}}{5.0 \times 10^9} + 1 \right) = 68.8 \text{ yr}$$

For comparison, if there were no growth in the rate of production, the 1000 billion tons of coal would last 200 years.

Example 3.6 makes an important point. If we simply divide the remaining amount of a resource by the current rate of production, we can get a misleading estimate of the remaining lifetime for that resource. If exponential growth is assumed, what would seem to be an abundant resource may actually be consumed surprisingly quickly. Continuing this example for world coal, let us perform a sensitivity analysis on the assumptions used. Table 3.2 presents the remaining lifetime of coal reserves assuming an initial production rate of 5.0 billion tons per year, for various estimates of total available supply and for differing rates of production growth.

Notice how the lifetime is quite insensitive to estimates of the total available resource when there is exponential growth in consumption. At 3 percent growth, for example, a total supply of 500 billion tons of coal would last for 46 years; if our supply is 4 times as large, 2000 billion tons, the resource only lasts another 39 years. Having 4 times as much coal translates into just an 85 percent increase in resource lifetime.

This is a good time to mention the important distinction between the *reserves* of a mineral and the ultimately producible *resources*. As shown in Figure 3.6,

TABLE 3.2   YEARS REQUIRED TO CONSUME ALL OF THE WORLD'S RECOVERABLE COAL RESERVES ASSUMING VARIOUS RESERVE ESTIMATES AND DIFFERING EXPONENTIALLY INCREASING PRODUCTION GROWTH RATES[a]

Growth rate (%)	500 Billion tons	1000 Billion tons	2000 Billion tons
0	100	200	400
1	69	110	160
2	55	80	109
3	46	65	85
4	40	55	71
5	35	48	61

[a] The 1986 production rate was 5.0 billion tons and reserves were estimated at 1000 billion tons.

Source: Energy Information Administration (1988).

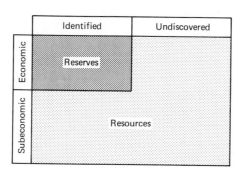

Figure 3.6 Classification of minerals. Reserves are a subcategory of resources.

reserves are quantities that can reasonably be assumed to exist and are producible with existing technology under present economic conditions. *Resources* include present reserves as well as deposits not yet discovered, or deposits that have been identified but are not recoverable under present technological and economic conditions.

As existing reserves are consumed, further exploration, advances in extraction technology, and higher acceptable prices may shift minerals into the reserves category. Estimating the lifetime of a mineral based on the available reserves rather than on the ultimately producible resources can therefore be misleading. U.S. oil reserves in 1970, for example, were estimated at 47 billion barrels, whereas in 1984 reserves were still 35 billion barrels in spite of a total production during that period of about 40 billion barrels.

A Symmetrical Production Cycle

There are many ways that we could imagine a complete production cycle to occur. The model of exponential growth until a resource is totally consumed that we just explored is just one example. It might seem an unlikely one, since the day before the resource collapses, the industry is at full production, and the day after, it is totally finished. It is a useful model, however, to dispel any myths about the possibility of long-term exponential growth in the consumption rate of any finite resource.

A more reasonable approach to estimating resource cycles was popularized by M. King Hubbert (1969). Hubbert argued that consumption would more likely follow a course that might begin with exponential growth while the resource is abundant and relatively cheap. But as new sources get harder to find, prices go up and substitutions begin to take some of the market. Eventually, consumption rates would peak and then begin a downward trend as the combination of high prices and resource substitutions would prevail. A graph of resource consumption versus time would therefore start at zero, rise, peak, and then decrease back to zero, with the area under the curve equalling the total resource consumed.

Hubbert suggested that a symmetrical production cycle that resembles a bell-shaped curve be used. One such curve is very common in probability theory,

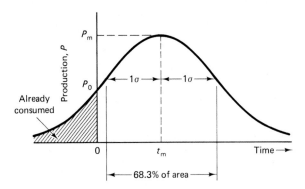

Figure 3.7 Resource production following a Gaussian distribution function.

where it is called the *normal* or *Gaussian* function. Figure 3.7 shows a graph of the function and identifies some of the key parameters used to define it.

The equation for the complete production cycle of a resource corresponding to Figure 3.7 is

$$P = P_m \exp\left[-\frac{1}{2}\left(\frac{t - t_m}{\sigma}\right)^2\right] \tag{3.14}$$

where P = the production rate of the resource

P_m = the maximum production rate

t_m = time at which the maximum production rate occurs

σ = standard deviation, a measure of the width of the bell-shaped curve

The parameter σ is the standard deviation of a normal density function, and in this application it has units of time. Within $\pm 1\sigma$ away from the time of maximum production, 68.3 percent of the production occurs; within $\pm 2\sigma$, 95 percent of the production occurs. Notice that with this bell-shaped curve, it is not possible to talk about the resource ever being totally exhausted. It is more appropriate to specify the length of time required for some major fraction of it to be used up. Hubbert uses 80 percent as his criterion, which corresponds to $\pm 1.3\sigma$ away from the year of maximum production.

We can find the total amount of the resource ever produced, Q_∞, by integrating (3.14):

$$Q_\infty = \int_{-\infty}^{\infty} P \, dt = \int_{-\infty}^{\infty} P_m \exp\left[-\frac{1}{2}\left(\frac{t - t_m}{\sigma}\right)^2\right] dt$$

which works out to

$$Q_\infty = \sqrt{2\pi}\sigma P_m \tag{3.15}$$

Equation 3.15 can be used to find σ if the ultimate amount of the resource ever to be produced, Q_∞, and the maximum production rate P_m can be estimated.

It is also interesting to find the length of time required to reach the maximum production rate. If we set $t = 0$ in (3.14), we find the following expression for the initial production rate P_0:

$$P_0 = P_m \exp\left[-\frac{1}{2}\left(\frac{t_m}{\sigma}\right)^2\right] \tag{3.16}$$

which leads to the following expression for the time required to reach maximum production:

$$t_m = \sigma \sqrt{2 \ln \frac{P_m}{P_0}} \tag{3.17}$$

Let us demonstrate the use of these equations by fitting a Gaussian curve to data for U.S. coal resources.

Example 3.7   U.S. Coal Production

Suppose ultimate total production of U.S. coal is four times the 1987 recoverable reserves which were estimated at 290×10^9 (short) tons. The U.S. coal production rate in 1987 was 0.9×10^9 tons/year. How long would it take to reach a peak production rate equal to 8 times the 1987 rate if a Gaussian production curve is followed?

Solution   Equation 3.15 gives us the relationship we need to find an appropriate σ:

$$\sigma = \frac{Q_\infty}{P_m\sqrt{2\pi}} = \frac{4 \times 290 \times 10^9 \text{ ton}}{8 \times 0.9 \times 10^9 \text{ ton/yr}\sqrt{2\pi}} = 64 \text{ yr}$$

A standard deviation of 64 years says that in a period of 128 years (2σ) about 68 percent of the coal would be consumed and in 166 years ($2 \times 1.3\sigma$) 80 percent would be gone. For comparison, at *current* consumption rates it would take 1030 years to consume 80 percent.

To find the time required to reach peak production, use (3.17)

$$t_m = \sigma \sqrt{2 \ln \frac{P_m}{P_0}} = 64\sqrt{2 \ln 8} = 130 \text{ yr}$$

The complete production curve is given by (3.14):

$$P = P_m \exp\left[-\frac{1}{2}\left(\frac{t - t_m}{\sigma}\right)^2\right] = 7.2 \times 10^9 \exp\left[-\frac{1}{2}\left(\frac{t - 130}{64}\right)^2\right] \text{ tons/yr}$$

which is plotted in Figure 3.8 along with the production curves that result from the maximum production rate reaching 16, 4, and 2 times the current production rates. The trade-offs between achieving high production rates versus making the resource last beyond the next century are readily apparent from this figure.

Hubbert's analysis of world oil production is presented in Figure 3.9 for two scenarios. One scenario is based on fitting the historical production record to a Gaussian curve; the other is based on a conservation scenario where oil production remains constant at the 1975 rate for as long as possible before a sudden decline takes place. With the bell-shaped curve, world oil consumption would

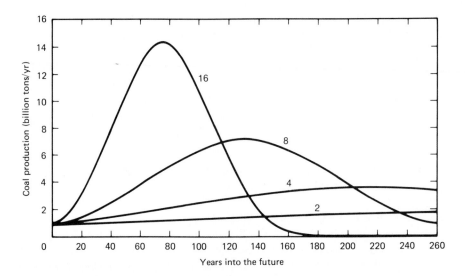

Figure 3.8 Gaussian curves fitted to U.S. production of coal. Ultimate production is equal to four times the 1987 reserves. The parameter is the ratio of the peak production rate to the initial (1987) production rate. At the 1987 production rate of 0.9 Gton/year, the resource would last 1290 years (see Example 3.7).

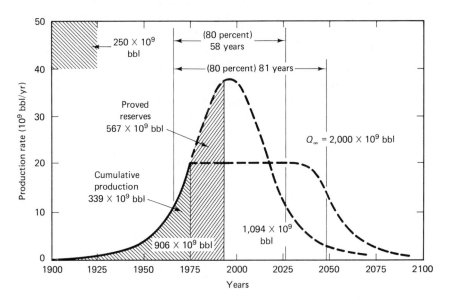

Figure 3.9 Two possible production rate curves for the world's ultimately recoverable crude oil resources. The Gaussian curve indicates peak production would be reached before the year 2000, while the "conservation" scenario, which holds production to the 1975 level as long as possible, extends the resource for only a few decades more (Hubbert, 1977).

peak just before the turn of the century; with the conservation scenario, supplies would last until about 2035. The actual production curve between 1975 and 1986 came very close to matching the conservation scenario.

With either scenario, it is clear that time is quickly running out on the cheapest, most convenient fossil fuel resource that we have. It would seem prudent to be developing alternatives to oil at the fastest possible rate, using what we have as efficiently as possible, and treating this crucial resource as merely our bridge to the future.

3.4 POPULATION GROWTH

The simple exponential growth model can serve us well for short-time horizons, but, obviously, even small growth rates cannot continue for long before environmental constraints limit further increases.

The typical growth curve for bacteria, shown in Figure 3.10, illustrates some of the complexities that natural biological systems often exhibit. The growth curve is divided into phases designated as lag, exponential, stationary, and death. The *lag phase,* characterized by little or no growth, corresponds to an initial period of time when bacteria are first inoculated into a fresh medium. After the bacteria have adjusted to their new environment, a period of rapid growth, the *exponential phase,* follows. During this time, conditions are optimal and the population doubles with great regularity. (Notice that the vertical scale in Figure 3.10 is logarithmic so that exponential growth produces a straight line.) As the bacterial food supply begins to be depleted, or as toxic metabolic products accumulate, the population enters the no-growth, or *stationary phase.* Finally, as the environment becomes more and more hostile, the *death phase* is reached and the population declines.

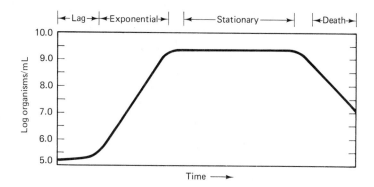

Figure 3.10   Typical growth curve for a bacterial population. (*Source:* Brock, *Biology of Microorganisms,* 2d ed., © 1974. Reprinted by permission of Prentice Hall, Inc., Englewood Cliffs, New Jersey.)

Logistic Growth

Population projections are quite often mathematically modeled with a *logistic* or S-shaped (*sigmoidal*) growth curve, such as the one shown in Figure 3.11. Such a curve has great intuitive appeal. It suggests an early exponential growth phase, while conditions for growth are optimal, followed by slower and slower growth as the population nears the carrying capacity of its environment. Biologists have successfully used logistic curves to model populations of many organisms, including protozoa, yeast cells, water fleas, fruitflies, pond snails, worker ants, and sheep (Southwick, 1976).

Mathematically, the logistic curve is derived from the following differential equation:

$$\frac{dN}{dt} = rN\left(1 - \frac{N}{K}\right) \tag{3.18}$$

where N is population size, r is a growth rate, and K is called the *carrying capacity* of the environment. Notice that when N is much less than K, the rate of change of population is proportional to population size. That is, population grows exponentially with a growth rate r. As N increases, the rate of growth slows down, and eventually, as N approaches K, growth stops altogether and the population stabilizes at a level equal to the carrying capacity. The factor $(1 - N/K)$ is often called the *environmental resistance*. As population grows, the resistance to further population growth continuously increases.

The solution to (3.18) is

$$N = \frac{K}{1 + e^{-r(t-t^*)}} \tag{3.19}$$

Note that t^* corresponds to the time at which $N = K/2$. Substituting $t = 0$ into (3.19) lets us solve for t^*:

$$t^* = \frac{1}{r}\ln\left(\frac{K}{N_0} - 1\right) \tag{3.20}$$

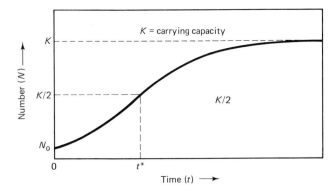

Figure 3.11 The logistic growth curve suggests a smooth transition from exponential growth to a steady-state population.

where N_0 is the population at time $t = 0$. In the usual application of (3.19), the population growth rate is known at $t = 0$, but this is not the same as the growth rate r. To find r, let us introduce another factor, R_0. Let R_0 = instantaneous rate of growth at $t = 0$. If we characterize the growth at $t = 0$ as exponential, then

$$\frac{dN}{dt}\Big|_{t=0} = R_0 N_0 \qquad (3.21)$$

But, from (3.18)

$$\frac{dN}{dt}\Big|_{t=0} = r N_0 \left(1 - \frac{N_0}{K}\right) \qquad (3.22)$$

so that, equating (3.21) with (3.22) yields

$$r = \frac{R_0}{1 - N_0/K} \qquad (3.23)$$

Equation 3.23 lets us use quantities that are known at $t = 0$, namely the population size (N_0) and the population growth rate (R_0), to find the appropriate growth factor r for (3.19). The following example demonstrates this process.

Example 3.8   Logistic Human Population Curve

Suppose the human population follows a logistic curve until it stabilizes at 15.0 billion. In 1986 the world's population was 5.0 billion and its growth rate was 1.7 percent. When would the population reach 7.5 billion? 14 billion?

Solution   We need to find r using (3.23)

$$r = \frac{R_0}{1 - N_0/K} = \frac{0.017}{1 - (5.0 \times 10^9)/(15.0 \times 10^9)} = 0.0255$$

The time required to reach 7.5 billion, or half of the final population size, can be found from (3.20):

$$t^* = \frac{1}{r} \ln\left(\frac{K}{N_0} - 1\right)$$

$$= \frac{1}{0.0255} \ln\left(\frac{15 \times 10^9}{5 \times 10^9} - 1\right) = 27 \text{ yr}$$

To determine the number of years that it will take to reach 14.0 billion, we need to solve (3.19) for t:

$$N = \frac{K}{1 + e^{-r(t-t^*)}} \qquad (3.19)$$

$$t = t^* - \frac{1}{r} \ln\left(\frac{K}{N} - 1\right)$$

$$= 27 - \frac{1}{0.0255} \ln\left(\frac{15}{14} - 1\right) = 130 \text{ yr}$$

Maximum Sustainable Yield

The logistic curve can also be used to introduce another useful concept in population biology called the *maximum sustainable yield* of an ecosystem. The maximum sustainable yield is the maximum rate that individuals can be harvested (removed) without reducing the population size. Imagine, for example, harvesting fish from a pond. If the pond is at its carrying capacity, there will be no population growth, so that any fish removed will reduce the population. Therefore, the maximum sustainable yield will correspond to some population size less than the carrying capacity. In fact, since yield is the same as dN/dt, the maximum yield will correspond to the point on the logistic curve where the slope is a maximum. If we set the derivative of the slope equal to zero, we can find that point. The slope of the logistic curve is given by (3.18):

$$\frac{dN}{dt} = rN \left(1 - \frac{N}{K} \right) \tag{3.18}$$

Setting the derivative equal to zero gives

$$\frac{d}{dt} \left(\frac{dN}{dt} \right) = r \frac{dN}{dt} - \frac{r}{K} \left(2N \frac{dN}{dt} \right) = 0$$

which yields

$$1 - \frac{2N}{K} = 0$$

$$N = \frac{K}{2} \tag{3.24}$$

That is, if population growth is logistic, then the maximum sustainable yield will be obtained when the population is half the carrying capacity.

Human Population Growth: Fertility and Mortality

The logistic growth equations just demonstrated are frequently used with some degree of accuracy and predictive capability for density dependent populations, but they are not often used to predict human population growth. More detailed information on fertility and mortality rates and population age composition are usually available for humans, increasing our ability to make population projections. Using such data enables us to ask such questions as the following: What if every couple had just two children? What if such replacement level fertility were to take 40 years to achieve? What would be the effect of reduced fertility rates on the ratios of elderly and children to workers in the population? The study of human population dynamics is known as *demography* and what follows is but a brief glimpse at some of the key definitions and most useful mathematics involved.

The simplest measure of fertility is the *crude birth rate*, which is just the number of live births per 1000 population in a given year. It is called crude because it does not take into account the fraction of the population that is physically capable of giving birth in any given year. The crude birth rate is typically in the range of 40 to 50 per 1000 per year for the poorest, least-developed countries of the world, while it is typically between 10 and 20 for the more developed ones. For the world as a whole, in 1989, the crude birth rate was 28 per thousand. That means the 5.2 billion people alive then would have had about $5.2 \times 10^9 \times (28/1000)$ $= 145 \times 10^6$ live births in that year. (These statistics, and all others in this section, are taken from data supplied by the Population Reference Bureau in a most useful annual publication called the *World Population Data Sheet.*)

The *total fertility rate* is the average number of children that would be born alive to a woman, assuming that current age-specific birth rates remain constant. It is a good measure of the average number of children each woman is likely to have during her lifetime. In the developed countries of the world in 1989, the total fertility rate was 1.9 children per female, while in the less developed countries (excluding China) it was 4.7.

The number of children that a woman must have, on the average, to replace herself with one daughter in the next generation, is called *replacement level fertility*. Replacement level fertility accounts for differences in the ratio of male to female births as well as child mortality rates. For example, in the United States, replacement level fertility is 2.11. At that fertility level, 100 women would bear, on the average, 211 children. Statistically, 108 of the children would be boys and 103 would be girls. About 3 percent of the girls would be expected to die before bearing children, leaving a net of 100 women in the next generation. Notice that the level of fertility that leads to replacement will vary from country to country due to different mortality rates. In many developing countries, replacement level fertility is approximately 2.7 children.

As will be demonstrated later, having achieved replacement level fertility does not necessarily mean a population has stopped growing. There may be a large enough fraction of the population already born, but not yet of child bearing age, that for several decades more births will occur in the population than there will be deaths. The continuation of population growth, in spite of replacement level fertility being achieved, is a phenomenon known as *population momentum.*

The simplest measure of mortality is the *crude death rate,* which is the number of deaths per 1000 population per year. Again, caution should be exercised in interpreting *crude* death rates since the age composition of the population is not accounted for. The United States and Guatemala, for example, both had crude death rates of 9 per 1000 in 1989, but that in no way indicates equivalent risks of mortality. In Guatemala only 3 percent of the population is over 65 years of age (and hence at greater risk of dying) while in the United States a much large fraction of the population, 12 percent, is over 65.

An especially important measure of mortality is the *infant mortality rate,* which is the number of deaths to infants (under 1 year of age) per 1000 live births in a given year. The infant mortality rate is one of the best indicators of poverty in a country. In the poorest countries of the world, infant mortality rates may be as high as 200, which means one child in five will not live to see its first birthday. In the most developed countries, infant mortality rates are typically around 9 per 1000 per year.

Example 3.9   Birth and Death Statistics

In 1989, with a population of 5.2 billion, the world had a crude birth rate of 28, a crude death rate of 10, and an infant mortality rate of 75. What fraction of the total deaths is due to infant mortality? If the world as a whole were able to care for its infants as well as they are cared for in most developed countries, resulting in an infant mortality rate of 9, how many infant deaths would be avoided each year?

Solution   To find the number of infant deaths each year, we must first find the number of live births and then multiply that by the fraction of those births that die within the first year:

Now:

$$\text{Infant deaths} = \text{Population} \times \text{Crude birth rate} \times \text{Infant mortality rate}$$

$$= 5.2 \times 10^9 \times (28/1000) \times (75/1000) = 10.9 \times 10^6 \text{ per year}$$

$$\text{Total deaths} = \text{Population} \times \text{Crude death rate}$$

$$= 5.2 \times 10^9 \times (10/1000) = 52 \times 10^6 \text{ per year}$$

$$\text{Fraction infants} = 10.9/52 = 0.21 = 21 \text{ percent}$$

Then:

$$\text{Infant deaths} = 5.2 \times 10^9 \times (28/1000) \times (9/1000) = 1.3 \times 10^6 \text{ per year}$$

$$\text{Avoided deaths} = (10.9 - 1.3) \times 10^6 = 9.6 \text{ million per year}$$

It is often argued that reductions in the infant mortality rate would eventually result in reduced birth rates as people began to gain confidence that their offspring would be more likely to survive. Hence, the reduction of infant mortality rates through such measures as better nutrition, cleaner water, and better medical care, is thought to be a key to population stabilization.

The difference between crude birth rate b and crude death rate d is called the *rate of natural increase* of the population. While it can be expressed as a rate per 1000 of population, it is more common to express it either as a decimal fraction or as a percentage rate. As a simple, but important equation,

$$r = b - d \tag{3.25}$$

where r is the rate of natural increase. If r is treated as a constant, then the exponential relationships developed earlier can be used. For example, in 1989 the crude birth rate for the world was 28 and the crude death rate was 10, so the rate of natural increase was $(28 - 10)/1000 = 18/1000 = 0.018 = 1.8$ percent. If this rate continues, then the world will double in population in about $70/1.8 = 39$ years.

While it is reasonable to talk in terms of a rate of natural increase for the world, it is often necessary to include effects of migration when calculating growth rates for an individual country. Letting *m* be the *net migration rate,* which is the difference between immigration (in-migration) and emigration (out-migration), we can rewrite the above relationship as follows:

$$r = b - d + m \tag{3.26}$$

At the beginning of the industrial revolution, it is thought that crude birth rates were around 40 per 1000 while death rates were around 35 per 1000 yielding a rate of natural increase of about 0.5 percent for the world. As parts of the world began to achieve the benefits of a better and more assured food supply, improved sanitation, and modern medicines, death rates began to drop. As economic and social development has proceeded in these countries, birth rates have fallen to the point where the developed countries of the world now have an average crude birth rate of about 15 and a crude death rate of about 9, equivalent to a growth rate of 0.6 percent. These countries have undergone what is referred to as the *demographic transition,* a transition from high birth and death rates to low birth and death rates, as shown in Figure 3.12.

The less-developed countries of the world have experienced a sizeable drop in death rates, especially during the last half century. Imported medicines and insecticides have contributed to a rather sudden death rate decline resulting in the 1989 level of 10 per 1000. Birth rates have not fallen nearly as fast; they are still up around 31, giving a growth rate of 2.1 percent for three-fourths of the world population. The rapid (and, historically speaking, extraordinary) population growth that the world is experiencing now is almost entirely due to a drop in death rates without a corresponding drop in birth rates. Many argue that decreases in fertility depend on economic growth, and many countries today face the danger that economic growth may not be able to exceed population growth. Such countries may be temporarily "stuck" in the middle of the demographic transition and be facing the risk that population equilibrium may ultimately occur along the least desirable path, through rising death rates.

Some of these data on fertility and mortality are presented in Table 3.3 for the world, for more developed countries, for China, and for the less-developed countries excluding China. China has been separated out because, in terms of

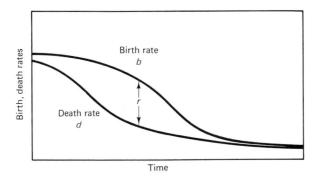

Figure 3.12 Stylized representation of the demographic transition as birth and death rates drop.

TABLE 3.3   SOME IMPORTANT POPULATION STATISTICS (1989)

Parameter	World	More developed countries	Less developed countries (non-China)	China
Population (millions)	5234	1206	2924	1104
(% of world)	100	23	56	21
Crude birth rate, b	28	15	35	21
Crude death rate, d	10	9	11	7
Natural increase, $r\%$	1.8	0.6	2.4	1.4
% Population under age 15	33	22	40	29
Total fertility rate	3.6	1.9	4.7	2.4
Infant mortality rate	75	15	93	44
Est. Population 2020 (million)	8330	1339	5468	1523
Population increase by 2020	3096	133	2544	419
Per capita GNP ($)	3330	12 070	820	300

Source: Population Reference Bureau, 1989.

population control, it is so different from the rest of the less-developed countries, and because its size so greatly influences the data.

Age Structure

A great amount of insight and predictive power can be obtained from a table or diagram representing the age composition of a population. A graphical presentation of the data, indicating numbers of people (or percentages of the population) in each age category, is called an *age structure* or a *population pyramid*. The shape of an age structure tells a great deal about the past and the near future of the population. Developing countries, for example, have pyramids that are roughly triangular in shape with each cohort larger than the cohort born before it, corresponding to a rapidly expanding population. An example pyramid is shown in Figure 3.13 for Morocco. It is not uncommon in such countries to have nearly half of the population younger than 15 years old. Even if replacement fertility is achieved in the near future, there are so many young women already born who will be having children that population size will continue to grow for many decades.

Figure 3.13 also shows a population pyramid for the United States in 1985. The sides are much more vertical, corresponding to a country that has reached a stage of slow growth. The pinched sides, corresponding to births in the 1930s, are people who have sometimes been called the "good times cohorts." Since there are so few of them, as they sought their educations, entered the job market, bought homes, and so on through their lives, the competition for those resources was less than it has been for people younger or older than them. On the other

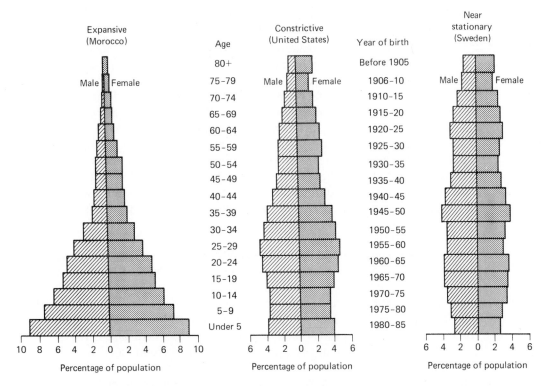

Figure 3.13   Three general profiles of age composition. (*Source:* Haupt and Kane, *Population Handbook,* © 1985. Reprinted by permission of Population Reference Bureau, Inc., Washington, D.C.)

hand, the baby boom children born in the late 1950s show up as a bulge in the age structure. As they went through school, resources were stretched to the limit, and as they left, schools closed behind them. Throughout their lives there will be more of them competing for the same things, and when they retire, there will be more of them to be supported by a smaller work force coming up behind them. The children born in the 1980s are an echo of the baby boomers, and another bulge is beginning to appear at the bottom of the age structure.

The third pyramid in Figure 3.13 is an example of a population that is nearly *stationary*. A stationary population is one that is constant in number; that is, it is not increasing or decreasing with time. Another important term that demographers use is population *stability*. A stable population is one that has constant birth and death rates so that the percentage of the population in any age category is unchanging. That is, a stable population has a pyramid whose shape does not change with time. A stable population may be growing or shrinking; it does not have to be stationary.

Human Population Projections

If the age structure for a population is combined with data on age-specific birth and death rates, it is possible to make realistic projections of future population size and composition. The techniques that we will explore now are especially useful for predicting the effects and implications of various changes in fertility patterns that might be imagined or advocated. For example, suppose we want to determine the effects of replacement level fertility. If replacement fertility were achieved today, how long would it take to achieve a stationary population? What would the age composition look like in the interim? What fraction of the population would need jobs? What fraction would be young and looking for work, and what fraction would be retired and expecting support? Or, suppose a policy is established that sets a target for maximum size of the population (as is the case for China): How many children would each couple need to have to achieve that goal and how would the number change with time?

Not long ago, the calculations required to develop these scenarios were tedious and best left to professional demographers. Now, however, with the widespread use of simple spreadsheet programs on personal computers, it is possible to work them out with relative ease.

The starting point in a population projection is the current age structure combined with mortality data obtained from *life tables*. Life tables are of particular use to insurance companies for predicting the average number of years of life remaining, as a function of the age of their clients, and they have become especially useful data sets for demographers. A life table is developed by applying a real population's age-specific death rates (the fraction of the people in a given age category who will die each year) to *hypothetical* stable and stationary populations having 100 000 live births per year, evenly distributed through the year, with no migration. As the 100 000 people added each year get older, their ranks are thinned in accordance with the age-specific death rates. It is then possible to calculate the numbers of people who would be alive within each age category in the following year.

Table 3.4 is an abridged version of a life table for the United States. These data, remember, are for a hypothetical population with 100 000 live births each year (and 100 000 deaths each year as well since this is a stationary population). The first column shows the age interval, x to $x + 5$ (e.g., 10–14 means people who have had their 10th birthday but not their 15th). The second column is the number of people who would be alive at any given time in the corresponding age interval, and is designated L_x, where x is the age at the beginning of the interval. The third column, L_{x+5}/L_x, is the ratio of the number of people in the next interval to the number in the current interval; it is the probability that someone aged x to $x + 5$ will live 5 more years (except in the case of those 80 and older, where the catchall category of 85+ years old modifies the interpretation).

If we assume that the age-specific death rates that were used to produce Table 3.4 remain constant, we can use the table to make future population projec-

TABLE 3.4   AGE DISTRIBUTION FOR A
HYPOTHETICAL, STATIONARY
POPULATION WITH 100 000 LIVE BIRTHS
PER YEAR[a]

Age interval x to $x + 5$	Number in interval L_x	$\dfrac{L_{x+5}}{L_x}$
0–4	494 285	0.9979
5–9	493 250	0.9989
10–14	492 699	0.9973
15–19	491 382	0.9951
20–24	488 972	0.9944
25–29	486 219	0.9940
30–34	483 310	0.9927
35–39	479 794	0.9898
40–44	474 877	0.9841
45–49	467 346	0.9745
50–54	455 448	0.9597
55–59	437 078	0.9381
60–64	410 020	0.9082
65–69	372 397	0.8658
70–74	322 425	0.8050
75–79	259 548	0.7163
80–84	185 922	0.9660
85 and over	179 601	0.0000

[a] Age-specific death rates are U.S. 1984 values.

Source: Abstracted from data in U.S. Dept. of
Health and Human Services (1987).

tions at 5-year intervals for all but the 0–4 age category. The 0–4 age category will
depend on fertility data to be discussed later.

If we let $P_x(0)$ be the number of people now in age category x to $x + 5$, and
$P_{x+5}(5)$ be the number in the next age category 5 years from now, then

$$P_{x+5}(5) = P_x(0) \frac{L_{x+5}}{L_x} \tag{3.27}$$

That is, 5 years from now, the number of people in the next 5-year age interval will
be equal to the number in the current interval times the probability of surviving,
(L_{x+5}/L_x) for the next 5 years.

For example, in the United States in 1985, there were 18.0 million people in
the age group 0–4 years; that is, $P_0(1985) = 18.0$ million. We would expect that in
1990 the number of people alive in the 5–9 age category would be

$$P_5(1990) = P_0(1985)(L_5/L_0)$$

$$= 18.0 \times 10^6 \times 0.9979 = 17.98 \text{ million}$$

Let's take what we have and apply it to the age composition of the United States in 1985 to predict as much of the structure as we can for 1990. This involves application of (3.27) to all of the categories, giving us a complete 1990 age distribution except for the 0- to 5-year-olds. The result is presented in Table 3.5.

To find the entry for ages 0–4 in 1990, we need to know something about birth rates. To do this calculation properly requires some sophistication since it requires manipulation of the statistics of child mortality during the first 5 years of life (see Keyfitz (1968) for details). Since our purpose here is to develop a tool for asking questions of the "what if" sort, rather than making actual population projections, we can use the following simple approach to estimating the number of children, ages 0–4 years, (P_0), to put into our age composition table:

$$P_0(5) = b_{15}P_{15}(0) + b_{20}P_{20}(0) + \cdots + b_{40}P_{40}(0) \tag{3.28}$$

where b_x is the number of the surviving children born per person in age category x to $x + 5$. Equation 3.28 has been written with the assumption that no children are born to individuals under 15 or over 45 years of age, though it could obviously have been written with more terms to include them. Notice that we can interpret the sum of the b_x's (Σb_x) to be the number of surviving children per person. Loosely stated, $2\Sigma b_x$ is the average number of children per family.

TABLE 3.5   1990 U.S. POPULATION PROJECTION BASED ON
THE 1985 AGE STRUCTURE (IGNORING IMMIGRATION)

Age interval x to $x + 5$	$\dfrac{L_{x+5}}{L_x}$	P_x (thousands) 1985	P_x (thousands) 1990
0–4	0.9979	18 020	$P_0(1990)$
5–9	0.9989	17 000	17 982
10–14	0.9973	16 068	16 981
15–19	0.9951	18 245	16 025
20–24	0.9944	20 491	18 156
25–29	0.9940	21 896	20 376
30–34	0.9927	20 178	21 765
35–39	0.9898	18 756	20 031
40–44	0.9841	14 362	18 564
45–49	0.9745	11 912	14 134
50–54	0.9597	10 748	11 609
55–59	0.9381	11 132	10 314
60–64	0.9082	10 948	10 443
65–69	0.8658	9 420	9 943
70–74	0.8050	7 616	8 156
75–79	0.7163	5 410	6 131
80–84	0.9660	3 312	3 875
85 and over	0.0000	2 113	3 199
		Total = 237 627	227 684 + $P_0(1990)$

Source: 1985 data from Vu (1985).

For example, if we want to investigate the effect of replacement fertility on population size and age structure, we need $\Sigma b_x = 1$, distributed among the age categories. Suppose we assume the following distribution:

$$b_{15} = 0 \qquad b_{20} = 0.3 \qquad b_{25} = 0.4 \qquad b_{30} = 0.3 \qquad b_{35} = b_{40} = 0$$

and apply it to the 1985 U.S. population to estimate the number of 0- to 4-year-olds in 1990.

$$P_0(1990) = 0.3 \times P_{20}(1985) + 0.4 \times P_{25}(1985) + 0.3 \times P_{30}(1985)$$

$$= 0.3 \times 20\ 491\ 000 + 0.4 \times 21\ 896\ 000 + 0.3 \times 20\ 178\ 000$$

$$= 20\ 959\ 000$$

With this 20 959 000 added to Table 3.5, the total population in the United States in 1990 would be 248 643 000. Notice that, in spite of hypothesizing replacement level fertility in 1985, the population in 1990 would have grown from 237.6 million to 248.6 million. This is an example of *population momentum*, a term used to describe the fact that a youthful age structure causes a population to continue growing for several decades after reaching replacement fertility, before it finally reaches stability.

Population Momentum

Let us begin with an extremely simplified example, that does not model reality very well, but that does keep the arithmetic manageable. Then we will apply the ideas to a real-world example.

Suppose we have an age structure with only three categories: 0–24 years, 25–49 years, and 50–74 years, subject to the following fertility and mortality conditions:

1. All births occur on the woman's 25th birthday.
2. All deaths occur at age 75.
3. The total fertility rate (TFR) is 4.0 for 25 years, then it drops instantaneously to 2.0.

Suppose a population of 5.0 billion people has an age structure at time $t = 0$ as shown in Figure 3.14a. Equal numbers of men and women will be assumed so that the number of children per person is half the total fertility rate. During the first 25 years, the 2.5 billion people who start in the 0–24 age category will all pass their 25th birthday and all of their children will be born. Since TFR is 4.0, these 2.5 billion individuals will bear 5.0 billion children (two per person). Those 2.5 billion people will now be in the 25–49 age category. Similarly, the 1.5 billion people who were 25–49 years old will now be ages 50–74. Finally, the 1.0 billion 50- to 74-year-olds will have all passed their 75th birthdays and, by the rules, they will all be dead. The age structure will be as shown in Figure 3.14b.

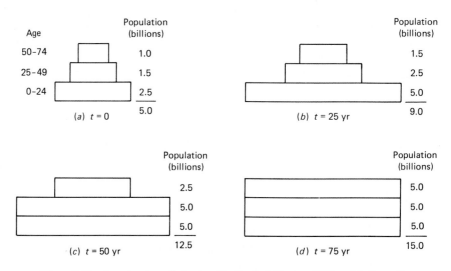

Figure 3.14   Age structure diagrams. For the first 25 years TFR = 4.0; thereafter it is at the replacement level of 2.0 (see text).

After the first 25 years have passed, TFR drops to the replacement level of 2.0 (one child per person). During the ensuing 25 years, the 5.0 billion 0- to 24-year-olds will have 5.0 billion children and, following the logic given above, the age structure at $t = 50$ years will be as shown in Figure 3.14c. With replacement level continuing, at $t = 75$ years the population stabilizes at 15 billion. A plot of population versus time is shown in Figure 3.15. Notice that the population stabilizes 50 years after replacement level fertility is achieved and that during that 50 years, it grows from 9 billion to 15 billion.

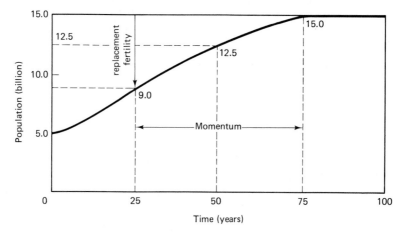

Figure 3.15   In the hypothetical example, it takes 50 years before population stabilizes after replacement level fertility begins.

Now let us add complexity by using an actual world population age structure with actual age specific mortality rates. To investigate the concept of population momentum, let us assume that replacement level fertility begins immediately. Suppose we assume the following birth rate factors, b_x:

$$b_{15} = 0.37 \qquad b_{20} = 0.40 \qquad b_{25} = 0.30 \qquad \text{with all other } b_x = 0$$

These factors sum to 1.07 children per person. This may seem a bit above replacement fertility but it is not. With world mortality rates being what they are, about 7 percent of the 0–5 age group will die before they reach their reproductive years, so we have added 7 percent to the sum of the b_x's.

The 1985 world age distribution and mortality rates are listed in Table 3.6 along with the assumed fertility factors. The projection is made using (3.27) and (3.28).

Example 3.10   Age Structure Calculations

Find the population of 0- to 4-year-olds and 5- to 9-year-olds in 1990 using the 1985 age structure, fertility, and mortality data given in Table 3.6.

Solution   The 0- to 4-year-olds can be found using (3.28):

$$P_0(1990) = b_{15}P_{15}(1985) + b_{20}P_{20}(1985) + b_{25}P_{25}(1985)$$

$$= 0.37 \times 495 + 0.40 \times 442 + 0.30 \times 384 = 475 \text{ million}$$

We can find the 5- to 9-year-olds with (3.27):

$$P_5(1990) = P_0(1985) \times (L_5/L_0) = 575 \times 0.971 = 558 \text{ million}$$

These numbers agree with those found in the 1990 column of Table 3.6.

From Table 3.6 (with the calculation carried a little further into the future), a graph of total population size can be drawn as shown in Figure 3.16. In spite of the impossible hypothesis of immediate replacement, the population still climbs

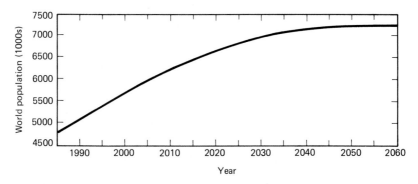

Figure 3.16   World population projection assuming immediate replacement fertility in 1985, showing effect of population momentum.

TABLE 3.6 POPULATION PROJECTION FOR THE WORLD ASSUMING IMMEDIATE REPLACEMENT FERTILITY (NUMBERS IN MILLIONS)

Age	L_{x+5}/L_x	Fertility	1985	1990	1995	2000	2005	2010	2015	2020	2025	2030	2035
0–4	0.971	0	575	475	518	542	560	536	522	519	546	551	542
5–9	0.991	0	530	558	461	503	526	544	521	507	504	530	535
10–14	0.991		523	525	553	457	499	522	539	516	502	499	526
15–19	0.988	0.37	495	518	521	548	453	494	517	534	512	498	495
20–24	0.986	0.40	442	489	512	514	542	448	488	511	528	505	492
25–29	0.986	0.30	384	436	482	505	507	534	441	481	504	520	498
30–34	0.983	0	348	379	430	475	498	500	527	435	475	497	513
35–40	0.978	0	288	342	372	422	467	489	491	518	428	467	488
40–45	0.973	0	237	282	335	364	413	457	479	481	506	418	456
45–49	0.962	0	222	231	274	326	354	402	445	466	468	493	407
50–54	0.947	0	196	214	222	264	313	341	387	428	448	450	474
55–59	0.922	0	173	186	202	210	250	297	323	366	405	424	426
60–64	0.881	0	139	160	171	186	194	230	273	297	338	374	391
65–69	0.825	0	106	122	141	151	164	171	203	241	262	297	329
70–74	1.273	0	82	87	101	116	124	136	141	167	199	216	245
75+	0	0	96	104	111	129	148	158	173	179	213	253	275
Total			4836	5108	5406	5713	6012	6258	6469	6646	6836	6993	7094

Source: Based on data in Vu (1985).

from 4.8 billion in 1985 to about 7.3 billion in 2060. Delaying the time to reach replacement results in much higher projections.

PROBLEMS

3.1. World population in 1850 has been estimated to have been about 1 billion. It reached 4 billion in 1975. Estimate the exponential rate of growth that would produce those numbers.

3.2. Assuming an initial population of 2 people, *about* how many doublings would it have taken to get to a world population of a little over 4 billion? Use the approximation that every 10 doublings multiplies the amount by about 1000. If it took *about* 2.2 million years to go from 2 people to 4+ billion, estimate the average doubling time. About what average exponential rate of growth would that correspond to? (You should be able to do this problem without a calculator, since only approximations are called for.)

3.3. Suppose we express the amount of land under cultivation as the product of four factors, each growing as follows: (1) the land required to grow a unit of food, −1 percent (due to greater productivity per unit of land); (2) the amount of food grown per calorie of food eaten by a human, +0.5 percent (with affluence, people consume more animal products, which greatly reduces the efficiency of land use); (3) the per capita calorie consumption, +0.1 percent; and, (4) the size of the population, +1.5 percent. At these rates, how long would it take to double the amount of cultivated land needed? At that time, how much less land would be required to grow a unit of food?

3.4. Suppose that world per capita energy demand increases at 1.5 percent per year, that fossil fuel emissions of carbon per unit of energy increase at 1 percent per year, and that world population grows at 1.5 percent per year. How long would it take before we are emitting carbon at twice the current rate?

3.5. Under the assumptions stated in Problem 3.4, if our initial rate of carbon emission is 5×10^9 tons C/year and if there are 700×10^9 tons of carbon in the atmosphere now, how long would it take to emit a total amount of carbon equal to the amount now contained in the atmosphere? If half of the carbon that we emit stays in the atmosphere (the other half being "absorbed" in other parts of the biosphere), how long would it take us to double atmospheric carbon concentrations by fossil fuel combustion?

3.6. World reserves of chromium are about 800 million tons and current usage is about 2 million tons per year. If growth in demand for chromium increases exponentially at a constant rate of 2.6 percent per year, how long would it take to use up current reserves? Suppose the total resource is 5 times current reserves. If the use rate continues to grow at 2.6 percent, how long would it take to use up the resource?

3.7. Suppose a Gaussian curve is used to approximate the production of chromium. If production peaks at 6 times its current rate of 2 million tons per year, how long would it take to reach that maximum if the total resource base is 4000 million tons (5 times the current reserves)? How long would it take to consume about 80 percent of the total resource?

3.8. Suppose we assume the following:

    a. Any chlorofluorocarbons (CFCs) released into the atmosphere remain in the atmosphere indefinitely.

    b. At current rates of release, the atmospheric concentration of fluorocarbons would double in 100 years. (*Hint:* What does that say about the ratio of the amount of CFC now present, Q, to the current rate of release, P_0?)

    c. Atmospheric release rates are, however, not constant but growing at 2 percent per year. How long would it take to double atmospheric CFC concentrations?

3.9. Bismuth-210 has a half-life of 4.85 days. If we start with 10 g of it now, how much would we have left in 7 days?

3.10. Suppose some sewage drifting down a stream decomposes with a reaction rate coefficient K equal to 0.2/day. What would be the "half-life" of this sewage? How much would be left after 5 days?

3.11. Assume that a population follows the simple logistic growth curve. Find the maximum sustainable yield as a function of carrying capacity K, the current population size N_0, and current growth rate R_0.

3.12. Suppose we stock a pond with 100 fish and note the population doubles every year for the first couple of years (with no harvesting) but after quite a number of years, the population stabilizes at what we think must be the carrying capacity of the pond, 2000 fish. Growth seems to have followed a logistic curve. Using the results of Problem 3.11, what would be the maximum sustainable fish yield from this pond?

3.13. Suppose human population growth follows a logistic curve until it stabilizes at 7.3 billion, starting with its 1985 population 4.84 billion and growth rate of 1.7 percent. When would the population reach 7 billion? (Compare this to the result in Figure 3.16.)

3.14. The following statistics are for India in 1985: population, 762 million; crude birth rate, 34; crude death rate, 13; infant mortality rate, 118 (rates are per thousand per year). Find (a) the fraction of the total deaths that are to infants less than 1 year old; (b) the "avoidable deaths" assuming any infant mortality above 10 could be avoided with better sanitation, food, and health care; (c) the annual increase in the number of people in India.

3.15. Consider a simplified age structure that divides a population into three groups: ages 0–24, with 3.0 million; 25–49, with 2.0 million; and 50–74, with 1.0 million. Suppose we impose the following simplified fertility and mortality constraints. All births are just before the 25th birthday only and no one dies until his or her 75th birthday, at which time they all die. Suppose we have replacement level fertility starting now. Draw the age structure in 25, 50, and 75 years. What is the total population size at each of these times?

3.16. Using the same initial population structure given in Problem 3.15, with all births just before the 25th birthday, draw the age structure in 25, 50, and 75 years under the following conditions: No deaths occur until the 50th birthday, at which time 20 percent die; the rest die on their 75th birthday. For the first 25 years, the total fertility rate is 4 (2 children/person), and thereafter, it is 2.

3.17. Consider simplified age structure shown in Figure P3.17. All births are just before the 20th birthday and all deaths are on the 60th birthday. Total population starts at

290 000 (half males, half females) and is growing at a constant rate of $3\frac{1}{2}$ percent per year.

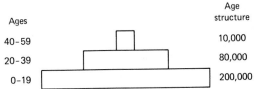

Figure P3.17

Draw the age structure in 20 years. If the total fertility rate is a single, constant value during those 20 years, what is it?

3.18. The age structure and survival data in the table are for China in 1980 (Harte, 1985). Suppose the birth factors (corresponding to a total fertility rate of 1.0) are as shown. Estimate the population of China in 1990.

Age	Population (millions)	$\dfrac{L_{x+10}}{L_x}$	b_x
0–9	235	0.957	0
10–19	224	0.987	0.25
20–29	182	0.980	0.25
30–39	124	0.964	0
40–49	95	0.924	0
50–59	69	0.826	0
60–69	42	0.633	0
70–79	24	0.316	0
80–89	6	0	0
Total	1001		

3.19. Use a spreadsheet to project the China population data given in Problem 3.18 out to the year 2030. What is the population at that time?

3.20. Use a spreadsheet to project the China population data given in Problem 3.18 out to the year 2030, but delay the births by one 10-year interval (that is, $b_{10} = 0$, $b_{20} = 0.25$, and $b_{30} = 0.25$). Compare the peak population in Problem 3.19 to that obtained by postponing births by 10 years.

REFERENCES

BROCK, T. D., 1974, *Biology of Microorganisms*, 2d ed., Prentice-Hall, Englewood Cliffs, NJ.

Energy Information Administration, 1988, *Annual Energy Review 1987*, DOE/EIA-0384(87), U.S. Department of Energy, Washington, DC.

HARTE, J., 1985, *Consider a Spherical Cow, A Course in Environmental Problem Solving,* William Kaufman, Los Altos, CA.

HAUPT, A., and T. T. KANE, 1985, *Population Handbook,* Population Reference Bureau, Inc., Washington, DC.

HUBBERT, M. K., 1969, in National Academy of Sciences, *Resources and Man,* Freeman, San Francisco, pp. 157–242.

HUBBERT, M. K., 1977, in Congressional Research Service, *Project Interdependence: U.S. and World Energy Outlook through 1990,* U.S. Government Printing Office, Washington, DC, pp. 632–644.

KEYFITZ, N., 1968, *Introduction to the Mathematics of Population,* Addison-Wesley, Reading, MA.

Population Reference Bureau, 1989, *1989 World Population Data Sheet,* Washington, DC.

SOUTHWICK, C. H., 1976, *Ecology and the Quality of our Environment,* 2d ed., Van Nostrand, New York.

U.S. Department of Health and Human Services, March 1987, *Vital Statistics of the United States, 1984, Life Tables,* Vol 2, Sec. 6, National Center for Health Statistics, Hyattsville, MD.

VU, M. T., 1985, *World Population Projections 1985, Short- and Long-Term Estimates by Age and Sex with Related Demographic Statistics,* World Bank, Johns Hopkins Press, Baltimore, 1985.

CHAPTER 4

Water
Pollution

When the well's dry, we know the worth of water.
(Ben Franklin, *Poor Richard's Almanac*)

4.1 INTRODUCTION

During the "ecology movement" of the early 1970s, our attention was focused on visible forms of air and water pollution. Lake Erie was pronounced dead, the Cuyahoga River was so polluted it caught on fire, and sewage from 50 million people across the country was discharged into our waterways with little or no treatment. The Clean Water Act of 1970 changed much of that. After $50 billion dollars worth of wastewater treatment grants, most cities have obtained modern sewage treatment facilities, Lake Erie is very much alive, and most of our rivers and lakes are approaching the goal of being "fishable and swimmable." The *National Pollutant Discharge Elimination System* (NPDES) now effectively controls discharges from most point sources of water pollution, greatly reducing the wanton disregard that many industries and municipalities showed in the past. In many ways, then, control of surface water pollution has been one of the success stories of the environmental movement of the 1970s and 1980s.

As we approach the end of the century, however, problems of both water quality and water quantity are returning to the forefront. Just as we seem to be reaching our goals for our lakes and rivers, we are beginning to find our beaches polluted with medical wastes and leftover sludge from wastewater treatment plants. Our traditional confidence in the quality of our drinking water has been seriously shaken as we find potential carcinogens in our groundwater and as we come to realize that we have been creating carcinogens by chlorinating our surface water. As well, the extended drought in the late 1980s, coupled with our growing awareness of the global climate change threat, has made the water quality problem seem even more serious as we begin to worry about the availability of enough water to meet our needs in the future.

After a brief introduction to water resources and water use in the United States, we will discuss the special characteristics and environmental problems associated with rivers, lakes, and groundwater.

4.2 WATER RESOURCES

Unusual Properties of Water

Water covers about 70 percent of the planet. It is so common that it is easy to assume that it is a typical liquid, much like any other. But, in fact, virtually every physical and chemical property is unusual when compared with other liquids, and these differences are essential to life as we know it. The following are some of these properties with comments on their importance to life (Berner and Berner, 1987).

Density. Water is the only common substance that expands when it freezes. In fact, a plot of density versus temperature shows a maximum density at 4 °C, which means that for temperatures above and below this point, water continuously becomes lighter and more buoyant. As a result, ice floats. If it did not, ice that would form on the surface of bodies of water would sink to the bottom, making it possible for lakes to freeze solid. The unusual density characteristics of water also lead to thermal stratification of lakes, affecting life in unusual ways, as will be described later in this chapter. The expansion of water as it freezes also contributes to the weathering of rocks by literally breaking them apart when water freezes in the cracks.

Melting and Boiling Points. If water were similar to other substances (such as H_2S, H_2Se, and H_2Te) it would boil at normal earth temperatures, thus existing mostly as a gas rather than a liquid or solid. It also has an unusually high difference in temperature between the melting point and boiling point, thus remaining a liquid over most of the globe. With only slightly different phase change temperatures, life on earth would be very different, if it could exist at all.

Specific Heat. The specific heat of water (4184 J/kg°C) is higher than it is for any other known liquid except ammonia. It is five times higher than the specific heat of most common heavy solids, such as rock and concrete. This property means that water heats and cools slower than almost anything else. This helps moderate climate near large bodies of water, and it also serves the important function of protecting life from rapid thermal fluctuations, which are often lethal.

Heat of Vaporization.   The heat required to vaporize water (2258 kJ/kg) is one of the highest of all liquids. This high heat of vaporization means that water vapor stores an unusually large amount of energy, energy that is released when the water vapor condenses. This property is important in distributing heat from one place on the globe to another and is a major factor affecting the earth's climate.

Water as a Solvent.   Water dissolves more substances than any other common solvent. As a result, it serves as an effective medium for both transporting dissolved nutrients to tissues and organs in living things as well as eliminating their wastes. Water also transports dissolved substances throughout the biosphere.

The Hydrologic Cycle

Almost all of the world's water (97.2 percent) is located in the oceans, but, as might be expected, the high concentration of salts renders the oceans virtually unusable as a source of water for municipal, agricultural, or most industrial needs. We do use the oceans, however, for thermal cooling of power plants and as a sink for much of our pollution. While desalination technologies exist that can remove the salts, the energy that would be required to produce significant quantities of fresh water in this way is prohibitive for all but the most extreme circumstances. Fortunately, the sun performs that service for us when it evaporates water and leaves the salts behind. Almost one-half of the sun's energy that strikes the earth's surface is converted to latent heat, removing water from wet surfaces by evaporation and from the leaves of plants by transpiration. The combination of processes, called *evapotranspiration,* requires an enormous amount of energy, equivalent to roughly 4000 times the rate at which humankind utilizes energy resources.

Evapotranspiration removes an amount of water about equivalent to a layer 1 m thick around the globe each year. About 88 percent of that is evaporation from the oceans, while the remaining 12 percent is evapotranspiration from the land. The resulting water vapor is transported by moving air masses and eventually condenses and returns to the earth's surface as precipitation. Over the oceans, there is more evaporation than precipitation, and over the land it is the other way around, more precipitation than evapotranspiration. The difference between precipitation and evapotranspiration on land is water that is returned to the oceans, both by stream flow and groundwater flow, as *runoff*. Figure 4.1 illustrates this simple concept of evapotranspiration, precipitation, and runoff as components of the *hydrologic cycle*. This representation of the hydrologic cycle is highly simplified, masking many of the complexities of timing and distribution. Snowfall may remain locked in polar ice for thousands of years; groundwater may emerge at the surface contributing to surface water flow and vice versa; droughts and floods attest to the erratic rates of precipitation; and our own activities, most

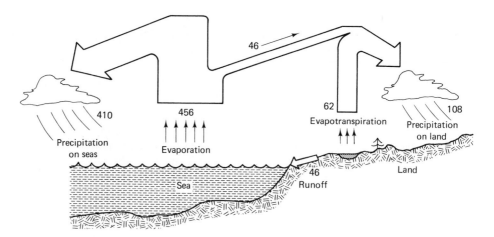

Figure 4.1 The hydrologic cycle. Units are 10^{12} m³/year. (Data from Budyko, 1974.)

importantly those contributing to global climate change, are likely to be in the process of significantly altering these balances.

As the data in Figure 4.1 indicate, nearly 60 percent of the precipitation falling on the earth's land masses is eventually returned to the atmosphere as evapotranspiration. Of the roughly 40 percent that does not evaporate, though, most collects on the surface, flowing into streams and rivers and emptying into the oceans, while some seeps into the soil to become underground water that slowly moves toward the seas. This combined groundwater and surface water runoff is a renewable supply of fresh water that can potentially be used year after year without ever depleting the freshwater resources of the world.

While the rates of evaporation, precipitation, and runoff are obviously important, the amounts of water stored in various locations and forms are also critical. It has already been mentioned that almost all of the world's water is contained in the oceans. The remainder is distributed as shown in Table 4.1, with the significant conclusion being that only about 0.3 percent of all water is available as fresh water for possible human use. That amount decreases by perhaps a factor of 100 if water that is inaccessible or too polluted is omitted.

Water Usage in the United States

In the United States, we pipe, pump, or divert an amount of fresh and saline water equivalent to about 40 percent of our runoff to meet our domestic, industrial, cooling water, rural, and irrigation needs. This 670 billion m³/year of water demand is about the same as the average flow of the Mississippi River.

At first glance, to be "using" an amount of water equal to 40 percent of our runoff might seem like we are already beginning to approach the physical limits of

TABLE 4.1 STOCKS OF WATER ON EARTH

Location	Amount (10^{15} m³)	Percentage of world supply
Oceans	1350	97.2
Icecaps and glaciers	29	2.09
Groundwater within 1 km	4.2	0.30
Groundwater below 1 km	4.2	0.30
Freshwater lakes	0.125	0.009
Saline lakes and inland seas	0.104	0.007
Soil water	0.067	0.005
Atmosphere	0.013	0.0009
Water in living biomass	0.003	0.0002
Average in stream channels	0.001	0.00007

Source: Harte (1985).

what is available. That figure is somewhat misleading, however. It is misleading, in part because some of that is saline water (about 16 percent, mostly used for power plant cooling), but mostly it is misleading because it does not account for the fact that water that is withdrawn and used for some purpose can usually be returned and reused again, provided it is not too polluted. It is important then, to distinguish between water *withdrawals,* water *returns,* and *consumptive* uses. Water that is withdrawn from a source is either consumed or returned:

$$\text{Withdrawals} = \text{Consumption} + \text{Returns} \qquad (4.1)$$

Water is said to be consumed when it is evaporated, transpired, incorporated into products or crops, consumed by people or livestock, or otherwise made unavailable for further reuse. About one-fourth of our water withdrawals are consumed, with most of that consumption resulting from evapotranspiration of irrigation water.

If we consider consumptive uses of water as the most critical indicator of approaching limits, it would appear that water resources in the United States will be more than adequate far into the future. Unfortunately, much of our annual runoff and water returns are not in the right place at the right time, or of high enough quality to be useful. As a result, water shortages are predicted for most regions of the country in the early part of the next century, with the western half of the United States expected to have the most difficult time meeting its needs.

Table 4.2 shows freshwater use in the United States broken down into irrigation, municipal (both residential and commercial), industrial (manufacturing and minerals production), and thermal electric power (steam plant cooling water). It includes amounts withdrawn, consumed, and returned. About one-fourth of U.S. freshwater needs are supplied from groundwater sources and the remaining three-fourths are derived from surface water. That distribution is very unevenly divided across the country. In certain states, the fraction of withdrawals derived from

TABLE 4.2   U.S. FRESHWATER USAGE (10^9 m³/year)

Usage	Withdrawals	Consumption	Returns
Irrigation	223 (48%)	122 (84%)	101
Municipal	40 (9%)	10 (7%)	30
Manufacturing/minerals	80 (17%)	11 (8%)	69
Thermal electric power	123 (26%)	2 (1%)	121
Totals	467 (100%)	145 (100%)	322
Groundwater withdrawals	113 (24%)		
Surface water withdrawals	354 (76%)		

Source: U.S. Water Resources Council (1978). Totals may not add due to rounding.

groundwater is much higher. In California, for example, it is closer to 50 percent. In other states it is nearly zero.

While water to cool power plants accounts for a significant fraction of freshwater withdrawals, essentially all of that water is returned. Moreover, it is re-

TABLE 4.3   EXAMPLES OF (FRESH AND SALINE) WITHDRAWALS TO SUPPLY VARIOUS END USES

Use	Average withdrawals	
	Liters	Gallons
Total average home use per person per day	**340**	**90**
Drinking water per person per day	2	0.5
Cooking per person per day	23	6
Watering lawn (per minute)	38	10
Toilet, per flush	19	5
Taking a shower (per minute)	8	2
Taking a bath	135	35
Washing machine per load	230	60
Total irrigation per person per day	**2 540**	**670**
One egg	150	40
Glass of milk	380	100
One pound of flour	285	75
One pound of rice	2 120	560
One pound of grain-fed beef	3 030	800
One pound of cotton	7 730	2 040
Total industrial and commercial per person per day	**4 520**	**1 190**
Cooling water per person per day	3 710	980
Refine one gallon of gasoline from crude oil	38	10
One Sunday newspaper	1 060	280
One pound of aluminum	3 790	1 000
One automobile	380 000	100 000

Source: Based on U.S. Geological Survey (1984).

turned in the immediate vicinity of the area from which it was withdrawn. The principal form of pollution associated with those returns is heat, which dissipates reasonably quickly to the atmosphere. Discounting those withdrawals changes the picture considerably making irrigation water for agriculture dominate. Excluding power plant cooling, irrigation accounts for 65 percent of freshwater withdrawals and 85 percent of consumption. In some areas, the predominant role of agriculture is even more dramatic. In California, for example, a state that has historically had an extremely difficult time meeting its needs for water, irrigation accounts for about 90 percent of both freshwater withdrawals and consumption (excluding cooling water). To help understand the significance of that number, consider the fact that during drought years, California municipal users have been asked to bear the brunt of water conservation efforts, in spite of the fact that they account for only a few percent of total water usage.

It is interesting to note the amounts of water withdrawn to supply us with various common commodities. Table 4.3 shows some examples. Notice that average home use is estimated at around 90 gallons of water per person per day, which is an almost insignificant fraction of the nearly 2000-gallon total water withdrawals per person per day for all our society's uses.

4.3 WATER POLLUTANTS

There are a number of ways to approach the problem of water pollution. We begin in this chapter by giving a brief description of the principal categories of water pollutants, and then we highlight some of the most common problems associated with various pollutants as they are discharged into groundwater, rivers, and lakes.

Oxygen-Demanding Wastes

One of the most important water quality parameters is the amount of dissolved oxygen (DO) present. As mentioned in Chapter 2, the saturated value of dissolved oxygen in water is modest, on the order of 8 to 15 mg/L, depending on temperature and salinity. Minimum amounts required for a healthy fish population may be as high as 5–8 mg/L for active species, such as trout, or as low as 3 mg/L for less desirable species, such as carp.

Oxygen-demanding wastes are substances that oxidize in the receiving body of water, reducing the amount of DO available. As DO drops, fish and other aquatic life are threatened and, in the extreme case, killed. In addition, as dissolved oxygen levels fall, undesirable odors, tastes, and colors reduce the acceptability of the water as a domestic supply and reduce its attractiveness for recreational uses. Oxygen-demanding wastes are usually biodegradable organic substances contained in municipal wastewaters or in effluents from certain industries, such as food processing and paper production. In addition, the oxidation of certain inorganic compounds may also contribute to the oxygen demand. Even naturally occurring organic matter, such as leaves and animal droppings, that find their way into surface water add to the DO depletion.

There are several measures of oxygen demand commonly used. The *chemical oxygen demand,* or COD, is the amount of oxygen needed to chemically oxidize the wastes, while the *biochemical oxygen demand,* or BOD, is the amount of oxygen required by microorganisms to biologically degrade the wastes. BOD has traditionally been the most important measure of the strength of organic pollution, and the amount of BOD reduction in a wastewater treatment plant is a key indicator of process performance. It will be discussed more fully in later sections.

Pathogens

It has long been known that contaminated water is responsible for the spread of many contagious diseases. In a famous study published in 1849, Dr. John Snow provided some of the earliest evidence of the relationship between human waste, drinking water, and disease. He noted that individuals who drank from a particular well on Broad Street in London were much more likely to become victims of a local cholera epidemic than those from the same neighborhood who drank from a different well. He not only found a likely source of the contamination, sewage from the home of a cholera patient, but he was able to effectively end the epidemic by simply removing the handle from the pump on the Broad Street well. It was not until later in the nineteenth century, however, when Pasteur and others convincingly established the germ theory of disease, that the role of pathogenic microorganisms in such epidemics was understood.

Pathogens are disease-producing organisms that grow and multiply within the host. Examples of pathogens associated with water include *bacteria* responsible for cholera, bacillary dysentery, typhoid, and paratyphoid fever; *viruses* responsible for infectious hepatitis and poliomyelitis; *protozoa,* which cause amebic dysentery and giardiasis; and *helminths,* or parasitic worms, which cause diseases such as schistosomiasis and dracontiasis (guinea worm). The intestinal discharges of an infected individual, a carrier, may contain billions of these pathogens, which, if allowed to enter the water supply, can cause epidemics of immense proportions. Carriers may not even necessarily exhibit symptoms of their disease, which makes it even more important to carefully protect all water supplies from any human waste contamination.

Not that long ago, even developed countries such as the United States experienced numerous epidemics of waterborne diseases such as typhoid and cholera. At the turn of the century, typhoid, for example, was killing approximately 28 000 Americans each year (US EPA, 1986). In one tragic incident in 1885, almost 90 000 people in Chicago died of typhoid or cholera when untreated sewage was drawn directly into the drinking water supply during a severe storm. It was only after the advent of chlorination, which began in the United States in

1908, that outbreaks of waterborne diseases such as these became rare. Even now, however, inadequate sanitation in developing countries continues to contribute to high rates of disease and death. The World Health Organization, for example, estimates that approximately 80 percent of all sickness in the world is attributable to inadequate water or sanitation. This includes roughly 10–20 million children each year who die of diarrheal diseases alone (Morrison, 1983).

Contaminated water caused by poor sanitation can lead to both waterborne and water-contact diseases. *Waterborne* diseases are those acquired by ingestion of pathogens not only in drinking water, but also from water that makes it into a person's mouth from washing food, utensils, and hands. In developing countries, such water is often taken from open wells or streams that are easily polluted. Simply enclosing a well and replacing the dirty rope and bucket with a less-easily contaminated handpump can dramatically reduce the incidence rate of these diseases. In the United States, most waterborne diseases are adequately controlled and we do not need to worry about water coming from the tap. However, as every backpacker knows these days, it is no longer safe to drink surface water from even the most sparkling stream because of the risk of giardiasis caused by the *Giardia lamblia* protozoa. *Giardia* cysts passed through the feces of carriers pose an unusual threat to surface water and even to municipal supply systems. They can be carried by wild animals as well as humans, may survive for months in the environment, and are not easily destroyed by chlorination.

Water-contact diseases do not even require that individuals ingest the water. Schistosomiasis (bilharzia) is the most common water-contact disease in the world, affecting approximately 200 million people. It is spread by free swimming larva in the water, called *cercaria,* that attach themselves to human skin, penetrate it, and enter the bloodstream. Cercaria mature in the liver into worms that lay masses of eggs on the walls of the intestine. When these eggs are excreted into water, they hatch and have only a few hours to find a snail host in which they develop into new cercaria. Cercaria excreted by the snails, then, have a few days to find another human host, continuing the cycle. Continuation of the cycle requires continued contamination by schistosomiasis carriers in waters that are still enough to allow snails to thrive. Unfortunately, development projects such as dams and irrigation canals, built in countries with poor sanitation, often lead to an increase in schistosomiasis by creating still water conditions in which the intermediate hosts, the snails, can multiply.

Water also plays an indirect role in other diseases common in developing countries. Insects that breed in water, or bite near water, are responsible for the spread of malaria, affecting some 160 million people and killing 1 million each year. Yellow fever, sleeping sickness, and river blindness are spread in this same way. Inadequate supplies of water for personal hygiene results in skin diseases such as scabies, leprosy, and yaws, as well as eye diseases such as trachoma and conjunctivitis. Table 4.4 summarizes some of these water-related problems.

TABLE 4.4   DISEASES RELATED TO POOR WATER SUPPLY AND SANITATION

Type	Spread by	Examples	Prevalence
Waterborne	Drinking water con- taminated by patho- gens, or washing hands, food, or utensils in contami- nated water.	Typhoid, cholera, dysentery, diarrhea, hepatitis, guinea worm disease.	6 million children under five die from diarrhea each year; 10–20 million chil- dren die each year from all types of diarrheal disease. 10 to 48 million annual guinea worm cases.
Water-contact	Invertebrates living in water which act as carriers (vectors).	Schistosomiasis (bil- harzia), leptospiro- sis, tularemia.	Over 200 million people infected worldwide with schistosomiasis.
Water-hygiene	Inadequate supplies of water for per- sonal hygiene.	Skin diseases: scabies, leprosy, yaws. Eye diseases: trachoma, conjunctivitis.	500 million infected with trachoma (blindness occurs in severe cases); prevalence of skin diseases approaches 80% of population in some areas.

Source: Adapted from Morrison (1983).

Nutrients

Nutrients are chemicals, such as nitrogen, phosphorus, carbon, sulfur, calcium, potassium, iron, manganese, boron, and cobalt, that are essential to the growth of living things. In terms of water quality, nutrients can be considered as pollutants when their concentrations are sufficient to allow excessive growth of aquatic plants, particularly algae. When nutrients stimulate the growth of algae, the at- tractiveness of the body of water for recreational uses, as a drinking water supply, and as a viable habitat for other living things can be adversely affected.

Nutrient enrichment can lead to blooms of algae which eventually die and decompose. Their decomposition removes oxygen from the water, potentially leading to levels of DO that are insufficient to sustain normal life forms. Algae and decaying organic matter add color, turbidity, odors, and objectionable tastes to water that are difficult to remove and that may greatly reduce its acceptability as a domestic water source. The process of nutrient enrichment, called *eutrophica- tion,* is an especially important one in lakes, and it is described more fully in Section 4.5.

Aquatic species require a long list of nutrients for growth and reproduction, but from a water quality perspective, the three most important ones are carbon, nitrogen, and phosphorus. Plants require relatively large amounts of each of these three nutrients and unless all three are available, growth will be limited. The nutrient that is least available relative to the plant's needs is called the *limiting nutrient*. This suggests that algal growth can be controlled by identifying and reducing the supply of that particular nutrient. Carbon is usually available from a number of natural sources including alkalinity, dissolved carbon dioxide from the atmosphere, and decaying organic matter, so it is not often the limiting nutrient. Rather, it is usually either nitrogen or phosphorus that controls algal growth rates. In general, seawater is most often limited by nitrogen, while freshwater lakes are most often limited by phosphorus (Welch, 1980).

Major sources of both nitrogen and phosphorus include municipal wastewater discharges, runoff from animal feedlots, and chemical fertilizers. In addition, certain bacteria and blue-green algae can obtain nitrogen directly from the atmosphere. These life forms are usually abundant in lakes that have high rates of biological productivity, making the control of nitrogen in such lakes extremely difficult. Certain forms of acid rain can also contribute nitrogen to lakes. While there are several special sources of nitrogen, the only unusual source of phosphorus is from detergents. When phosphorus is the limiting nutrient in a lake that is experiencing an algal problem, it is especially important to limit the nearby use of phosphate in detergents.

Not only is nitrogen capable of contributing to eutrophication problems, but when found in drinking water a particular form of it can pose a serious public health threat. Nitrogen in water is commonly found in the form of nitrate (NO_3), which is itself not particularly dangerous. However, certain bacteria commonly found in the intestinal tract of infants can convert nitrates to highly toxic nitrites (NO_2). Nitrites have a greater affinity for hemoglobin in the bloodstream than does oxygen, and when they replace that needed oxygen, a condition known as *methemoglobinemia* results. The resulting oxygen starvation causes a bluish discoloration of the infant; hence, it is commonly referred to as the "blue baby" syndrome. In extreme cases the victim may die from suffocation. Usually after the age of about 6 months, the digestive system of a child is sufficiently developed that this syndrome does not occur.

Salts

Water naturally accumulates a variety of dissolved solids, or *salts,* as it passes through soils and rocks on its way to the sea. These salts typically include such cations as sodium, calcium, magnesium, and potassium, and anions such as chloride, sulfate, and bicarbonate. While a careful analysis of salinity would result in a list of the concentrations of the primary cations and anions, a simpler, more commonly used measure of salinity is the concentration of *total dissolved solids* (TDS). As a rough approximation, *fresh* water can be considered to be water with less than 1500 mg/L TDS; *brackish* waters may have TDS values up to 5000 mg/L;

and, *saline* waters are those with concentrations above 5000 mg/L (Tchobanoglous and Schroeder, 1985). Seawater contains 30 000–34 000 mg/L TDS.

The concentration of dissolved solids is an important indicator of the usefulness of water for various applications. Drinking water, for example, has a recommended maximum contaminant level for TDS of 500 mg/L. Livestock can tolerate higher concentrations. Upper limits for stock water concentrations quoted by the U.S. Geological Survey (1985) include poultry at 2860 mg/L, pigs at 4290 mg/L, and beef cattle at 10 100 mg/L. Of greater importance, however, is the salt tolerance of crops. As the concentration of salts in irrigation water increases above 500 mg/L, the need for careful water management to maintain crop yields becomes increasingly important. With sufficient drainage to keep salts from accumulating in the soil, up to 1500 mg/L TDS can be tolerated by most crops with little loss of yield (Frederick and Hanson, 1982), but at concentrations above 2100 mg/L, water is generally unsuitable for irrigation except for the most salt tolerant of crops.

All naturally occurring water has some amount of salt in it. In addition, many industries discharge high concentrations of salts, and urban runoff may contain large amounts in areas where salt is used to keep ice from forming on roads in the winter. While such human activities may increase salinity by adding salts to a given volume of water, it is more often the opposite process, the removal of fresh water by evaporation, that causes salinity problems. When water evaporates, the salts are left behind, and since there is less remaining fresh water to dilute them, their concentration increases.

Irrigated agriculture, especially in arid areas, is always vulnerable to an accumulation of salts due to this evapotranspiration on the cropland itself. The salinity is enhanced by the increased evaporation in storage reservoirs that typically accompany irrigation projects. In addition, irrigation drainage water may dissolve more salt from the soils with which it comes in contact, further increasing its salinity. As a result, irrigation drainage water is always higher in salinity than the supply water and with every reuse, its salt concentration increases even more. In rivers that are heavily used for irrigation, the salt concentration progressively increases downstream as the volume of water available to dilute salts decreases due to evaporation and diversions, and as the salt load increases due to salty drainage water returning from irrigated lands. As an example, Table 4.5 shows decreasing flows and increasing TDS for the Rio Grande as it travels from New Mexico to Texas.

It has been estimated that about one-third of the irrigated lands in the western part of the United States have a salinity problem that is increasing with time, the most seriously affected regions being in the Lower Colorado River Basin and the west side of the San Joaquin Valley in California (Frederick and Hanson, 1982). Salinity problems are also having major impacts on irrigated lands in Iraq, Pakistan, India, Mexico, Argentina, Mali, and North Africa, among others. The collapse of ancient civilizations, such as those that once flourished in the Fertile

TABLE 4.5 MEAN ANNUAL DISCHARGE AND TOTAL
DISSOLVED SOLIDS ALONG THE RIO GRANDE AS IT
TRAVELS THROUGH NEW MEXICO AND TEXAS

Station	Discharge (10^6 m^3)	Dissolved solids (mg/L)
Otowi Bridge, NM	1.33	221
San Marcial, NM	1.05	449
Elephant Butte Outlet, NM	0.97	478
Caballo Dam, NM	0.96	515
Leasburg Dam, NM	0.92	551
El Paso, TX	0.65	787
Fort Quitman, TX	0.25	1691

Source: Skogerboe and Law (1971).

Crescent in what is now Iraq, is now often attributed to the demise of irrigated agriculture caused by accumulating salt (Reisner, 1986).

One way to approach the salt problem is with desalination. Desalination technologies are available, but the energy required makes the economics of desalting unattractive in almost all circumstances. More commonly, the problem of accumulating salts in soils is dealt with by simply applying more irrigation water than is actually needed by the crops to flush the salts away. This increases costs, wastes water, which may not be abundantly available in the first place, and unless adequate drainage is available, it increases the likelihood that a rising water table will cause waterlogging, thus drowning plant roots in salt-laden water. Providing adequate drainage can be an expensive and challenging task involving extensive on-farm subsurface drainage systems coupled with a central drain and disposal system. Since irrigation return water contains not only salts but fertilizers and pesticides as well, finding an acceptable method of disposal is difficult.

Thermal Pollution

A large steam-electric power plant requires an enormous amount of cooling water. A typical nuclear plant, for example, warms about 40 m^3/s of cooling water by 10 °C as it passes through the plant's condenser. If that heat is released into a local river or lake, the resulting rise in temperature can dramatically affect life in the vicinity of the thermal plume.

There are some circumstances when warmed water might be considered beneficial. Within certain limits, thermal additions can promote fish growth, and fishing may actually be improved in the vicinity of a power plant. On the other hand, sudden changes in temperature caused by periodic plant outages, both planned and unanticipated, can make it difficult for the local ecology to ever acclimate. For some species, such as trout and salmon, any increase in temperature is undesirable.

As water temperature increases, two factors combine to make it more difficult for aquatic life to get sufficient oxygen to meet its needs. The first results from the fact that metabolic rates tend to increase with temperature, generally by about a factor of 2 for each 10 °C rise in temperature. This causes an increase in the amount of oxygen required by organisms. At the same time, the available supplies of dissolved oxygen are reduced both because waste assimilation is quicker, drawing down DO at a faster rate, and because the amount of DO that the water can hold decreases with temperature. Thus, as temperatures increase, the demand for oxygen goes up while the amount of DO available goes down.

Heavy Metals

In some contexts, the definition of a *metal* is based on physical properties. Metals are characterized by high thermal and electrical conductivity, high reflectivity and metallic luster, strength, and ductility. From a biological perspective, however, it is more common to use a broader definition that says a metal is an element that will give up one or more electrons to form a cation in an aqueous solution. With this latter definition, there are about 80 elements that can be called metals.

The term *heavy metal* is less precisely defined. In chemical terms it can refer to metals with specific gravity greater than about 4 or 5, but more often, the term is simply used to denote metals that are toxic. The list of toxic metals includes aluminum, arsenic, beryllium, bismuth, cadmium, chromium, cobalt, copper, iron, lead, manganese, mercury, nickel, selenium, strontium, thallium, tin, titanium, and zinc. Some of these metals, such as chromium and iron, are essential nutrients in our diets, but in higher doses are extremely toxic.

Metals may be inhaled, as is often the case with lead, for example, and they may be ingested. How well they are absorbed in the body depends somewhat on the particular metal in question. Some metal salts, such as those formed with lead, tin, and cadmium, are poorly absorbed, while the salts of others such as arsenic and thallium are almost completely absorbed. The most important route for the elimination of metals is via the kidneys. In fact, kidneys can be considered to be complex filters whose primary purpose is to eliminate toxic substances from the body. The kidneys contain millions of excretory units called nephrons, and chemicals that are toxic to the kidneys are called *nephrotoxins*. Cadmium, lead, and mercury are examples of nephrotoxic metals. Metals have a range of adverse impacts on the body, including nervous system and kidney damage, creation of mutations, and induction of tumors.

Pesticides

The term *pesticide* is used to cover a range of chemicals that kill organisms that humans consider undesirable and includes the more specific categories of insecticides, herbicides, rodenticides, and fungicides. There are three main groups of synthetic organic insecticides: *organochlorines* (also known as *chlorinated hydrocarbons*), *organophosphates,* and *carbamates.* In addition, a number of herbicides, including the chlorophenoxy compounds 2,4,5-T (which contains the impurity dioxin, which is one of the most potent toxins known) and 2,4-D are common water pollutants.

The most well-known organochlorine pesticide is DDT (dichlorodiphenyltrichloroethane) which has been widely used to control insects that carry such diseases such as malaria, typhus, and plague. By contributing to the control of these diseases, DDT is credited with saving literally millions of lives worldwide. In spite of its more recent reputation as a dangerous pesticide, in terms of human toxicity DDT is considered to be relatively safe. It was its impact on food chains, rather than human toxicity, that led to its ban. Organochlorine pesticides, such as DDT, have two properties that cause them to be particularly disruptive to food chains. They are very *persistent,* which means they last a long time in the environment before being broken down into other substances, and they are quite *soluble* in lipids, which means they easily accumulate in fatty tissue.

The accumulation of organochlorine pesticides in fatty tissue means that organisms at successively higher trophic levels in a food chain are consuming food that has successively higher concentrations of pesticide. At the top of the food chain, body concentrations of these pesticides are the highest, and it is there that organochlorine toxicity has been most recognizable. Birds, for example, are high on the food chain and it was the adverse effect of DDT on their reproductive success that focused attention on this particular pesticide. DDT interferes with calcium metabolism in birds, resulting in eggs with shells that are too thin to support the weight of the parent. The resulting difficulty to reproduce has been shown to affect a number of species, including peregrine falcons, bald eagles, ospreys, and brown pelicans.

This phenomenon in which the concentration of a chemical increases at higher levels in the food chain is known as *biomagnification* or *bioconcentration.* In Chapter 5, quantitative estimates of the bioconcentration factor for a number of important chemicals, including organochlorine pesticides, will be given (Table 5.10).

Other widely used organochlorines included methoxychlor, chlordane, heptachlor, aldrin, dieldrin, endrin, endosulfan, and Kepone. Animal studies have shown dieldrin, heptachlor, and chlordane produce liver cancers, and aldrin, dieldrin, and endrin have been shown to cause birth defects in mice and hamsters. Workers' exposure to Kepone in a manufacturing plant in Virginia showed severe neurological damage and the plant was ultimately closed. Given the ecosystem disruption, the toxicity, and the biological resistance to these pesticides that many insect species have developed, organochlorines have largely been replaced with organophosphates and carbamates.

The organophosphates, such as parathion, malathion, diazinon, TEPP (tetraethyl prophosphate), and dimethoate, are effective against a wide range of insects and they are not persistent. However, they are much more toxic than the organochlorines that they have replaced. They are rapidly absorbed through the skin, lungs, and gastrointestinal tract and, hence, unless proper precautions are taken, they are very hazardous to those who use them. Humans exposed to excessive amounts have shown a range of symptoms including tremor, confusion, slurred speech, muscle twitching, and convulsions.

Popular carbamate pesticides include propoxur, carbaryl, and aldicarb. Acute human exposure to carbamates has led to a range of symptoms, such as nausea, vomiting, blurred vision, and in extreme cases, convulsions.

Volatile Organic Compounds

Volatile organic compounds (VOCs) are among the most commonly found contaminants in groundwater. They are often used as solvents in industrial processes and a number of them are either known or suspected carcinogens or mutagens. Their volatility means they are not often found in concentrations above a few micrograms per liter in surface waters, but in groundwater their concentrations can be hundreds or thousands of times higher. Their volatility also suggests the most common method of treatment, which is to aerate the water to encourage them to vaporize.

Five VOCs are especially toxic and their presence in drinking water is cause for special concern. The most toxic of the five is *vinyl chloride* (chloroethylene). It is a known human carcinogen used primarily in the production of polyvinyl chloride resins. *Tetrachloroethylene* is used as a solvent, as a heat transfer medium, and in the manufacture of chlorofluorocarbons. It causes tumors in animals but there is inadequate evidence to call it a human carcinogen. Of the five, it is the one most commonly found in groundwater. *Trichloroethylene* (TCE) is a solvent that was quite commonly used to clean everything from electronics parts to jet engines and septic tanks. It is a suspected carcinogen and it is among the most frequently found contaminants in groundwater. *1,2-Dichloroethane* is a metal degreaser that is also used in the manufacture of a number of products including vinyl chloride, tetraethyllead, fumigants, varnish removers, and soap compounds. Though it is not a known carcinogen, high levels of exposure are known to cause injury to the central nervous system, liver, and kidneys. It is also a common groundwater contaminant that is quite soluble, making it one of the more difficult to remove by air stripping. *Carbon tetrachloride* was a common household cleaning agent that is now more often used in grain fumigants, fire extinguishers, and solvents. It is very toxic if ingested; only a few milliliters can produce death. It is relatively insoluble in water, however, and so it is only occasionally found in contaminated groundwater.

4.4 SURFACE WATER QUALITY: RIVERS AND STREAMS

Surface water is obviously highly susceptible to contamination. It has historically been the most convenient sewer for industry and municipalities alike, while at the same time it is the source of the majority of our water for all purposes. One particular category of pollutants, oxygen-demanding wastes, has been such a pervasive surface-water problem, affecting both moving water and still water, that it will be given special attention.

Biochemical Oxygen Demand (BOD)

When biodegradable organic matter is released into a body of water, microorganisms, especially bacteria, feed on the wastes, breaking it down into simpler organic and inorganic substances. When this decomposition takes place in an *aerobic* environment, that is, in the presence of oxygen, the process produces nonobjectionable, stable end products such as carbon dioxide (CO_2), sulfate (SO_4), orthophosphate (PO_4), and nitrate (NO_3). A simplified representation of aerobic decomposition is given by the following:

$$\text{Organic matter} + O_2 \xrightarrow{\text{microorganisms}} CO_2 + H_2O + \text{New cells} + \text{Stable products}$$

When insufficient oxygen is available, the resulting *anaerobic* decomposition is performed by completely different microorganisms. They produce end products that can be highly objectionable, including hydrogen sulfide (H_2S), ammonia (NH_3), and methane (CH_4). Anaerobic decomposition can be represented by the following:

$$\text{Organic matter} \xrightarrow{\text{microorganisms}} CO_2 + CH_4 + \text{New cells} + \text{Unstable products}$$

The methane produced is physically stable, biologically degradable, and a potent greenhouse gas. When emitted from bodies of water it is often called swamp gas. It is also generated in the anaerobic environment of landfills, where it is sometimes collected and used as an energy source.

The amount of oxygen required by microorganisms to oxidize organic wastes aerobically is called the biochemical oxygen demand (BOD). BOD may have various units, but most often it is expressed in milligrams of oxygen required per liter of wastes (mg/L) or the equivalent g/m^3.

Five-Day BOD Test

The total amount of oxygen that will be required for biodegradation is an important measure of the impact that a given waste stream will have on the receiving body of water. While we could imagine a test in which the oxygen required to completely degrade a sample of waste would be measured, such a test would require an extended period of time (several weeks), making it impractical. As a result, it has become standard practice simply to measure and report the oxygen demand over a shorter, restricted period of 5 days, realizing that the ultimate demand is considerably higher.

The 5-day BOD, or BOD_5, is the total amount of oxygen consumed by microorganisms during the first 5 days of biodegradation. In its simplest form, a

BOD_5 test would involve putting a sample of waste into a stoppered bottle, measuring the concentration of dissolved oxygen in the sample at the beginning of the test and again 5 days later. The difference in DO would be the 5-day BOD. Light must be kept out of the bottle to keep algae from adding oxygen by photosynthesis and the stopper is used to keep air from replenishing DO that has been removed by biodegradation. To standardize the procedure, the test is run at a fixed temperature of 20 °C. Since the oxygen demand of typical waste is several hundred milligrams per liter, and since the saturated value of DO for water at 20 °C is only 9.1 mg/L, it is usually necessary to dilute the sample to keep final DO above zero. If during the 5 days the DO drops to zero, then the test is invalid, since more oxygen would have been removed had more been available.

The 5-day BOD of a diluted sample is given by

$$BOD_5 = \frac{DO_i - DO_f}{P} \qquad (4.2)$$

where DO_i = the initial dissolved oxygen of the diluted wastewater
DO_f = the final DO of the diluted wastewater
P = the dilution fraction = $\dfrac{\text{Volume of wastewater}}{\text{Volume of wastewater plus dilution water}}$

A standard BOD bottle holds 300 mL, so P is just the volume of wastewater divided by 300 mL.

Example 4.1   Unseeded 5-Day BOD Test

A 10.0-mL sample of sewage mixed with enough water to fill a 300-mL bottle has an initial DO of 9.0 mg/L. To help assure an accurate test, it is desirable to have at least a 2.0-mg/L drop in DO during the 5-day run, and the final DO should be at least 2.0 mg/L. For what range of BOD_5 would this dilution produce the desired results?

Solution   The dilution fraction is $P = 10/300$. To get at least a 2.0-mg/L drop in DO, using (4.2)

$$BOD_5 \geq \frac{DO_i - DO_f}{P} = \frac{2.0 \text{ mg/L}}{10/300} = 60 \text{ mg/L}$$

and to assure at least 2.0 mg/L of DO remaining after 5 days requires that

$$BOD_5 \leq \frac{9.0 - 2.0 \text{ mg/L}}{10/300} = 210 \text{ mg/L}$$

So this dilution will be satisfactory for BOD_5 values between 60 and 210 mg/L.

So far we have assumed that the dilution water added to the waste sample has no BOD of its own, which would be the case if pure water were added. In some cases it is necessary to seed the dilution water with microorganisms to assure there is an adequate bacterial population to carry out the biodegradation. In such cases, to find the BOD of the waste itself, it is necessary to subtract the oxygen demand caused by the seed from the demand in the mixed sample of waste and dilution water.

To be able to sort out the effect of seeded dilution water from the waste itself, two BOD bottles must be prepared, one containing just the seeded dilution water and the other containing the mixture of both the wastewater and seeded dilution water. The change in DO in the bottle containing just seeded dilution water (called the "blank"), as well as the change in DO in the mixture are then noted. The oxygen demand of the waste itself (BOD_w) can then be determined as follows:

$$BOD_m \, V_m = BOD_w \, V_w + BOD_d \, V_d \tag{4.3}$$

where BOD_m = BOD of the mixture of wastewater and seeded dilution water
BOD_w = BOD of the wastewater alone
BOD_d = BOD of the seeded dilution water alone
V_w = the volume of wastewater in the mixture
V_d = the volume of dilution water in the mixture
V_m = the volume of the mixture = $V_d + V_w$

As before, let P equal the fraction of the mixture that is wastewater = V_w/V_m, so that $(1 - P)$ is the fraction of the mixture which is seeded dilution water = V_d/V_m. Rearranging (4.3) gives

$$BOD_w = BOD_m \frac{V_m}{V_w} - BOD_d \left(\frac{V_d}{V_w} \times \frac{V_m}{V_m} \right) \tag{4.4}$$

where the last term has been multiplied by unity (V_m/V_m). A slight rearrangement of (4.4) yields

$$BOD_w = \frac{BOD_m}{V_w/V_m} - BOD_d \frac{V_d/V_m}{V_w/V_m} \tag{4.5}$$

Substituting the definitions of P and $(1 - P)$ into Eq. 4.5 gives

$$BOD_w = \frac{BOD_m - BOD_d(1 - P)}{P} \tag{4.6}$$

Since $BOD_m = DO_i - DO_f$ and $BOD_d = B_i - B_f$

where B_i = initial DO in the seeded dilution water (blank)
B_f = final DO in the seeded dilution water

our final expression for the BOD of the waste itself is thus

$$BOD_w = \frac{(DO_i - DO_f) - (B_i - B_f)(1 - P)}{P} \tag{4.7}$$

Example 4.2   A Seeded BOD Test

A test bottle containing just seeded dilution water has its DO level drop by 1.0 mg/L in a 5-day test. A 300-mL BOD bottle filled with 15 mL of wastewater and the rest seeded dilution water (sometimes expressed as a dilution of 1 : 20) experiences a drop of 7.2 mg/L in the same time period. What would be the 5-day BOD of the waste?

Solution   The dilution factor P is

$$P = \frac{15}{300} = 0.05$$

Using (4.7), the 5-day BOD of the waste would be

$$BOD_5 = \frac{7.2 - 1.0(1 - 0.05)}{0.05} = 125 \text{ mg/L}$$

Modeling BOD as a First-Order Reaction

It is often assumed that the rate of decomposition of organic wastes is proportional to the amount of waste available. If we let L_t represent the amount of oxygen demand left after time t, then, assuming a first-order reaction, we can write

$$\frac{dL_t}{dt} = -kL_t \tag{4.8}$$

where k = BOD reaction rate constant (time^{-1})

The solution to (4.8) is

$$L_t = L_0 e^{-kt} \tag{4.9}$$

where L_0 is the *ultimate carbonaceous oxygen demand*. It is the total amount of oxygen required by microorganisms to oxidize the carbonaceous portion of the waste to simple carbon dioxide and water. (Later we will see there is an additional demand for oxygen associated with the oxidation of nitrogen compounds.) The ultimate carbonaceous oxygen demand is the sum of the amount of oxygen consumed by the waste in the first t days (BOD_t), plus the amount of oxygen remaining to be consumed after time t. That is,

$$L_0 = BOD_t + L_t \tag{4.10}$$

Combining (4.9) and (4.10) gives us

$$BOD_t = L_0(1 - e^{-kt}) \tag{4.11}$$

A graph of Eqs. 4.9 and 4.11 is given in Figure 4.2.

Example 4.3   Estimating L_0 from BOD$_5$

The dissolved oxygen in an unseeded sample of diluted waste having an initial DO of 9.0 mg/L is measured to be 3.0 mg/L after 5 days. The dilution factor P is 0.030, and the reaction rate constant k is 0.22/day.

a. What is the 5-day BOD of the waste?

b. What would be the ultimate carbonaceous BOD?

c. What would be the remaining oxygen demand after 5 days?

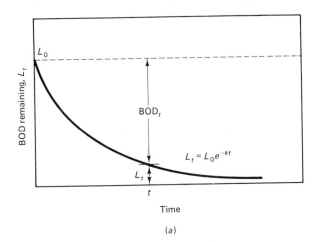

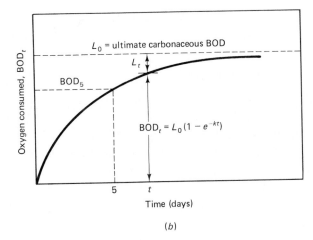

Figure 4.2 Idealized carbonaceous oxygen demand: (*a*) the BOD remaining as a function of time, and (*b*) the oxygen consumed.

Solution

a. From (4.2), the oxygen demand in the first 5 days is

$$\text{BOD}_5 = \frac{\text{DO}_i - \text{DO}_f}{P} = \frac{9.0 - 3.0}{0.030} = 200 \text{ mg/L}$$

b. We can find the ultimate BOD by rearranging Eq. 4.11:

$$L_0 = \frac{\text{BOD}_5}{1 - e^{-kt}} = \frac{200}{1 - e^{-0.22 \times 5}} = 300 \text{ mg/L}$$

c. After 5 days, 200 mg/L of oxygen demand out of the total 300 mg/L would have already been used. The remaining oxygen demand would therefore be 100 mg/L.

The BOD Reaction Rate Constant, k

The BOD reaction rate constant, k, is a factor that indicates the rate of biodegradation of wastes. As k increases, the rate at which dissolved oxygen is used increases, although the ultimate amount required, L_0, does not change. The reaction rate will depend on a number of factors, including the nature of the waste itself (some, such as simple sugars and starches, degrade easier than others, like cellulose), the ability of the available microorganisms to degrade the wastes in question (it may take some time for a healthy population of organisms to be able to thrive on the particular waste in question), and the temperature (as temperatures increase, so does the rate of biodegradation).

Some typical values of the BOD reaction rate constant, at 20 °C, are given in Table 4.6. Notice that raw sewage has a higher rate constant than either well-treated sewage or polluted river water. This is because raw sewage contains a larger proportion of easily degradable organics that exert their oxygen demand quite quickly, leaving a remainder that decays more slowly.

The rate of biodegradation of wastes increases with increasing temperature. To account for these changes, the reaction rate constant k is often modified using the following equation:

$$k = k_{20}\theta^{(T-20)} \tag{4.12}$$

where k_{20} is the reaction rate constant at the standard 20 °C laboratory reference temperature, and k is the reaction rate at a different temperature T (expressed in °C). θ is a temperature coefficient that is itself somewhat temperature dependent. The most commonly used value for θ is 1.047, which will be sufficiently accurate for our purposes.

TABLE 4.6 TYPICAL VALUES FOR THE BOD RATE
CONSTANT[a] k AT 20 °C

Sample	k (day^{-1})
Raw sewage	0.35–0.70
Well-treated sewage	0.10–0.25
Polluted river water	0.10–0.25

[a] Note these rate constants, as well as all others in this chapter, assume exponentials that are to the base e. Some sources use rate constants to the base 10.

Source: Based on Davis and Cornwell (1985).

Example 4.4 Temperature Dependence of BOD$_5$

In Example 4.3 the wastes had an ultimate BOD equal to 300 mg/L. At 20 °C, the 5-day BOD was 200 mg/L and the reaction rate constant was 0.22/day. What would be the 5-day BOD of this waste at 25 °C?

Solution   First we will adjust the reaction rate constant with Eq. 4.12, using a value of θ equal to 1.047:

$$k = k_{20}\theta^{(T-20)} = 0.22(1.047)^{(25-20)} = 0.277/\text{day}$$

So, from (4.11),

$$\text{BOD}_5 = L_0(1 - e^{-k5}) = 300(1 - e^{-0.277 \times 5}) = 225 \text{ mg/L}$$

Notice that the ultimate BOD is not temperature dependent but the 5-day BOD is. The same total amount of oxygen is required, but as temperatures increase, it gets used sooner.

Nitrification

So far, it has been assumed that the only oxygen demand is associated with the biodegradation of the carbonaceous portion of the wastes. There is a significant additional demand, however, caused by the oxidation of nitrogen compounds that we will now briefly describe.

Nitrogen is the critical element required for protein synthesis and, hence, is essential to life. When living things die, or excrete waste products, nitrogen that was tied to complex organic molecules is converted to ammonia by bacteria and fungi. Then, in aerobic environments nitrite bacteria (*Nitrosomonas*) convert ammonia to nitrite (NO_2^-), and nitrate bacteria (*Nitrobacter*) convert nitrite to nitrate (NO_3^-). This process, called *nitrification,* can be represented with the following two reactions:

$$2NH_3 + 3O_2 \xrightarrow{\textit{Nitrosomonas}} 2NO_2^- + 2H^+ + 2H_2O \qquad (4.13)$$

$$2NO_2^- + O_2 \xrightarrow{\textit{Nitrobacter}} 2NO_3^- \qquad (4.14)$$

Nitrification is just one part of the biogeochemical cycle for nitrogen, which is shown in Figure 4.3. As is suggested there, nitrogen exists in many forms as it moves through the biosphere. In the atmosphere it is principally in the form of molecular nitrogen (N_2), with a small but important fraction being nitrous oxide (N_2O). (Nitrous oxide is a greenhouse gas that will be considered again in Chapter 8.) Nitrogen in the form of N_2 is unusable by plants and must first be transformed into either ammonia (NH_3) or nitrate (NO_3^-) in the process called *nitrogen fixation.* Nitrogen fixation occurs during electrical storms when N_2 oxidizes, combines with water, and is rained out as HNO_3. Certain bacteria and blue-green algae are also capable of fixing nitrogen. Under anaerobic conditions, certain denitrifying bacteria are capable of reducing NO_3 back into NO_2 and N_2, completing the nitrogen cycle.

While the entire nitrogen cycle obviously is important, our concern in this section is with the nitrification process itself in which organic nitrogen is converted to ammonia, nitrite, and nitrate. As illustrated in Figure 4.4, the process proceeds sequentially and it is a matter of days before the rate of oxidation of ammonia is sufficient to create a significant oxygen demand. To distinguish this

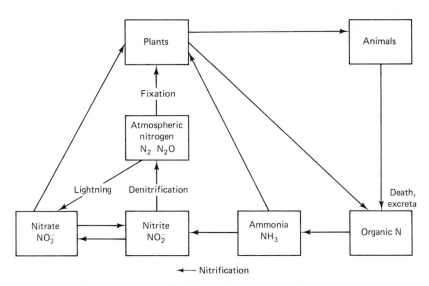

Figure 4.3   A simplified biogeochemical cycle for nitrogen.

nitrogenous biochemical oxygen demand from the *carbonaceous biochemical oxygen demand,* the two are sometimes specified separately as NBOD and CBOD, respectively.

Figure 4.5 illustrates these two demands as they might be exerted for typical municipal wastes. Notice the NBOD does not normally begin to exert itself for at

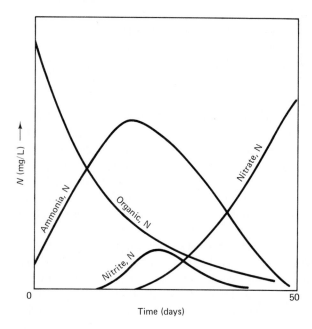

Figure 4.4   Changes in nitrogen forms in polluted water under aerobic conditions. (*Source:* Sawyer and McCarty, *Chemistry for Environmental Engineering,* 3d ed., © 1978. Reprinted by permission of McGraw-Hill, Inc.)

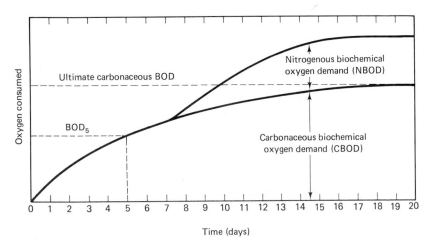

Figure 4.5 Illustrating the carbonaceous and nitrogenous biochemical oxygen demand.

least 5–8 days, so most 5-day BOD tests are not affected by nitrification. In fact, since nitrification would complicate the interpretation of test results, it is now an accepted practice to modify wastes in a way that will inhibit nitrification during that 5-day period.

A stoichiometric analysis of (4.13) and (4.14) allows us to quantify the oxygen demand associated with nitrification, as the following example illustrates.

Example 4.5 Nitrogenous Oxygen Demand

Some domestic wastewater has 30 mg/L of nitrogen either in the form of organic nitrogen or ammonia. Assuming that very few new cells of bacteria are formed during the nitrification of the waste (that is, that the oxygen demand can be found from a simple stoichiometric analysis of the nitrification reactions given above), find

a. The ultimate nitrogenous oxygen demand.

b. The ratio of the ultimate NBOD to the concentration of nitrogen in the waste.

Solution

a. Combining the two nitrification reactions (4.13) and (4.14) yields

$$NH_3 + 2O_2 \longrightarrow NO_3^- + H^+ + H_2O$$

The molecular weight of NH_3 is 17 (14 + 3 × 1) and the molecular weight of O_2 is 32 (2 × 16). The above reaction indicates that one g-mole of NH_3 (17 g) requires two g-moles of O_2 (2 × 32 = 64 g). Since 17 g of NH_3 contains 14 g of N, and the concentration of N is 30 mg/L, we can find the final, or ultimate, NBOD:

$$NBOD = 30 \text{ mgN/L} \times \frac{17 \text{ g NH}_3}{14 \text{ g N}} \times \frac{64 \text{ g O}_2}{17 \text{ g NH}_3} = 137 \text{ mgO}_2/L$$

    b. The oxygen demand due to nitrification divided by the concentration of nitro-
       gen in the waste is

$$\frac{137 \text{ mgO}_2/\text{L}}{30 \text{ mgN}/\text{L}} = 4.57 \text{ mgO}_2/\text{mgN}$$

The total concentration of organic and ammonia nitrogen in wastewater is
known as the *total Kjeldahl nitrogen,* or TKN. As was demonstrated in the above
example, the nitrogenous oxygen demand can be estimated by multiplying the
TKN by 4.57. This is a result worth noting:

$$\text{Ultimate NBOD} \approx 4.6 \times \text{TKN} \tag{4.15}$$

Since untreated domestic wastewaters typically contain approximately 15–
50 mg/L of TKN, the oxygen demand caused by nitrification is considerable,
ranging from roughly 70 to 230 mg/L (Tchobanoglous and Schroeder 1985). For
comparison, typical raw sewage has an ultimate carbonaceous oxygen demand of
250–350 mg/L.

Other Measures of Oxygen Demand

In addition to the CBOD and NBOD measures already presented, there are
two other indicators that are sometimes used to describe the oxygen demand of
wastes. These are the *theoretical oxygen demand* (ThOD) and the *chemical oxygen demand* (COD).

As was described in Chapter 2, the theoretical oxygen demand is the amount
of oxygen required to completely oxidize a particular organic substance, as calculated from simple stoichiometric considerations. The result will be slightly higher
than the measured biochemical oxygen demand since actual biodegradation does
not completely oxidize all of the carbon to CO_2 (some of it is incorporated into
new bacterial cells). This measure is of limited usefulness in practice since it
presupposes a particular, single pollutant with a known chemical formula.

Some organic materials, such as cellulose, phenols, benzene, and tannic
acid, resist biodegradation. Others, such as pesticides and various industrial
chemicals, are nonbiodegradable because they are toxic to microorganisms. The
chemical oxygen demand, COD, is a measured quantity that does not depend
either on the ability of microorganisms to degrade the waste or on knowledge of
the particular substances in question. In a COD test, a strong chemical oxidizing
agent is used to oxidize the organics rather than relying on microorganisms to do
the job. The COD test is much quicker than a BOD test, taking only a matter of
hours. However, it does not distinguish between the oxygen demand that will
actually be felt in a natural environment due to biodegradation, and the chemical
oxidation of inert organic matter. Nor does it provide any information on the rate
at which actual biodegradation will take place. The measured value of COD is
higher than BOD, though for easily biodegradable matter the two will be quite
similar. In fact, the COD test is sometimes used as a way to estimate the ultimate
BOD.

The Effect of Oxygen-Demanding Wastes on Rivers

The amount of dissolved oxygen in water is one of the most commonly used indicators of a river's health. As DO drops below 4 or 5 mg/L, the forms of life that can survive begin to be reduced. In the extreme case, when anaerobic conditions exist, most higher forms of life are killed or driven off. Noxious conditions, including floating sludges, bubbling, odorous gases, and slimy fungal growths, then prevail.

A number of factors affect the amount of DO available in a river. Oxygen-demanding wastes remove DO; plants add DO during the day, but remove it at night; and the respiration of organisms living in sediments removes oxygen. Tributaries draining into the river bring their own oxygen supplies that mix with those of the main river. In the summer, rising temperatures reduce the solubility of oxygen, while lower flows reduce the rate at which oxygen enters the water from the atmosphere. To properly model all of these effects and their interactions is an exceedingly difficult task. A simple analysis, however, can provide insight into the most important parameters that affect DO. We should remember, however, that our results are only a first approximation to reality.

The simplest model of the oxygen resources in a river focuses on two key processes: the removal of oxygen by microorganisms during biodegradation, and the replenishment of oxygen through reaeration at the surface of the river. In this simple model, it is assumed that there is a continuous discharge of waste at a given location on the river. As the water and wastes flow downriver, it is assumed that they are uniformly mixed at any given cross section of river, and it is assumed that there is no dispersion of wastes in the direction of flow. That is, *plug flow* conditions are assumed.

Deoxygenation

The rate of deoxygenation at any point in the river is assumed to be proportional to the BOD remaining at that point. That is,

$$\text{Rate of deoxygenation} = k_d L_t \tag{4.16}$$

where k_d = the deoxygenation rate constant (day^{-1})

L_t = the BOD remaining t (days) after the wastes enter the river (mg/L)

The deoxygenation rate constant k_d is often assumed to be the same as the (temperature adjusted) BOD rate constant k obtained in a standard, laboratory BOD test. For deep, slowly moving rivers, this seems to be a reasonable approximation, but for turbulent, shallow, rapidly moving streams, the approximation is less valid. Such streams have deoxygenation constants that can be significantly higher than the values determined in the laboratory.

Substituting Eq. 4.9, which gives BOD remaining at any time t, into Eq. 4.16 gives

$$\text{Rate of deoxygenation} = k_d L_0 e^{-k_d t} \tag{4.17}$$

where L_0 is the BOD of the mixture of streamwater and wastewater at the point of discharge. Assuming complete and instantaneous mixing,

$$L_0 = \frac{Q_w L_w + Q_r L_r}{Q_w + Q_r} \qquad (4.18)$$

where

L_0 = initial BOD of the mixture of streamwater and wastewater (mg/L)

L_r = ultimate BOD of the river just upstream of the point of discharge (mg/L)

L_w = ultimate BOD of the wastewater (mg/L)

Q_r = volumetric flow rate of the river just upstream of the discharge point (m³/s)

Q_w = volumetric flow rate of wastewater (m³/s)

Example 4.6 Downstream BOD

A municipal wastewater treatment plant serving a city of 200 000 discharges 1.10 m³/s of treated effluent having an ultimate BOD of 50.0 mg/L into a stream that has a flow of 8.70 m³/s and a BOD of its own equal to 6.0 mg/L (see Figure 4.6). The deoxygenation constant k_d is 0.20/day.

a. Assuming complete and instantaneous mixing, estimate the ultimate BOD of the river just downstream from the outfall.

b. If the stream has constant cross section so that it flows at a fixed speed equal to 0.30 m/s, estimate the BOD of the stream at a distance 30 000 m downstream.

Solution

a. The BOD of the mixture of effluent and stream water can be found using (4.18):

$$L_0 = \frac{1.10 \text{ m}^3/\text{s} \times 50.0 \text{ mg/L} + 8.70 \text{ m}^3/\text{s} \times 6.0 \text{ mg/L}}{(1.10 + 8.70) \text{ m}^3/\text{s}} = 10.9 \text{ mg/L}$$

b. At a speed of 0.30 m/s, the time required for the waste to reach a distance 30 000 m downstream would be

$$t = \frac{30\ 000 \text{ m}}{0.30 \text{ m/s}} \times \frac{\text{hr}}{3600 \text{ s}} \times \frac{\text{day}}{24 \text{ hr}} = 1.16 \text{ days}$$

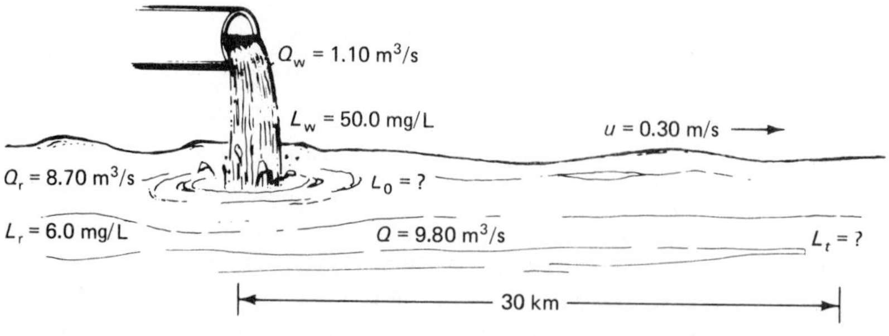

$Q_w = 1.10 \text{ m}^3/\text{s}$

$L_w = 50.0 \text{ mg/L}$ $u = 0.30 \text{ m/s} \longrightarrow$

$Q_r = 8.70 \text{ m}^3/\text{s}$ $L_0 = ?$

$L_r = 6.0 \text{ mg/L}$ $Q = 9.80 \text{ m}^3/\text{s}$ $L_t = ?$

$\longmapsto$ 30 km $\longmapsto$

Figure 4.6 Finding the BOD of the mixture of wastewater and river in Example 4.6.

so, the BOD remaining at that point, 30 km downstream, would be

$$L_t = L_0 e^{-k_d t} = 10.9 \; e^{-(0.2/d \times 1.16 d)} = 8.7 \; \text{mg/L}$$

Reaeration

The rate at which oxygen is replenished is assumed to be proportional to the difference between the actual DO in the river at any given location, and the saturated value of dissolved oxygen. This difference is called the oxygen deficit D:

$$\text{Rate of reaeration} = k_r \, D \tag{4.19}$$

where k_r = reaeration constant (time^{-1})
 $\quad D$ = dissolved oxygen deficit = $(\text{DO}_s - \text{DO})$
 $\quad \text{DO}_s$ = saturated value of dissolved oxygen $\qquad\qquad$ (4.20)
 $\quad \text{DO}$ = actual dissolved oxygen at a given location in the river

The reaeration constant k_r is very much dependent on the particular conditions in the river. A fast moving, shallow stream will have a much higher reaeration constant than a sluggish stream or a pond. Many attempts have been made to empirically relate key stream parameters to the reaeration constant, with the most commonly used formulation being the following (O'Connor and Dobbins, 1958):

$$k_r = \frac{3.9 u^{1/2}}{H^{3/2}} \tag{4.21}$$

where k_r = reaeration coefficient at 20 °C (day^{-1})
 $\quad u$ = average stream velocity (m/s)
 $\quad H$ = average stream depth (m)

Typical values of the reaeration constant k_r for various bodies of water are given in Table 4.7. Adjustments to the reaeration rate constant for temperatures other than 20 °C can be made using (4.12) but with a temperature coefficient θ equal to 1.024.

TABLE 4.7 TYPICAL REAERATION CONSTANTS FOR VARIOUS WATER BODIES

Water body	Range of k_r at 20 °C (base e) (day^{-1})
Small ponds and backwaters	0.10–0.23
Sluggish streams and large lakes	0.23–0.35
Large streams of low velocity	0.35–0.46
Large streams of normal velocity	0.46–0.69
Swift streams	0.69–1.15
Rapids and waterfalls	>1.15

Source: Tchobanoglous and Schroeder (1985).

TABLE 4.8 SOLUBILITY OF OXYGEN IN WATER (mg/L) AT 1 atm PRESSURE

Temperature (°C)	Chloride concentration in water (mg/L)			
	0	5000	10 000	15 000
0	14.62	13.73	12.89	12.10
5	12.77	12.02	11.32	10.66
10	11.29	10.66	10.06	9.49
15	10.08	9.54	9.03	8.54
20	9.09	8.62	8.17	7.75
25	8.26	7.85	7.46	7.08
30	7.56	7.19	6.85	6.51

Source: Thomann and Mueller (1987).

The solubility of oxygen in water DO_s was first introduced in Chapter 2, where it was noted that the saturated value of dissolved oxygen varies with temperature, barometric pressure, and salinity. Table 4.8 gives representative values of the solubility of oxygen in water at various temperatures and chloride concentrations.

Both the wastewater that is being discharged into a stream, and the stream itself, are likely to have some oxygen deficit. If we assume complete mixing of the two, we can calculate the initial deficit of the polluted river using a weighted average based on their individual concentrations of dissolved oxygen:

$$D_0 = DO_s - \frac{Q_w DO_w + Q_r DO_r}{Q_w + Q_r} \qquad (4.22)$$

where D_0 = initial oxygen deficit of mixture of river and wastewater

DO_s = saturated value of DO in water at the temperature of the river

DO_w = DO in the wastewater

DO_r = DO in the river just upstream of the wastewater discharge point

Example 4.7 Initial Oxygen Deficit

The wastewater in Example 4.6 has a dissolved oxygen concentration of 2.0 mg/L and a discharge rate of 1.10 m³/s. The river into which it is being discharged has DO equal to 8.3 mg/L, a flow rate of 8.70 m³/s, and a temperature of 20 °C. Assuming complete and instantaneous mixing, estimate the initial dissolved oxygen deficit of the mixture of wastewater and river water just downstream from the discharge point.

Solution The initial amount of dissolved oxygen in the mixture of waste and river would be

$$DO = \frac{1.10 \text{ m}^3/\text{s} \times 2.0 \text{ mg/L} + 8.70 \text{ m}^3/\text{s} \times 8.3 \text{ mg/L}}{(1.10 + 8.70) \text{ m}^3/\text{s}} = 7.6 \text{ mg/L}$$

The saturated value of DO at 20 °C is given in Table 4.8 as 9.09 mg/L, so the initial deficit would be

$$D_0 = 9.09 \text{ mg/L} - 7.6 \text{ mg/L} = 1.5 \text{ mg/L}$$

The Oxygen-Sag Curve

The deoxygenation caused by microbial decomposition of wastes and oxygenation by reaeration are competing processes that are simultaneously removing and adding oxygen to a stream. Combining the two equations (4.17) and (4.19) yields the following expression for the rate of increase of the oxygen deficit:

Rate of increase of the deficit = Rate of deoxygenation − Rate of oxygenation

$$\frac{dD}{dt} = k_d L_0 e^{-k_d t} - k_r D \tag{4.23}$$

which has the solution

$$D = \frac{k_d L_0}{k_r - k_d} (e^{-k_d t} - e^{-k_r t}) + D_0 e^{-k_r t} \tag{4.24}$$

Equation 4.24 is the classic *Streeter–Phelps oxygen-sag equation* first described in 1925. In this equation, t represents the time of travel in the stream from the point of discharge to the point in question downstream.

For the special case where $k_r = k_d$, the solution to (4.23) becomes

$$D = (k_d L_0 t + D_0)e^{-k_d t} \tag{4.25}$$

If the stream has constant cross-sectional area, and it is traveling at a speed u, then time and distance downstream are related by

$$x = ut \tag{4.26}$$

where x = distance downstream
  u = stream speed
  t = elapsed time between discharge point and distance x downstream

and (4.24) can be rewritten as

$$D = \frac{k_d L_0}{k_r - k_d} (e^{-k_d x/u} - e^{-k_r x/u}) + D_0 e^{-k_r x/u} \tag{4.27}$$

Subtracting the oxygen deficit, given by (4.24) or (4.27), from the saturation value DO_s gives DO as a function of time or distance downstream. A plot of this DO is given in Figure 4.7. As can be seen, there is a stretch of river immediately downstream of the discharge point ($x = 0$ or $t = 0$) where the decomposition process withdraws oxygen at a faster rate than reaeration can replace it, causing DO to drop. DO reaches a minimum at a particular location and time, called the *critical point*. At this point, the rate of deoxygenation equals the rate of reaeration. Beyond the critical point, reaeration exceeds deoxygenation and the stream naturally recovers.

The location of the critical point and the corresponding minimum value of DO are of obvious importance. It is at this point where stream conditions are at their worst. Setting the derivative of the oxygen deficit equal to zero and solving

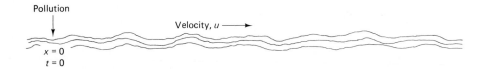

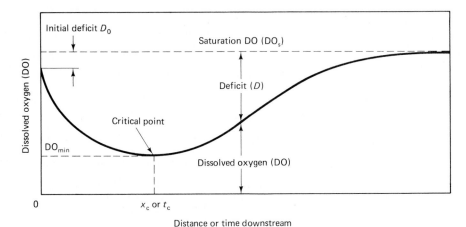

Figure 4.7 Streeter–Phelps oxygen-sag curve.

for the critical time yields

$$t_c = \frac{1}{k_r - k_d} \ln \left\{ \frac{k_r}{k_d} \left[1 - \frac{D_0(k_r - k_d)}{k_d L_0} \right] \right\} \tag{4.28}$$

The maximum deficit can then be found by substituting the value obtained for the critical time t_c into (4.24).

Example 4.8 The Streeter–Phelps Oxygen-Sag Curve

Just below the point where a continuous discharge of pollution mixes with a river, the BOD is 10.9 mg/L and DO is 7.6 mg/L. The river and waste mixture has a temperature of 20 °C, a deoxygenation constant of 0.20/day, an average flow speed of 0.30 m/s, and an average depth of 3.0 m. (In other words, this is just a continuation of the problem started in Examples 4.6 and 4.7.)

a. Find the time and distance downstream at which the oxygen deficit is a maximum.

b. Find the minimum value of DO.

Solution From Table 4.8, the saturation value of DO at 20 °C is 9.1 mg/L, so the initial deficit is

$$D_0 = 9.1 - 7.6 = 1.5 \text{ mg/L}$$

To estimate the reaeration constant, we can use the O'Connor and Dobbins relationship given in (4.21):

$$k_r = \frac{3.9 u^{1/2}}{H^{3/2}} = \frac{3.9(0.30)^{1/2}}{(3.0)^{3/2}} = 0.41/\text{day}$$

a. Using (4.28) we can find the time at which the deficit is a maximum

$$t_c = \frac{1}{k_r - k_d} \ln \left\{ \frac{k_r}{k_d} \left[1 - \frac{D_0(k_r - k_d)}{k_d L_0} \right] \right\}$$

$$= \frac{1}{(0.41-0.20)} \ln \left\{ \frac{0.41}{0.20} \left[1 - \frac{1.5(0.41 - 0.20)}{0.20 \times 10.9} \right] \right\} = 2.67 \text{ days}$$

and the critical distance downstream would be

$$x_c = ut_c = 0.30 \text{ m/s} \times 3600 \text{ s/hr} \times 24 \text{ hr/day} \times 2.67 \text{ days} = 69\,300 \text{ m}$$

which is about 43 miles.

b. The maximum deficit can be found from (4.24):

$$D = \frac{k_d L_0}{k_r - k_d} (e^{-k_d t} - e^{-k_r t}) + D_0 e^{-k_r t}$$

$$= \frac{0.20 \times 10.9}{0.41 - 0.20} (e^{-0.20 \times 2.67} - e^{-0.41 \times 2.67}) + 1.5 e^{-0.41 \times 2.67}$$

$$= 3.1 \text{ mg/L}$$

so the minimum value of DO will be the saturation value minus this maximum deficit:

$$DO_{min} = (9.1 - 3.1) \text{mg/L} = 6.0 \text{ mg/L}$$

In the above example, the lowest value of DO was found to be 6.0 mg/L, an amount sufficient for most forms of aquatic life. Had the minimum DO been unacceptably low, a trial and error process could have been used to determine the reduction in wastewater BOD required to achieve a more desirable level. This can be somewhat tedious by hand, but it is a simple exercise on a computer.

The effect of temperature on the oxygen-sag curve is particularly important. As temperature increases, the rate of deoxygenation increases, while the solubility of oxygen decreases. The combination of these effects causes the critical point downstream to be reached sooner. The minimum value of DO will be lower, as well, as is shown in Figure 4.8 (drawn for the example problem at 20 and 30 °C). Thus, a stream that may have sufficient DO in colder months, may have an unacceptable deficit in the warmer months of summer. Figure 4.8 also illustrates the potential adverse impact associated with thermal pollution caused by power plants. A river that might have been able to accept a certain sewage load without adverse effects could have unacceptably low oxygen levels when a power plant is added.

The oxygen-sag curve is affected by a number of other factors besides those already discussed. Deoxygenation, for example, increases along stretches of a river where sludge that has accumulated along the bottom causes its own oxygen demand. Algae can add DO during the daytime hours while photosynthesis is occurring, but at night its respiration removes DO. The net effect is a diurnal variation in DO that can lead to elevated levels of DO in the late afternoon and

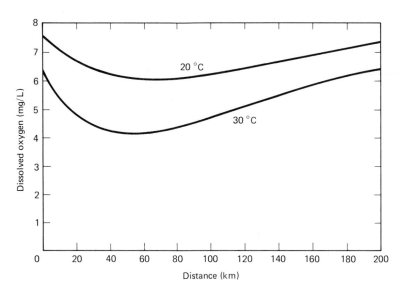

Figure 4.8   At higher temperatures, the minimum value of DO is lower and occurs closer to the source.

depressed concentrations at night. For a lake or a slow-moving stream that is
already overloaded with BOD and choked with algae, it is not unusual for respiration to cause offensive, anaerobic conditions late at night, while appearing to be
fine throughout the day. Another confounding factor not included in our simplified analysis is nitrification. This can cause a second dip later in the oxygen-sag
curve as its effects are felt. Properly accounting for these complicating factors
requires an extensive survey of stream conditions and a more detailed, more
complex model than was presented here.

4.5 WATER QUALITY IN LAKES AND RESERVOIRS

Eutrophication

All lakes are subject to a natural aging process known as *eutrophication*. It is
caused by the gradual accumulation of silt and organic matter in the lake. A
young lake is characterized by a low nutrient content and low plant productivity.
Such *oligotrophic* ("few foods") lakes gradually acquire nutrients from their
drainage basins, which enables increased aquatic growth. Over time, the increased biological productivity causes the water to become murky with phytoplankton, while decaying organic matter depletes the available oxygen. The lake
becomes *eutrophic* ("well fed"). As the accumulating silt and organic debris
causes the lake to get shallower and warmer, more plants take root along the
shallow edges, and the lake slowly transforms into a marsh or bog.

While such eutrophication is a natural process that may take thousands of years, it is possible to greatly accelerate the rate of change through human activities. Such cases are called *cultural eutrophication*. Municipal wastewater, industrial wastes, and runoff from fertilized agricultural lands add nutrients that stimulate algal growth and degrade water quality. Algal blooms die and decay, causing unsightly, odorous clumps of rotting debris and diminished oxygen. Oxygen depletion drives out fish, particularly those that must live in the colder, more oxygen-depleted regions near the bottom of the lake. If anaerobic conditions are created, hydrogen sulfide is formed and metals such as iron and manganese, which are normally tied up as precipitates in sediments, are dissolved and released into the lake.

Lakes and reservoirs, being convenient bodies of surface water, are potentially vulnerable to a host of other pollution problems besides eutrophication. However, unlike streams, which can transport pollutants "away," the slowness with which pollutants can be dispersed in a lake makes a lake less likely to be subject to blatant contamination by municipal or industrial point sources. However, lakes are often exposed to nonpoint sources, including urban and agricultural runoff that may contain toxic chemicals, as well as oxygen-demanding wastes and nutrients. They are also vulnerable to problems of acid rain.

Controlling Factors in Eutrophication

There are many factors that control the rate of production of algae, including the availability of sunlight to power the photosynthetic reactions and the concentration of nutrients required for growth.

The amount of light available is related to the transparency of the water, which is in turn a function of the level of eutrophication. An oligotrophic lake, such as Lake Tahoe, may have enough sunlight to allow significant rates of photosynthesis to take place at a depth of 100 m or more, while eutrophic lakes may be so murky that photosynthesis is restricted to a thin layer of water very near the surface. The top layer of water in a lake, where plants produce more oxygen by photosynthesis than they remove by respiration, is called the *euphotic zone*. Below that lies the *profundal zone*. The transition between the two zones is designated the *light compensation level*. The light compensation level corresponds roughly to a depth at which light intensity is about 1 percent of full sunlight.

While the amount of sunlight available can be a restricting factor in algal growth, it is not something that we could imagine controlling as a way to slow eutrophication. Since it is nutrient stimulation by human activity that causes cultural eutrophication in the first place, it certainly makes more sense to attempt to control eutrophication by restricting the supply of nutrients. The list of the nutrients that we might consider restricting, however, is long. It includes carbon, nitrogen, phosphorus, sulfur, calcium, magnesium, potassium, sodium, iron, manganese, zinc, copper, boron, and perhaps others. Fortunately, the problem is made manageable by focusing on just a single nutrient, usually either phosphorus or nitrogen.

Justus Liebig, in 1840, first formulated the idea that "growth of a plant is dependent on the amount of foodstuff that is presented to it in minimum quantity." This has come to be known as *Liebig's law of the minimum*. In essence, this law states that algal growth will be limited by the nutrient that is least available relative to its needs; therefore, the quickest way to control eutrophication would be to identify the *limiting nutrient* and reduce its concentration.

Liebig's law also implies that reductions in a nonlimiting nutrient will not provide effective control unless its concentration can be reduced to the point where it becomes the limiting nutrient. Thus, reducing phosphorus loading (for example, by eliminating phosphorus in detergents) will have much less effect in a lake that is nitrogen limited than in one that is phosphorus limited. It is important to note, however, that in eutrophic lakes the dominant species of algae are often blue-green (*Cyanophyta*). These are able to obtain nitrogen directly from the atmosphere. Thus, the only practical method of eutrophication control is to focus on phosphorus, whether or not it is initially the limiting nutrient.

To help illustrate the relative amounts of nitrogen and phosphorus that are required for algal growth, consider the following frequently used representation of algal photosynthesis (Stumm and Morgan, 1981):

$$106CO_2 + 16NO_3^- + HPO_4^{2-} + 122H_2O + 18H^+ \longrightarrow$$

$$C_{106}H_{263}O_{110}N_{16}P + 138O_2 \qquad (4.29)$$

Using a simple stoichiometric analysis, the ratio of the weights of nitrogen to phosphorus in this algae would be

$$\frac{N}{P} = \frac{16 \times 14}{1 \times 31} = 7.2$$

For a first approximation then, it takes about 7 times more nitrogen than phosphorus to produce a given amount of algae. Accounting for variations in plant stoichiometry, however, N/P ratios in a body of water over 20 generally indicate that phosphorus is the limiting nutrient, whereas N/P ratios of 5 or less reflect nitrogen limited systems (Thomann and Mueller, 1987).

Sawyer (1947) suggests that phosphorus concentrations in excess of 0.015 mg/L and nitrogen concentrations above 0.3 mg/L are sufficient to cause blooms of algae. These are in line with more recent estimates that suggest that 0.010 mg/L of phosphorus are acceptable while 0.020 mg/L are excessive (Vollenweider, 1975).

A Simple Phosphorus Model

In Chapter 1, the mass balance approach was introduced as a way to model certain idealized environmental problems. We can apply that technique to the analysis of a phosphorus-limited eutrophication problem. Suppose we want to estimate the phosphorus concentration that would be expected in a completely mixed lake under steady-state conditions, given some combination of phosphorus sources and sinks. By comparing the calculated phosphorus level with phosphorus concentrations that are generally considered acceptable, we would be able to estimate the amount of phosphorus control needed to prevent a eutrophication problem.

   An idealized lake model is shown in Figure 4.9. Phosphorus can enter the lake from a variety of sources, including the incoming streamflow, runoff from adjacent lands, and industrial or municipal point sources. It is removed by both settling into sediments and by flowing out with the stream flow leaving the lake.

   If we assume that the lake is well mixed, and that steady-state conditions prevail, we can write the following phosphorus balance:

$$\text{Rate of addition of P} = \text{Rate of removal of P}$$

$$S = QP + v_s AP \tag{4.30}$$

where S = the rate of addition of phosphorus from all sources (g/s)
   P = the concentration of phosphorus (g/m^3)
   Q = the stream outflow rate (m^3/s)
   v_s = the phosphorus settling rate (m/s)
   A = the surface area of the lake (m^2)

which results in a steady-state concentration of

$$P = \frac{S}{Q + v_s A} \tag{4.31}$$

The settling rate v_s is an empirically determined quantity that is difficult to predict with any confidence. Based on work by a number of authors, Thomann and Mueller (1987) suggest that lakes have a settling rate of approximately 3–30 m/year.

Example 4.9   Phosphorus Loading in a Lake

   A lake with surface area equal to 80×10^6 m^2 is fed by a stream having an average flow of 15.0 m^3/s and an average total phosphorus concentration of 0.010 mg/L. In addition, treated effluent from a wastewater treatment plant adds 0.20 m^3/s of flow having 5.0 mg/L total phosphorus. The phosphorus settling rate is estimated at 10 m/year.

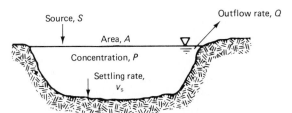

Source, S Outflow rate, Q

Area, A

Concentration, P

Settling rate, v_s

Figure 4.9   Well-mixed lake phosphorus mass balance.

a. Estimate the average total phosphorus concentration.
b. What rate of phosphorus removal at the wastewater treatment plant would be required to keep the concentration of phosphorus in the lake at an acceptable level of 0.010 mg/L?

Solution

a. The phosphorus loading from the incoming stream is

$$S_s = 15.0 \text{ m}^3/\text{s} \times 0.010 \text{ mg/L} \times \frac{1 \text{ g/m}^3}{\text{mg/L}} = 0.15 \text{ g/s}$$

and from the wastewater treatment plant it is

$$S_w = 0.20 \text{ m}^3/\text{s} \times 5.0 \text{ mg/L} \times \frac{1 \text{ g/m}^3}{\text{mg/L}} = 1.0 \text{ g/s}$$

for a total loading of $S = 0.15 + 1.0 = 1.15$ g/s. The flow rate out of the lake (neglecting evaporation) would be the sum of the inlet stream flow and the effluent flow

$$Q = 15.0 \text{ m}^3/\text{s} + 0.20 \text{ m}^3/\text{s} = 15.2 \text{ m}^3/\text{s}$$

The estimated settling rate is

$$v_s = \frac{10 \text{ m/yr}}{365 \text{ day/yr} \times 24 \text{ hr/day} \times 3600 \text{ s/hr}} = 3.17 \times 10^{-7} \text{ m/s}$$

Using (4.31), the steady-state concentration of total phosphorus would be

$$P = \frac{S}{Q + v_s A} = \frac{1.15 \text{ g/s}}{15.2 \text{ m}^3/\text{s} + 3.17 \times 10^{-7} \text{ m/s} \times 80 \times 10^6 \text{ m}^2}$$

$$= 0.028 \text{ g/m}^3 = 0.028 \text{ mg/L}$$

which is above the 0.010 mg/L suggested for an acceptable concentration.
b. To reach 0.010 mg/L, the total phosphorus loading must be

$$S = P(Q + v_s A)$$

$$= 0.010 \text{ g/m}^3 (15.2 \text{ m}^3/\text{s} + 3.17 \times 10^{-7} \text{ m/s} \times 80 \times 10^6 \text{ m}^2) = 0.41 \text{ g/s}$$

Of that, the amount that the wastewater treatment plant could contribute would be

$$S_w = (0.41 - 0.15) \text{ g/s} = 0.26 \text{ g/s}$$

Since S_w is now 1.0 g/s, there is a need for 74 percent phosphorus removal.

It must be emphasized that the model presented here depends on a number of assumptions that are not particularly valid in a real lake. For example, due to a phenomenon called *thermal stratification* that will be described in the next section, the assumption that the lake is well mixed is usually reasonable only during certain times of the year and in certain parts of the lake. The assumption of steady state ignores the dynamic behavior of lakes as weather and seasons change. We have also assumed that phosphorus is the controlling nutrient and that a simple measure of its concentration will provide an adequate indication of the potential for eutrophication. Finally, the model assumes a constant phosphorus settling rate that does not depend on such factors as the fraction of incoming phosphorus that is in particulate form, the movement of phosphorus both to sediments and from sediments, and physical parameters of the lake such as the ratio of lake volume to streamflow (the hydraulic detention time). In spite of these limitations, however, the model does illustrate some key parameters affecting eutrophication, and it does provide a reasonable introduction to more complex models that are in actual use and that do provide quite useful results. For further discussion of these issues, see, for example, Thomann and Mueller (1987).

Thermal Stratification

As we have seen, nutrients stimulate algal growth, and the subsequent death and decay of that algae can lead to oxygen depletion. This oxygen depletion problem is worsened by certain physical characteristics of lakes that we will now consider.

Figure 4.10 shows the relative density of water as a function of its temperature. One of the most unusual properties of water is the fact that its density does not monotonically decrease with increasing temperatures. Instead, it has a maximum point at 4 °C, as shown. One result of this density maximum is that ice floats because the water surrounding it is slightly warmer and denser. If water were like other liquids, freezing surface water would sink and it would be possible for lakes to freeze solid from the bottom up. Fortunately, this is not the case.

Above 4 °C the density of water decreases with temperature. As a result, a lake warmed by the sun during the summer will tend to have a layer of warm water floating on top of the denser, colder water below. Conversely, in the winter, if the lake's surface drops below 4 °C, it will create a layer of cold water that floats on top of the less dense, 4 °C water below. These density differences between surface water and the water nearer to the bottom inhibit vertical mixing in the lake. This leads to a very stable layering effect known as *thermal stratification*.

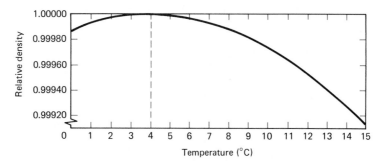

Figure 4.10   The density of water reaches a maximum at 4 °C.

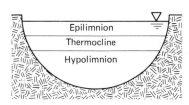
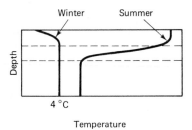

Figure 4.11 Thermal stratification of a lake showing winter and summer stratification temperature profiles.

Figure 4.11 shows the stratification that typically occurs in a deep lake, in the temperate zone, during the summer. In the upper layer, known as the *epilimnion,* the warm water is completely mixed by the action of wind and waves, causing an almost uniform temperature profile. The thickness of the epilimnion varies from lake to lake and from month to month. In a small lake it may be only a meter or so deep, whereas in large lakes it may extend down to 20 m or more. Below the epilimnion is a transition layer called the *thermocline,* or *metalimnion,* in which the temperature drops rather quickly. Most swimmers have experienced a thermocline's sudden drop in temperature when diving down into a lake. Below the thermocline is a region of cold water called the *hypolimnion.*

In terms of mixing, summer stratification creates essentially two separate lakes: a warm lake (epilimnion) floating on top of a cold lake (hypolimnion). The separation is quite stable, and becomes increasingly so as the summer progresses. Once summer stratification begins, the lack of mixing between the layers causes the epilimnion, which is absorbing solar energy, to warm even faster, creating an even greater density difference. This difference is enhanced in a eutrophic lake since in such lakes most, if not all, of the absorption of solar energy occurs in the epilimnion.

As the seasons progress and winter approaches, the temperature of the epilimnion begins to drop and the marked stratification of summer begins to disappear. Sometime in the fall, perhaps along with a passing storm that stirs things up, the stratification will disappear, the temperature will become uniform with depth, and complete mixing of the lake becomes possible. This is called the *fall overturn*. Similarly, in climates that are cold enough for the surface to drop below 4 °C, there will be a winter stratification, followed by a *spring overturn* when the surface warms up enough to allow complete mixing once again.

Stratification and Dissolved Oxygen

Dissolved oxygen, one of the most important water quality parameters, is greatly affected by both eutrophication and thermal stratification. Consider, for example, two different stratified lakes, one oligotrophic and one eutrophic. In both lakes, the waters of the epilimnion can be expected to be rich in DO since oxygen is readily available from reaeration and photosynthesis. The hypolimnion, on the other hand, is cut off from the oxygen-rich epilimnion by stratification. The only source of oxygen in the hypolimnion will be the result of photosynthesis that will occur only if the water is clear enough to allow the euphotic zone to extend below the thermocline. That is, the hypolimnion of the clear, oligotrophic lake at least has the possibility of having a source of oxygen, while that of the eutrophic lake does not.

In addition, the eutrophic lake is rich in nutrients and organic matter. Algal blooms suddenly appear and die off, leaving rotting algae that washes onto the beaches or sinks to the bottom. The rain of organic debris into the hypolimnion leads to increased oxygen demands there. Thus, not only is there inherently less oxygen available in the hypolimnion, but there is also more demand for oxygen due to decomposition, especially if the lake is eutrophic. Once summer stratification sets in, DO in the hypolimnion will begin dropping, driving fish out of the colder bottom regions of the lake and into the warmer, more oxygen-rich, surface waters. As lakes eutrophy, fish that require cold water for survival are the first victims. In the extreme case, the hypolimnion of a eutrophic lake can become anaerobic during the summer, as is suggested in Figure 4.12.

During the fall and spring overturns, which may last for several weeks, the lake's waters become completely mixed. Nutrients from the bottom are distributed throughout the lake and oxygen from the epilimnion becomes mixed with the oxygen-poor hypolimnion. The lake, in essence, takes a deep breath of air.

In winter, demands for oxygen decrease as metabolic rates decrease, while at the same time the ability of water to hold oxygen increases. Thus, even though winter stratification may occur, its effects tend not to be as severe as those in the summer. If ice forms, however, both reaeration and photosynthesis may cease to provide oxygen, and fish may die.

Acidification of Lakes

As was indicated in Chapter 2, all rainfall is naturally somewhat acidic. As was demonstrated there, pure water in equilibrium with atmospheric carbon dioxide forms a weak solution of carbonic acid (H_2CO_3) with a pH of about 5.6. As a result, acid rain is usually defined to be precipitation with pH less than 5.6. It is

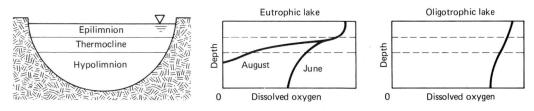

Figure 4.12   DO curves for eutrophic and oligotrophic lakes during summer thermal stratification.

not unusual for rainfall in the northeastern United States to have pH between 4.0 and 5.0 and for acid fogs in southern California to have pH less than 3.0. There is little question that such pH values are associated with anthropogenic emissions of sulfur and nitrogen oxides formed during the combustion of fossil fuels, as will be discussed in Chapter 7. Since some sulfur oxides are actually particles that can settle out of the atmosphere without precipitation, the popular expression "acid rain" is more correctly described as *acid deposition*.

The effects of acid deposition on materials, terrestrial ecosystems, and aquatic ecosystems are still only partially understood, but some features are emerging quite clearly. Acids degrade building materials, especially limestone, marble (a form of limestone), various commonly used metals such as galvanized steel, and certain paints. In fact, the increased rate of weathering and erosion of building surfaces and monuments was one of the first indications of adverse impacts from acid rain. Terrestrial ecosystems, especially forests, seem to be experiencing considerable stress due to acid deposition, with reductions in growth and increased mortality being especially severe in portions of the eastern United States, Canada, and northern Europe. It is the impact of acidification on aquatic ecosystems, however, that has focused our attention on the problem of acid rain.

Bicarbonate Buffering

Aquatic organisms are very sensitive to pH. Most are severely stressed if pH drops below 5.5 and very few are able to survive when pH falls below 5.0. Moreover, as pH drops, certain toxic minerals such as aluminum, lead, and mercury, which are normally insoluble and, hence, relatively harmless, enter solution and can be lethal to fish and other organisms.

It is important to note, however, that adding acid to a solution may have little or no effect on pH, depending on whether or not the solution has *buffers*. Buffers are substances capable of neutralizing added hydrogen ions. The available buffering of an aquatic ecosystem is a function not only of the chemical characteristics of the lake itself, but also of nearby soils through which water percolates as it travels from land to the lake. Thus, information on the pH of precipitation alone, without taking into account the chemical characteristics of the receiving body of water and surrounding soils, is a poor indicator of the potential effect of acid rain on an aquatic ecosystem.

Most lakes are buffered by bicarbonate (HCO_3^-), which is related to carbonic acid (H_2CO_3) by the reaction

$$H_2CO_3 \quad \rightleftharpoons \quad H^+ + HCO_3^- \qquad (4.32)$$

We have already encountered this reaction in Chapter 2, where the carbonate system was first described. Some bicarbonate results from the dissociation of carbonic acid, as is suggested in (4.32), and some comes from soils. Consider what happens to a lake containing bicarbonate when hydrogen ions (acid) are added. As reaction (4.32) suggests, some of the added hydrogen ions will react with bicarbonate to form neutral carbonic acid. To the extent that this occurs, the addition of hydrogen ions does not show up as an increase in hydrogen ion concentration. Thus, since pH is a measure of the hydrogen ion concentration, adding acid to a solution containing bicarbonate will have only minor impacts on pH because bicarbonate is a buffer.

Notice that the reaction of hydrogen ions with bicarbonate removes bicarbonate from solution. Unless there is some source of bicarbonate replacement its concentration can decrease as more acid is added, to the point where there is so little left that it is no longer capable of buffering the solution. At that point additional acid causes very rapid decreases in pH, as is suggested in Figure 4.13. For pH values above 6.3, the bicarbonate acts as an effective buffer, and there is very little change in pH as acid is added. As the pH drops below this point, the bicarbonate buffering is rapidly depleted, and, as is shown in the figure, lakes having a pH in the range of 5.0 to 6.0 are very sensitive to small changes in acid. Below pH 5.0, the lakes are unbuffered and chronically acidic (Wright, 1984).

The implications of Figure 4.13 are worth emphasizing. Acid precipitation may have little or no effect on a lake's pH up until the point where the natural buffering is exhausted. Continued exposure to acidification beyond that point can result in a rapid drop in pH that can be disastrous to the ecosystem.

In a Norwegian study, 684 lakes were categorized by their water chemistry into bicarbonate lakes, transition lakes, and acid lakes, and for each category observations of fish populations were made (Wright, 1984; Henriksen, 1980). Four fish-status categories were chosen: barren (indicative of long-term, chronic failures in fish reproduction), sparce populations (which reflect occasional failures), good populations, and overpopulated. The results, shown in Figure 4.14, clearly show the correlation between fish populations and lake acidification.

Importance of the Local Watershed

If there is a source of bicarbonate to replace that which is removed during acidification, then the buffering ability of bicarbonate can be extensive. Consider the ability of calcareous soils, which are rich in calcium carbonate ($CaCO_3$), to

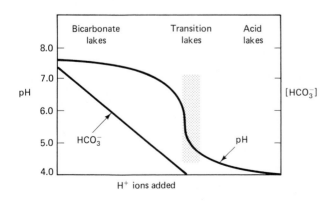

Figure 4.13 Bicarbonate buffering strongly resists acidification until pH drops below 6.3. As more H^+ ions are added, pH decreases rapidly (after Henriksen, 1980).

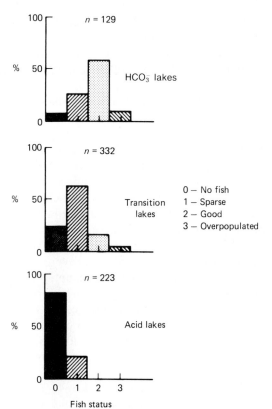

Figure 4.14　Frequency histograms of fish status for 684 Norwegian lakes categorized as bicarbonate, transition, or acid lakes (Wright, 1984).

provide buffering:

$$H^+ + CaCO_3 \longrightarrow Ca^{2+} + HCO_3^- \qquad (4.33)$$

As this reaction suggests, added hydrogen ions will react with $CaCO_3$ to form more bicarbonate. By removing those added hydrogen ions, calcium carbonate acts as a buffer. In addition, the bicarbonate that is formed can replenish the lake's natural bicarbonate buffers. Calcareous lakes, which have an abundance of calcium carbonate, are thus invulnerable to acidification. One way to temporarily mitigate the effects of acidification is to copy this natural phenomenon by artificially treating vulnerable lakes with limestone ($CaCO_3$). This approach is being vigorously pursued in a number of countries impacted by acid rain, including Sweden, Norway, Canada, and the United States (Shepard, 1986).

The ability of nearby soils to buffer acid deposition is an extremely important determinant of whether or not a lake will be subject to acidification. Soil systems derived from calcareous rock, for example, are better able to assimilate acid deposition than soils derived from granite bedrock. The depth and permeability of soils are also important characteristics. Thin, relatively impermeable soils provide little contact between soil and runoff, which reduces the ability of natural soil buffers to affect acidity. The size and shape of the watershed itself also affect a lake's vulnerability. Steep slopes and a small watershed area create conditions in which the runoff has little time to interact with soil buffers, increasing the likelihood of lake acidification. Even the type of vegetation growing in the watershed can affect acidification. Rainwater that falls through a forest canopy interacts with natural materials in the trees, such as sap, and its pH is affected. Deciduous foliage tends to decrease acidity, while water dripping from evergreen foliage tends to be more acidic than the rain itself.

In other words, to a large extent, the characteristics of the watershed itself will determine the vulnerability of a lake to acid. The most vulnerable lakes will be in areas with shallow soil of low permeability, granite bedrock, a steep watershed, and a predominance of conifers. Using factors such as these enables scientists to predict areas where lakes are potentially most sensitive to acid rain. One such prediction, using bedrock geology as the criterion, produced the map shown in Figure 4.15 (Galloway and Cowling, 1978). A comparison between Figure 7.14, which shows the average pH of rainfall across North America, and this figure, illustrating potentially sensitive regions, shows striking overlaps. Large areas in the northeastern portion of the continent, especially those near the Canadian border, have the unfortunate combination of vulnerable soils and very acid precipitation.

Mobilization of Aluminum

While the carbonate buffering system is the most important one at pH values above 6, for lower pH other compounds become increasingly important. At pH below 5, aluminum may provide the dominant buffering effect in the soil (USEPA, 1984). Aluminum is the third most abundant element in the earth's outer crust, but it rarely occurs in solution in any significant concentration unless the pH is very low, in which cases reactions such as the following occur. For the clay mineral kaolinite ($Al_2Si_2O_5(OH)_4$),

$$Al_2Si_2O_5(OH)_4 + 6H^+ \longrightarrow 2Al^{3+} + 2Si(OH)_4 + H_2O \qquad (4.34)$$

and for the commonly occurring gibbsite ($Al(OH)_3$),

$$Al(OH)_3 + 3H^+ \longrightarrow Al^{3+} + 3H_2O \qquad (4.35)$$

In each of these reactions, the removal of hydrogen ions provides the buffering capacity.

Of even greater importance than the buffering ability of aluminum is the resulting liberation of highly toxic Al^{3+} ions that enter solution. Aluminum that is normally bound up in soil minerals, and hence is harmless to living organisms, is leached from the soil by acid deposition and moves with runoff into streams and lakes. The toxic action of aluminum on fish seems to be a combination of effects, but the most important is its ability to cause mucus clogging of the gills, leading to respiratory distress and death. It has been found to be toxic to fish at concentra-

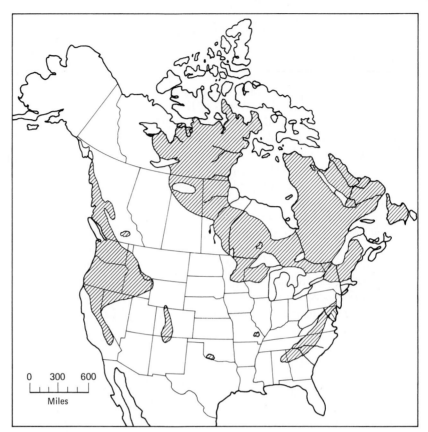

Figure 4.15 Regions in North America containing lakes that would be sensitive to potential acidification by acid precipitation (shaded areas), based on bedrock geology (USEPA, 1984; based on Galloway and Cowling, 1978).

tions as low as 0.1 mg/L and it has been implicated in fish kills in field observations, field experiments, and laboratory studies (USEPA, 1984). Especially troubling is the fact that aluminum can cause fish kills at a moderate value of pH that would, by itself, not be considered harmful.

4.6 GROUNDWATER

Groundwater is the source of about one-third of this country's drinking water. In the western part of the United States, it supplies closer to half of the total public supply. In rural areas, almost all of the water supply comes from groundwater, and more than one-third of our 100 largest cities depend on it for at least part of their supply. Historically, groundwater has been considered to be so safe to drink that many water companies deliver it untreated to their customers. We are quickly learning, however, that some of our groundwater is becoming contaminated with hazardous substances from hundreds of thousands of legally constructed and operated landfills, surface impoundments, and septic systems, as well as illegal and uncontrolled hazardous waste sites. Once contaminated, groundwater is difficult, if not impossible, to restore. It has already become necessary to shut down thousands of drinking water wells across the country.

Aquifers

Underground water, or *subsurface water,* occurs in two zones (Figure 4.16), distinguished by whether or not water fills all of the cracks and pores between particles of soil and rock. The *unsaturated* zone that lies just beneath the land surface is characterized by crevices that contain both air and water. Water in the unsaturated zone, called *vadose water,* is essentially unavailable for use. That is, it cannot be pumped, though plants certainly use soil water that lies near the surface. In the *saturated* zone, all spaces between soil particles are filled with water. Water in the saturated zone is called *groundwater,* and the upper boundary of the saturated zone is called the *water table.* There is a transition region between these two zones called the *capillary fringe,* where water rises into small cracks as a result of the attraction between water and rock surfaces. There is some controversy about whether to include capillary water in the saturated or unsaturated zone with many now recommending that it be given its own designa-

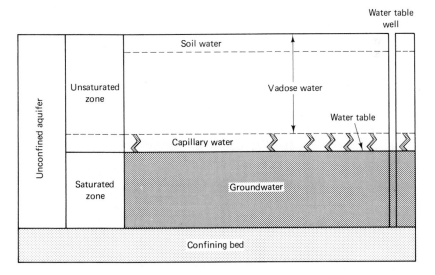

Figure 4.16 An unconfined aquifer is made up of saturated and unsaturated zones. A well penetrating the saturated zone would have water at the level of the water table.

tion called the *tension-saturated zone*. For our purposes such distinctions are unimportant.

Figure 4.16 illustrates an *unconfined aquifer* situated above a *confining bed*. An aquifer is a saturated geologic layer that is permeable enough to allow water to flow fairly easily through it, while a confining bed, or, as it is sometimes called, an *aquitard* or an *aquiclude*, is a relatively impermeable layer that greatly restricts the movement of groundwater. The two terms are not precisely defined, and are often used in a relative sense. A well drilled into the saturated zone of an unconfined aquifer will have water at atmospheric pressure at the level of the water table. It is not unusual to have a local impermeable layer in the midst of an unsaturated zone, above the main body of groundwater. Downward percolating water is trapped above this layer, creating a *perched water table* as shown in Figure 4.17.

Groundwater also occurs in *confined aquifers*, which are aquifers sandwiched between two aquitards. Water in a confined aquifer can be under pressure so that a well drilled into it may have water naturally rising above the upper surface of the aquifer, in which case it is called an *artesian well*. A line drawn at the level to which water would rise in an artesian well defines a surface called the *piezometric surface* or the *potentiometric surface*. In some cases, enough pressure may exist to force water to rise above ground level and flow without pumping, in which case it is called a *flowing artesian well*. Figure 4.17 shows these distinctions.

The amount of water that can be stored in a saturated aquifer depends on the *porosity* of the soil or rock which makes up the aquifer. Porosity (η) is defined to be the ratio of the volume of voids (openings) to the total volume of material:

$$\text{Porosity}(\eta) = \frac{\text{Volume of voids}}{\text{Total volume}} \qquad (4.36)$$

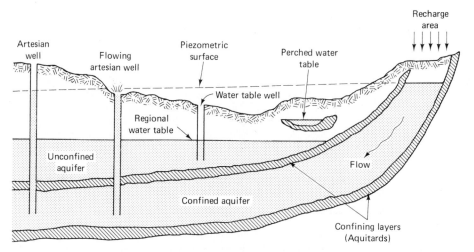

Figure 4.17   A confined aquifer and artesian wells.

While porosity describes the water-bearing capacity of a geologic formation, it is not a good indicator of the total amount of water that can be removed from that formation. Some water will always be retained as a film on rock surfaces and in very small cracks and openings. The volume of water that can actually be drained from an *unconfined* aquifer per unit of area per unit decline in the water table is called the *specific yield,* or the *effective porosity.* Representative values of porosity and specific yield for selected materials are given in Table 4.9.

It should be noted that some of the volume of water removed from an aquifer during pumping is due to expansion of the water as its pressure is reduced during removal, and some is due to compression in the aquifer as the water is removed. For an unconfined aquifer, these effects are negligible and the amount of water that can be released from the aquifer is virtually the same as the specific yield. For confined aquifers, however, these factors have to be accounted for and so a different term, the *storage coefficient,* is used as the measure of water released per unit of aquifer area per unit decline in head.

Example 4.10   Specific Yield

For an aquifer of sand having characteristics as given in Table 4.9, what volume of water would be stored in a saturated column with cross-sectional area equal to 1.0 m^2 and depth 2.0 m? How much water could be extracted from that volume?

Solution   The volume of material is 1.0 m^2 × 2.0 m = 2.0 m^3 so the volume of water stored would be

$$\text{Volume of water} = \text{Porosity} \times \text{Volume of material}$$

$$= 0.35 \times 2.0 \text{ m}^3 = 0.7 \text{ m}^3$$

The amount that could be removed would be

$$\text{Yield} = \text{Specific yield} \times \text{Volume of material} = 0.25 \times 2.0 \text{ m}^3 = 0.5 \text{ m}^3$$

Hydraulic Gradient

The slope of the water table in an unconfined aquifer, or of the piezometric surface in a confined aquifer, is called the *hydraulic gradient.* It is measured in the direc-

TABLE 4.9   REPRESENTATIVE VALUES OF POROSITY AND SPECIFIC YIELD

Material	Porosity (%)	Specific yield (%)
Clay	45	3
Sand	35	25
Gravel	25	22
Gravel and sand	20	16
Sandstone	15	8
Dense limestone and shale	5	2
Quartzite, granite	1	0.5

Source: Linsley et al. (1975).

tion of the steepest change. The gradient is important because groundwater flows in the direction of the gradient and at a rate proportional to the gradient. To determine the gradient, it is useful to introduce the notion of hydraulic head as shown in Figure 4.18. The *elevation head* at a well, for example, is the vertical distance from some reference datum plane (usually taken to be sea level) and the bottom of the well. The *pressure head* is the distance from the bottom of the well to the water level in the well. The sum of the two is the *total head* and has dimensions of length such as "meters of water" or "feet of water."

In Figure 4.18, if we assume that the groundwater flows in the direction shown in the plane of the page, the gradient would be

$$\text{Hydraulic gradient} = \frac{\text{Change in head}}{\text{Horizontal distance}} = \frac{\Delta h}{L} \qquad (4.37)$$

or, in a microscopic sense,

$$\text{Hydraulic gradient} = \frac{dh}{dL} \qquad (4.38)$$

While Figure 4.18 suggests that knowing the head at only two wells is sufficient to determine the gradient, that is only true if the direction of groundwater flow happens to be directly from one well to the other. In the more general case, three wells are required to determine the gradient, and even then the solution is valid only in the vicinity of those wells.

The following simple graphical procedure can be used to determine the gradient using the locations and total head for three nearby wells (Heath, 1983):

1. Draw a line between the two wells with the highest and lowest head and divide that line into equal intervals. Identify the location on the line where the head is equal to the head of the third (intermediate head) well.

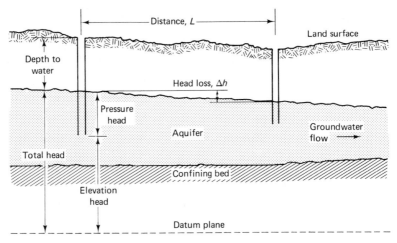

Figure 4.18   Various definitions of head in an unconfined aquifer. If groundwater flow is in the plane of the page, then the gradient is $\Delta h/L$.

2. Draw a line between the third (intermediate head) well and the spot on the above line that corresponds to the head at the third well. This is an *equipotential line,* meaning that the head anywhere along the line should be constant. Groundwater flow will be in a direction perpendicular to this line.

3. Draw a line perpendicular to the equipotential line through the well with the lowest (or the highest) head. This is a *flow line,* which means groundwater flow is in a direction parallel to this line.

4. Determine the gradient as the difference in head between the head on the equipotential and the head at the lowest (or highest) well, divided by the distance from the equipotential line to that well.

Figure 4.19 illustrates this procedure using data from the following example.

Example 4.11   Determining the Hydraulic Gradient from Three Wells

Two wells are drilled 200 m apart along an east–west axis. The west well has a total head of 30.2 m and the east well has a 30.0-m head. A third well located 100 m due south of the east well has a total head of 30.1 m. Find the magnitude and direction of the hydraulic gradient.

Solution   The locations of the wells are as indicated in Figure 4.19*a*. In Figure 4.19*b*, a line has been drawn between the wells with the highest (west, 30.2 m) and lowest (east, 30.0 m) head, and the location along that line with head equal to the

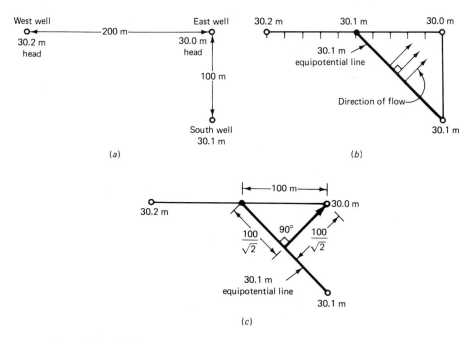

Figure 4.19   Finding the gradient from the location and total head of three wells in Example 4.11: (*a*) well plan, (*b*) equipotential, (*c*) gradient.

intermediate well (south, 30.1 m) has been indicated. The line corresponding to a
30.1-m equipotential has been drawn.

In Figure 4.19c, a perpendicular to the equipotential line has been drawn
through the east well. The direction of the gradient is thus at a 45° angle from the
south–west. The distance between the equipotential and the east well is easily determined by geometry to be $L = 100/\sqrt{2}$. So the gradient is

$$\text{Hydraulic gradient} = \frac{(30.1 - 30.0)\ \text{m}}{100/\sqrt{2}\ \text{m}} = 0.00141$$

Notice that the gradient is dimensionless.

Groundwater Flow: Darcy's Law

The basic equation governing groundwater flow was first formulated by the
French hydraulic engineer, Henri Darcy, in 1856. Based on laboratory experiments in which he studied the flow of water through sand filters, Darcy concluded
that flow rate Q is proportional to the cross sectional area A times the hydraulic
gradient (dh/dL):

$$Q = KA\frac{dh}{dL} \tag{4.39}$$

In (4.39), which is known as *Darcy's law* for flow through porous media, the
constant of proportionality K is called the *hydraulic conductivity* or the *coefficient
of permeability*. The hydraulic conductivity has units of length divided by time; in
metric units it is usually expressed in meters per day, in American units it is
traditionally given as gallons per day per square foot (also known as *meinzer
units*). Some approximate values of average hydraulic conductivity are given
Table 4.10, though it should be appreciated that these values are very rough.
Conductivities can easily vary over several orders of magnitude for any given
category of material, depending on differences in particle orientation and shape as
well as relative amounts of silt and clay that might be present.

TABLE 4.10 APPROXIMATE VALUES OF HYDRAULIC
CONDUCTIVITY

	Conductivity	
Material	gpd/ft²	m/day
Clay	0.01	0.0004
Sand	1 000	40
Gravel	100 000	4000
Gravel and sand	10 000	400
Sandstone	100	4
Dense limestone and shale	1	0.04
Quartzite, granite	0.01	0.0004

Source: Linsley et al. (1975).

Darcy's law assumes linearity between flow rate and hydraulic gradient, which is a valid assumption in most, but not all, circumstances. It breaks down when flow is turbulent, which may occur in the immediate vicinity of a pumped well. It is also invalid when water flows through extremely fine-grained materials, such as colloidal clays, and it should be used only when the medium is fully saturated with water.

Aquifers that have the same hydraulic conductivity throughout are said to be *homogeneous,* while those in which hydraulic conductivity differs from place to place are *heterogeneous.* Not only may hydraulic conductivity vary from place to place within the aquifer, but it may also depend on the direction of flow. It is quite common, for example, to have higher hydraulic conductivities in the horizontal direction than in the vertical. Aquifers that have the same hydraulic conductivity in any flow direction are said to be *isotropic,* while those in which conductivity depends on direction are *anisotropic.* While it is mathematically convenient to assume aquifers are both homogeneous and isotropic, they rarely, if ever, are.

Example 4.12   Flow through an Aquifer

A confined aquifer 20.0 m thick has two monitoring wells spaced 500.0 m apart along the direction of groundwater flow. The difference in water level in the wells is 2.0 m (the difference in piezometric head). The hydraulic conductivity is 50.0 m/day. Estimate the rate of flow per meter of distance perpendicular to the flow.

Solution   Figure 4.20 summarizes the data. The gradient is

$$\frac{\Delta h}{\Delta L} = \frac{2.0 \text{ m}}{500.0 \text{ m}} = 0.004$$

Using Darcy's law, with an arbitrary aquifer width of 1 m, yields

$$Q = KA \frac{\Delta h}{\Delta L}$$

$$= 50.0 \text{ m/day} \times 20.0 \text{ m} \times 1.0 \text{ m} \times 0.004 = 4.0 \text{ m}^3/\text{day per meter of width}$$

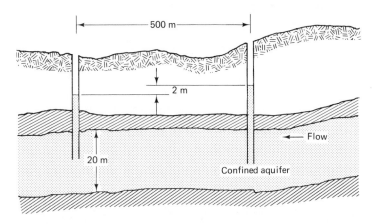

Figure 4.20   An example problem of flow through a confined aquifer.

Flow Velocity

It is often very important to be able to estimate the rate at which groundwater is moving through an aquifer, especially when a toxic plume exists upgradient from a water supply well. If we combine the usual relationship between flow rate, velocity, and cross-sectional area:

$$Q = Av \qquad (4.40)$$

with Darcy's law, we can solve for velocity:

$$v = \frac{Q}{A} = \frac{KA(dh/dL)}{A} = K\frac{dh}{dL} \qquad (4.41)$$

The velocity given in (4.41) is known as the *Darcy velocity*. It is not a "real" velocity in that it, in essence, assumes that the full cross-sectional area A is available for water to flow through. Since much of the cross-sectional area is made up of solids, the actual area through which all of the flow takes place is much smaller, and as a result, the *real groundwater velocity is considerably faster than the Darcy velocity*.

As suggested in Figure 4.21, consider the cross section of an aquifer to be made up of voids and solids, with A representing the total cross-sectional area, and A' being the area of voids filled with water. Letting v' be the actual *average linear velocity*, we can rewrite (4.40) as

$$Q = Av = A'v' \qquad (4.42)$$

Solving for v' and introducing an arbitrary length of aquifer, L, gives

$$v' = \frac{Av}{A'} = \frac{ALv}{A'L} = \frac{\text{Total volume} \times v}{\text{Void volume}} \qquad (4.43)$$

But, recall the ratio of void volume to total volume is just the porosity η, introduced in (4.36). Therefore, the actual average linear velocity through the aquifer is the Darcy velocity divided by porosity:

$$v' = \frac{\text{Darcy velocity}}{\text{Porosity}} = \frac{v}{\eta} \qquad (4.44)$$

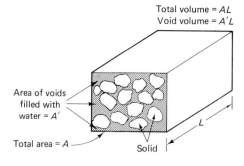

Total volume = AL
Void volume = $A'L$

Area of voids filled with water = A'

Total area = A

Solid

L

Figure 4.21 The cross-sectional area available for flow, A', is less than the overall cross-sectional area of the aquifer.

or, using (4.41),

$$v' = \frac{K(dh/dL)}{\eta} \tag{4.45}$$

Example 4.13   A Groundwater Plume

Suppose the aquifer in Example 4.12 has become contaminated upgradient of the two wells. Consider the upgradient well as a *monitoring* well whose purpose is to provide early detection of the approaching plume to help protect the second, drinking-water well. How long after the monitoring well is contaminated would you expect the drinking-water well to be contaminated? Make the following three assumptions (each of which will be challenged later):

1. Ignore dispersion or diffusion of the plume (that is, it does not spread out)
2. Assume the plume moves at the same speed as the groundwater
3. Ignore the "pulling" effect of the drinking-water well.

The aquifer has a porosity of 35 percent.

Solution   The Darcy velocity is given by (4.41):

$$v = K\frac{dh}{dL} = 50.0 \text{ m/d} \times 0.004 = 0.20 \text{ m/day}$$

The average linear velocity is the Darcy velocity divided by porosity:

$$v' = \frac{0.20 \text{ m/day}}{0.35} = 0.57 \text{ m/day}$$

so the time to travel the 500.0-m distance would be

$$t = \frac{500.0 \text{ m}}{0.57 \text{ m/day}} = 877 \text{ days} = 2.4 \text{ yr}$$

As this example illustrates, groundwater moves very slowly.

Dispersion and Retardation

A few comments on the assumptions made in Example 4.13 are in order. The first assumption was that there would be no dispersion, so the contamination would move forward with a sharp front, the so-called *plug-flow* case. In reality, contamination moving through an aquifer takes many different pathways to get from one point to another. It does not arrive all at once at a given location downgradient. This effect is easily demonstrated in the laboratory by establishing a steady-state flow regime in a column packed with a homogeneous granular material and then introducing a continuous stream of a nonreactive tracer, as shown in Figure 4.22a. If the tracer had no dispersion, a plot of concentration versus distance down the column at any given time t_1 would show a sharp jump at the distance $x_1 = v' t_1$, where v' is the average linear velocity. Instead, the front arrives smeared out, as shown in Figure 4.22b. Figure 4.22c shows the effect of disper-

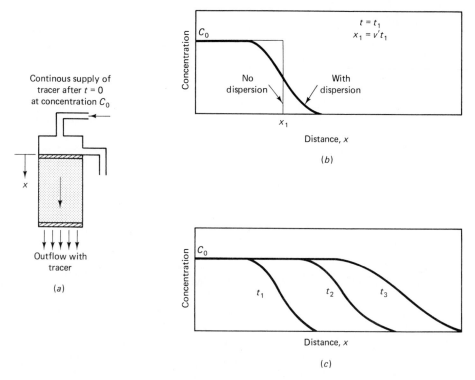

Figure 4.22   Longitudinal dispersion of a tracer passing through a column: (*a*) the column with steady flow and a continuous supply of tracer after *t* = 0; (*b*) the tracer concentration at some instant of time with and without dispersion; (*c*) the increased smearing of the tracer front with time, $t_1 < t_2 < t_3$. (*Source:* Freeze/Cherry, *Groundwater,* © 1979, p. 390. Reprinted by permission of Prentice Hall, Inc., Englewood Cliffs, New Jersey.)

sion at different instants of time ($t_1 < t_2 < t_3$) as the tracer passes through the column.

While the column experiment described in Figure 4.22 illustrates longitudinal dispersion, there is also dispersion normal to the main flow. Longitudinal dispersion is stronger than transverse dispersion. This tends to lead to plumes that are elliptical in shape, such as those shown in Figure 4.23.

A second assumption used in Example 4.13 was that the contamination would move at the same speed as the groundwater, which may or may not be the case in reality. Some contaminants are adsorbed onto soil particles, which reduces the overall solute flow rate relative to that of the groundwater itself. This reduction in speed is called the *retardation factor,*

$$\text{Retardation factor} = R = \frac{v'}{v_s} \geq 1 \tag{4.46}$$

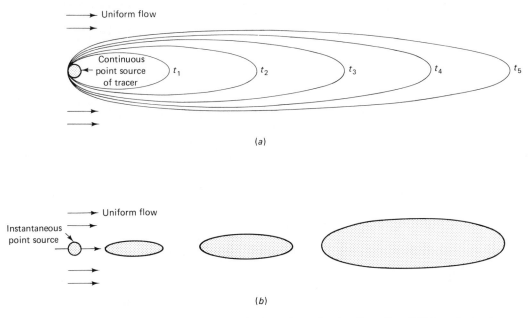

Figure 4.23 Spreading of a tracer in a two-dimensional uniform flow field in an isotropic sand: (*a*) continuous tracer feed with step function initial condition; (*b*) instantaneous point source. (*Source:* Freeze/Cherry, *Groundwater*, © 1979, p. 394. Reprinted by permission of Prentice Hall, Inc., Englewood Cliffs, New Jersey.)

where v_s is the average linear velocity of the retarded constituent, and v' is the average linear velocity of the groundwater. Retardation experiments are easy to perform using a column such as that shown in Figure 4.22, but difficult to perform in an actual aquifer.

In one such in situ experiment (Roberts et al., 1986), a number of organic solutes, including carbon tetrachloride (CTET) and tetrachloroethylene (PCE), were injected into the groundwater, along with a chloride tracer, which is assumed to move at the same rate as the groundwater itself. The position of the three plumes roughly 21 months later, as indicated by two-dimensional contours of depth-averaged concentrations, are shown in Figure 4.24. As can be seen, the center of the chloride plume has moved roughly 60 m away from the point of injection ($x = 0$, $y = 0$), the CTET has moved about 25 m, and the PCE has moved only a bit over 10 m.

Using actual plume measurements, such as are shown in Figure 4.25, Roberts et al. determined retardation factors for CTET and PCE, as well as for bromoform (BROM), dichlorobenzene (DCB), and hexachlorethane (HCE). As shown in Figure 4.25, retardation factors are not constants, but instead appear to increase over time and eventually reach a steady-state value.

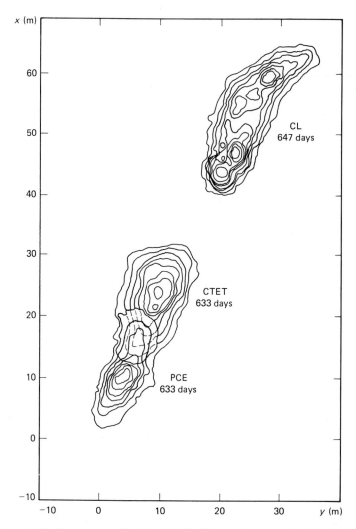

Figure 4.24 Plume separation for chloride (CL), carbon tetrachloride (CTET), and tetrachloroethylene (PCE) 21 months after injection into an actual aquifer. (*Source:* Roberts, Goltz, and Mackay, 1986. "A Natural Gradient Experiment on Solute Transport in a Sand Aquifer, 3, Retardation Estimates and Mass Balances for Organic Solutes." *Water Resources Research* 22(13):2047–2058, © by the American Geophysical Union.)

Cone of Depression

Another assumption made in Example 4.13 was that pumping water from the aquifer would not affect the hydraulic gradient. That is, in fact, not the case, since there must be a gradient toward the well to provide flow to the well. Moreover, the faster the well is pumped, the steeper the gradient will be in the vicinity of the

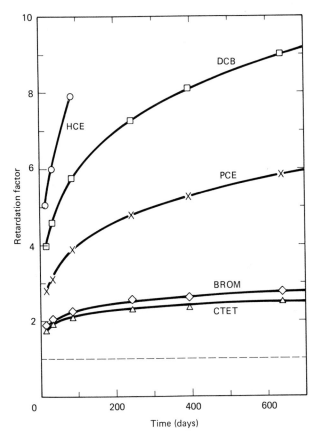

Figure 4.25 Retardation factors for carbon tetrachloride (CTET), tetrachloroethylene (PCE), bromoform (BROM), dichlorobenzene (DCB), and hexachloroethane (HCE). (*Source:* Roberts, Goltz, and Mackay, 1986. "A Natural Gradient Experiment on Solute Transport in a Sand Aquifer, 3, Retardation Estimates and Mass Balances for Organic Solutes." *Water Resources Research* 22(13):2047–2058, © by the American Geophysical Union.)

well. When a well is pumped, the water table in an unconfined aquifer or the piezometric surface for a confined aquifer, forms a *cone of depression* in the vicinity of the well, such as is shown in Figure 4.26.

If we make enough simplifying assumptions, we can use Darcy's law to derive an expression for the shape of the cone of depression, as follows. We will assume that pumping has been steady for a long enough time that the shape of the cone is no longer changing; that is, we assume equilibrium conditions. In addition, we will assume that the original water table is horizontal. If we also assume that the drawdown is small relative to the depth of the aquifer and that the well draws from the entire depth of the aquifer, then the flow to the well is horizontal and radial. Under these conditions, and using Figure 4.27, we can write Darcy's law for flow passing through a cylinder of radius r, depth h, and hydraulic gradient dh/dr as

$$Q = KA\,\frac{dh}{dr} = K\,2\pi rh\,\frac{dh}{dr} \qquad\qquad (4.47)$$

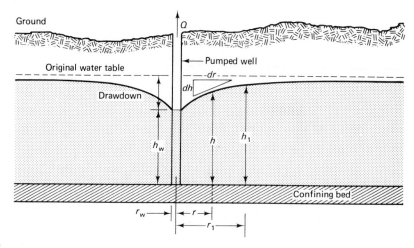

Figure 4.26   Cone of depression in an unconfined aquifer.

where the cross-sectional area of an imaginary cylinder around the well is $2\pi rh$, K is the hydraulic conductivity, and dh/dr is the slope of the water table at radius r. The flow through the cylinder toward the well equals the rate at which water is being pumped from the well, Q. Rearranging Eq. 4.47 into an integral form yields

$$\int_r^{r_1} Q\,\frac{dr}{r} = 2\pi K \int_h^{h_1} h\,dh \tag{4.48}$$

Notice how the limits have been set up. The radial term is integrated between two arbitrary values, r and r_1, corresponding to heads h and h_1. Integrating (4.48) gives

$$Q \ln \frac{r_1}{r} = \pi K(h_1^2 - h^2) \tag{4.49}$$

so,

$$Q = \frac{\pi K(h_1^2 - h^2)}{\ln(r_1/r)} \tag{4.50}$$

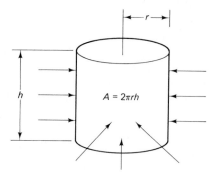

Figure 4.27   Flow to the well passes through a cylinder of area $A = 2\pi rh$.

Equation 4.50 for an unconfined aquifer can be used in several ways. It can be used to estimate the shape of the cone of depression for a given pumping rate, or it can be rearranged and used to determine the aquifer hydraulic conductivity K. To estimate K, two observation wells capable of measuring heads h and h_1 are set up at distances r and r_1 from a well pumping at a known rate Q. The data obtained can then be used in (4.50) as the following example illustrates. Actually, since equilibrium conditions take fairly long to be established, and since more information about the characteristics of the aquifer can be obtained with a transient study, this method of obtaining K is not commonly used. A similar, but more complex, model based on a transient analysis of the cone of depression is used (see, for example, Freeze and Cherry, 1979).

Example 4.14   Determining K from the Cone of Depression

Suppose a well 0.30 m in diameter has been pumped at a rate of 6000 m³/day for a long enough time that steady-state conditions apply. An observation well located 30.0 m from the pumped well has been drawn down by 1.0 m and another well at 100.0 m is drawn down by 0.50 m. The well extends completely through an unconfined aquifer 30.0 m thick.

    a. Determine the hydraulic conductivity K.

    b. Estimate the drawdown at the well.

Solution   It helps to put the data onto a drawing, as has been done in Figure 4.28.

    a. Rearranging (4.50) for K and then inserting the quantities from the figure gives

$$K = \frac{Q \ln(r_1/r)}{\pi(h_1^2 - h^2)}$$

$$= \frac{6000 \text{ m}^3/\text{day } \ln(100.0/30.0)}{\pi[(29.5)^2 - (29.0)^2] \text{ m}^2} = 78.6 \text{ m/day}$$

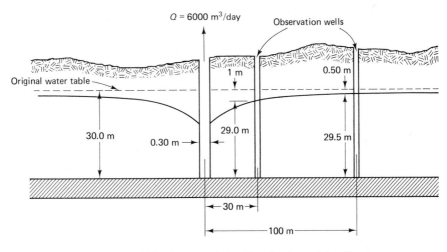

Figure 4.28   An example problem for determining K.

b. To estimate the drawdown, let $r = r_w = 0.30/2 = 0.15$ m, and let us use the first
    observation well for $r_1 = 30$ m and $h_1 = 29$ m. Using (4.50) to find the head at
    the outer edge of the well, h_w gives

$$Q = \frac{\pi K(h_1^2 - h_w^2)}{\ln(r_1/r_w)} = \frac{\pi(78.6 \text{ m/day})(29.0^2 - h_w^2) \text{ m}^2}{\ln(30.0/0.15)} = 6000 \text{ m}^3/\text{day}$$

Solving for h_w yields

$$h_w = 26.7 \text{ m}$$

so the drawdown would be $30.0 - 26.7 = 3.3$ m.

Equation 4.50 was derived for an unconfined aquifer. The derivation for a
confined aquifer is similar, but now the height of the cylinder at radius r, through
which water flows to reach the well, is a constant equal to the thickness of the
aquifer, B. Also, as shown in Figure 4.29, the cone of depression now appears in
the piezometric surface.

With the same assumptions as were used for the unconfined aquifer, we can
write Darcy's law as

$$Q = K2\pi r B \frac{dh}{dr} \tag{4.51}$$

which integrates to

$$Q = \frac{2\pi KB(h_1 - h)}{\ln(r_1/r)} \tag{4.52}$$

Control of Groundwater Plumes

The most common way to protect a drinking water well that may have an approaching plume of contaminated groundwater is to use some combination of
extraction wells and *injection wells*. Extraction wells are used to lower the water

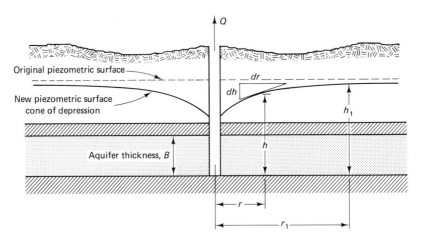

Figure 4.29 Cone of depression for a confined aquifer.

table (or piezometric surface), creating a hydraulic gradient that draws the plume to the wells. Injection wells raise the water table and push the plume away. Through careful design of the location and pumping rates of such wells, the hydraulic gradient can be manipulated in such a way that plumes can be kept away from drinking water wells and drawn toward extraction wells. Extracted, contaminated groundwater can then be treated and either reinjected back into the aquifer, reused, or released into the local surface water system.

Figure 4.30 shows the effect on the piezometric surface of two wells located near each other. In Figure 4.30a, the two wells are both extracting water from the aquifer, which lowers the piezometric surface over a wider distance than is obtained with a single well. And, in Figure 4.30b, one well is an extraction well, drawing down the surface, and the other is an injection well, raising the surface. Groundwater between the two wells is driven toward the extraction well.

Manipulating the hydraulic gradient to control and remove a groundwater plume is called *hydrodynamic control*. The well field used to create hydrodynamic control of a contaminant plume can be anything from a single extraction well, properly placed and pumping at the right rate (as will be described in the next section), to a complex array of extraction wells and injection wells. One way to try to protect a ''production well'' (e.g., a local drinking water well) from an approaching plume is to place a row of injection wells between the plume and the

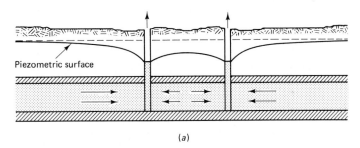

(a)

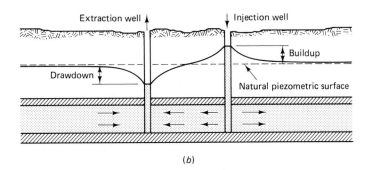

(b)

Figure 4.30 Manipulating the hydraulic gradient with multiple wells. (a) Two extraction wells broaden the capture region; (b) an injection well can push groundwater toward an extraction well.

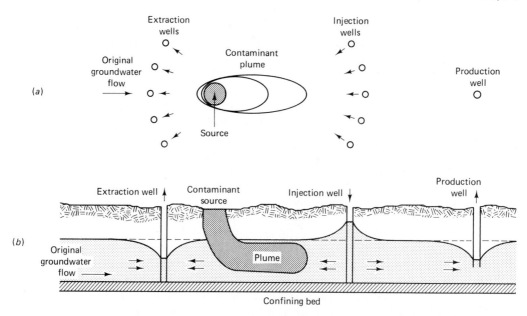

Figure 4.31 Hydrodynamic control with injection wells and extraction wells to protect a production well. (*a*) plan view; (*b*) cross section.

production well, as shown in Figure 4.31. Another row of extraction wells upgradient from the plume can help reverse the original direction of groundwater flow, increasing the control. A hydrodymanic control scheme of the sort illustrated can stop plume migration and protect the production well, but it would need to be coupled with a treatment system to eliminate the plume itself.

A more common approach to dealing with a groundwater contamination problem is to work with the natural gradient and place the extraction wells ahead of the plume, as shown in Figure 4.32. The rate at which the plume moves toward the extraction wells can be increased even more by treating the extracted groundwater (see Chapter 6) and reinjecting it back into the aquifer upgradient of the plume. Since rehabilitating an aquifer by pumping out the contaminants can take many years, this combination of extraction and injection wells is especially important, as it can shorten that time considerably.

Hydrodynamic control of groundwater plumes is an effective way to protect production wells. Construction costs are relatively low, and if the original well field is not sufficient to adequately control the plume, it is always possible to add additional wells when necessary. In addition, the environmental disturbance on the surface is minimal. On the negative side, the operation and maintenance costs can be high since the wells must be pumped for many years.

An alternative approach to protecting production wells is to literally surround the plume with a wall of impermeable material that extends from the surface down to the aquitard. A number of such *physical containment* schemes involving

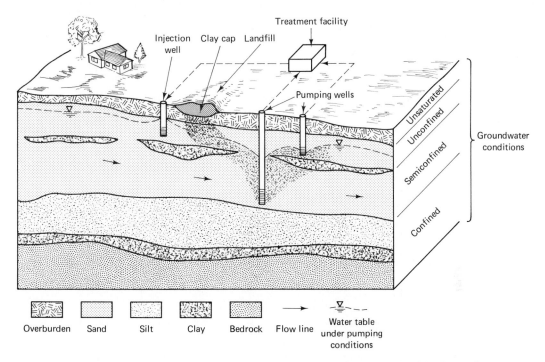

Figure 4.32 Aquifer rehabilitation by extraction, treatment, and reinjection (USEPA, 1990).

different materials and construction techniques are possible. Perhaps the most common is a *slurry cutoff wall* in which a narrow trench, about 1 to 2 m wide, is dug around the plume and then backfilled with a relatively impermeable mixture of soil and bentonite. Figure 4.33 shows some construction details of a 1-km-long (perimeter), 1-m-wide trench dug around a hazardous waste plume at a Superfund site in San Jose. The slurry wall extends through two aquifers, penetrating the clay aquitard beneath the lower, or B aquifer, for a total depth that ranges from 20 to 40 m. The slurry wall keeps the plume from migrating off-site while other remediation measures are applied to clean up the aquifer.

Groundwater Flow Nets and Capture-Zone Curves

One way to visualize the effect of extraction and injection wells is with a *flow net*. A flow net is a plan-view grid made of two kinds of lines: *equipotential lines,* which are contour lines drawn through points in the aquifer with equal head; and *flow lines,* which represent the horizontal direction in which the groundwater flows. Flow lines always cross equipotential lines at right angles.

A simple example of a flow net, drawn for a single extraction well in a region with no local natural hydraulic gradient, is given in Figure 4.34. In this case the

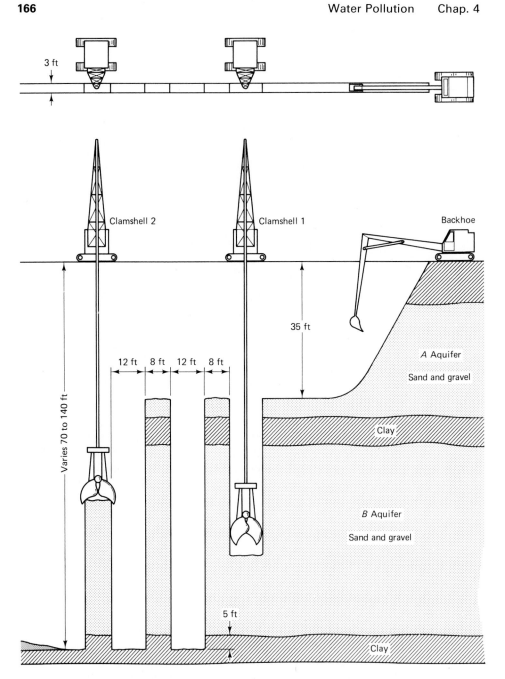

Figure 4.33 Schematic of slurry wall construction. (*Source:* Courtesy of Canonie Environmental Services.)

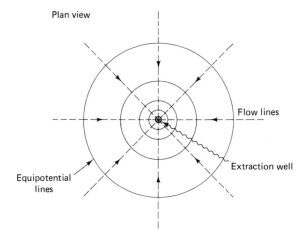

Figure 4.34 A flow net for an extraction well in an aquifer with no natural hydraulic gradient.

cone of depression is symmetrical around the well, and contour lines representing points with equal head appear as a series of circles extending outward from the well. Flow lines appear as a series of radiating lines that converge on the well. If an extraction well is located in an aquifer that has its own hydraulic gradient, then the cone of depression will no longer be symmetrical as it was in Figure 4.34. Instead, the flow lines for this more realistic case will be as shown in Figure 4.35, where some flow lines converge on the extraction well and some do not. The region within which all of the flow lines converge on the extraction well is called the *capture zone*. Groundwater contained within the capture zone will eventually find its way to the well and be pumped out of the aquifer. Flow lines outside of the

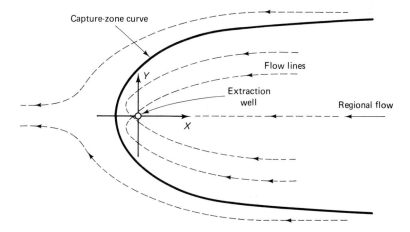

Figure 4.35 A single extraction well located at $X = 0$, $Y = 0$, in an area with a regional flow from right to left. The capture zone is the region in which all flow lines converge on the extraction well.

capture zone may be bent toward the well, but the regional flow associated with the natural hydraulic gradient is strong enough to carry that groundwater past the well.

Javandel and Tsang (1986) have developed the use of capture-zone type curves as an aid to the design of extraction well fields for aquifer cleanup. Their analysis is based on an assumed ideal aquifer, that is, one that is homogeneous, isotropic, uniform in cross section, and infinite in width. Also, it is either confined, or unconfined with an insignificant drawdown relative to the total thickness of the aquifer. They assume extraction wells that extend downward through the entire thickness of the aquifer and are screened to extract uniformly from every level. These restrictive assumptions describe circumstances that are unlikely to ever be precisely true in any real situation; nonetheless, the resulting analysis does give considerable insight into the main factors that affect more realistic, but complex, models.

For a single extraction well located at the origin of the coordinate system shown in Figure 4.35, Javandel and Tsang derive the following relationship between the x and y coordinates of the envelope surrounding the capture zone:

$$y = \pm \frac{Q}{2Bv} - \frac{Q}{2\pi Bv} \tan^{-1} \frac{y}{x} \tag{4.53}$$

where B is the aquifer thickness, v is the Darcy velocity (assumed parallel to, and in the direction of, the negative x axis), and Q is the pumping rate from the well. Equation 4.53 can be rewritten in terms of the angle ϕ (radians), where

$$\tan \phi = \frac{y}{x} \tag{4.54}$$

so that, for $0 \le \phi \le 2\pi$,

$$y = \frac{Q}{2Bv} \left(1 - \frac{\phi}{\pi} \right) \tag{4.55}$$

Equation 4.55 makes it easy to predict some important parameters of the capture zone. For example, as x approaches infinity, $\phi = 0$ and $y = Q/2Bv$, which sets the maximum total width of the capture zone at $2(Q/2Bv) = Q/Bv$. For $\phi = \pi/2$, that is, $x = 0$, y becomes equal to $Q/4Bv$. Thus, the width of the capture zone along the y axis is $Q/2Bv$, which is only half as broad as it is far from the well. These relationships are illustrated in Figure 4.36.

The width of the capture zone is directly proportional to the pumping rate Q and inversely proportional to the product of the regional (without the effect of the well) Darcy flow velocity v and the aquifer thickness B. Higher regional flow velocities therefore require higher pumping rates to capture the same area of plume. Usually there will be some maximum pumping rate determined by the acceptable amount of drawdown at the well, which restricts the size of the capture zone. Assuming that the aquifer characteristics have been determined and the plume boundaries defined, one way to use capture-zone-type curves is to first

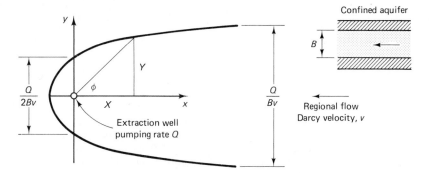

Figure 4.36   Capture-zone-type curve for a single extraction well located at the origin in an aquifer with regional flow velocity v and thickness B, and pumping at the rate Q. (Based on Javandel and Tsang, 1986.)

draw the curve corresponding to the maximum acceptable pumping rate. Then by superimposing the plume onto the capture zone curve (drawn to the same scale), it can be determined whether or not a single well will be sufficient and, if it is, where the well can be located. Figure 4.37 suggests the approach and the following example illustrates its use for a very idealized plume.

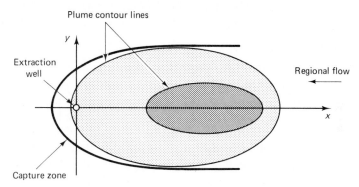

Figure 4.37   Superimposing the plume onto a capture-zone-type curve for a single extraction well.

Example 4.15   A Single Extraction Well

Consider a confined aquifer having thickness 20.0 m, hydraulic conductivity 1.0×10^{-3} m/s, and a regional hydraulic gradient equal to 0.002. The maximum pumping rate has been determined to be 0.004 m³/s. The aquifer has been contaminated and, for simplicity, consider the plume to be rectangular, with width 80.0 m. Locate a single extraction well so that it can totally remove the plume.

Solution   Let us first determine the regional Darcy velocity:

$$v = K \frac{dh}{dx} = 1.0 \times 10^{-3} \text{ m/s} \times 0.002 = 2.0 \times 10^{-6} \text{ m/s}$$

Now find the critical dimensions of the capture zone. Along the y axis, its width is

$$\frac{Q}{2Bv} = \frac{0.004 \text{ m}^3/\text{s}}{2 \times 20.0 \text{ m} \times 2.0 \times 10^{-6} \text{ m/s}} = 50.0 \text{ m}$$

and, at an infinite distance upgradient, the width of the capture zone is

$$\frac{Q}{Bv} = 100.0 \text{ m}$$

So, the 80.0-m-wide plume will fit within the capture zone if the well is located some distance downgradient. Using Figure 4.38 as a guide, we can determine the distance x that must separate the plume from the well. From (4.55), with $y = 40.0$ m,

$$y = \frac{Q}{2Bv} \left(1 - \frac{\phi}{\pi} \right) = 40.0 = 50 \left(1 - \frac{\phi}{\pi} \right)$$

so the angle, in radians to the point where the plume just touches the capture zone, is

$$\phi = 0.2\pi \text{ radians}$$

and, from the figure,

$$x = \frac{y}{\tan \phi} = \frac{40.0}{\tan(0.2\pi)} = 55.0 \text{ m}$$

so the extraction well should be placed 55.0 m ahead of the oncoming plume and directly in line with it.

The single-well solution found in Example 4.15 is not a particularly good one. The extraction well is far downgradient from the plume, which means a large volume of clean groundwater must be pumped before any of the contaminated plume even reaches the well. That can add years of pumping time and raise total costs considerably before the aquifer is rehabilitated.

A better solution would involve more extraction wells placed closer to the head of the plume. Javandel and Tsang have derived capture-zone-type curves for a series of n optimally placed wells, each pumping at the same rate Q, lined up

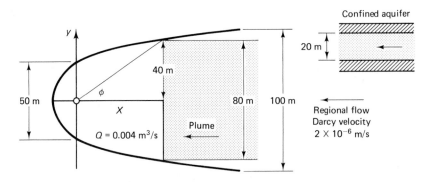

Figure 4.38 Example problem with single extraction well.

along the y axis. Optimality is defined to be the maximum spacing between wells that will still prevent any flow from passing between them. That distance for two wells has been determined to be $Q/\pi Bv$. If the wells are any farther apart than this, some of the flow can pass between them and not be captured. With this optimal spacing, the two wells will capture a plume as wide as Q/Bv along the y axis and as wide as $2Q/Bv$ a long distance upgradient from the wells, as shown in Figure 4.39a. Analogous parameters for the case of three optimally spaced wells are given in Figure 4.39b.

A general equation for the positive half of the capture-zone-type curve for n optimally spaced wells arranged symmetrically along the y axis is

$$y = \frac{Q}{2Bv}\left(n - \frac{1}{\pi}\sum_{i=1}^{n}\phi_i\right) \qquad (4.56)$$

where ϕ_i is the angle between a horizontal line through the ith well and a spot on the capture-zone curve.

To demonstrate use of these curves, let us redo Example 4.15, but this time we will use two wells.

Example 4.16   Capture Zone for Two Wells

Consider the same plume that was described in Example 4.15, that is, it is rectangular with width 80.0 m, in a confined aquifer with thickness $B = 20.0$ m, and Darcy velocity $v = 2.0 \times 10^{-6}$ m/s.

   a. If two optimally located wells are aligned along the leading edge of the plume, what minimum pumping rate Q would assure complete plume capture? How far apart should the wells be?

   b. If the plume is 1000 m long and the aquifer porosity is 0.30, how long would it take to pump an amount of water equal to the volume of water contained in the plume? Notice that it would take longer to pump out the whole plume since there will be some uncontaminated groundwater removed with the plume and we are ignoring retardation.

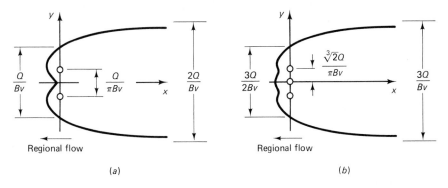

Figure 4.39 Capture-zone-type curves for optimally spaced wells along the y axis, each pumping at the rate Q: (a) two wells; (b) three wells.

Solution

a. The plume width along the y axis (also the leading edge of the plume) is 80.0 m so, from Figure 4.39,

$$\frac{Q}{Bv} = \frac{Q}{20.0 \text{ m} \times 2.0 \times 10^{-6} \text{ m/s}} = 80.0 \text{ m}$$

$$Q = 0.0032 \text{ m}^3/\text{s} \quad \text{(each)}$$

From Figure 4.39, the optimal spacing between two wells is given as

$$\text{Separation} = \frac{Q}{\pi Bv}$$

$$= \frac{0.0032 \text{ m}^3/\text{s}}{\pi \times 20.0 \text{ m} \times 2.0 \times 10^{-6} \text{ m/s}} = 25.5 \text{ m}$$

These dimensions are shown in Figure 4.40.

b. The volume of water contained in the plume is the porosity times the aquifer volume

$$V = 0.30 \times 80.0 \text{ m} \times 20.0 \text{ m} \times 1000 \text{ m} = 480\ 000 \text{ m}^3$$

At a total pumping rate of $2 \times 0.0032 \text{ m}^3/\text{s} = 0.0064 \text{ m}^3/\text{s}$, it would take

$$t = \frac{480\ 000 \text{ m}^3}{0.0064 \text{ m}^3/\text{s} \times 3600 \text{ s/hr} \times 24 \text{ hr/day} \times 365 \text{ day/yr}} = 2.4 \text{ yr}$$

to pump a volume of water equal to that contained in the plume. Again note, however, that the time required to pump the actual plume would be considerably greater than 2.4 years, depending on several factors, such as retardation and the fraction of the water pumped that is actually from the plume.

Since real aquifers are so much more complex than the ideal ones considered in this chapter, designing a well field is a much more difficult task than has been presented here. Interested readers are referred to more advanced texts such as those by Gupta (1989) or Willis and Yeh (1987).

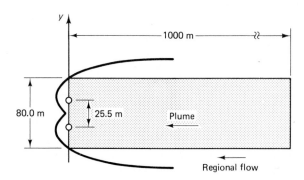

Figure 4.40 Example problem with two extraction wells.

PROBLEMS

4.1. In a standard 5-day BOD test:
  a. Why is the BOD bottle stoppered?
  b. Why is the test run in the dark (or in a black bottle)?
  c. Why is it usually necessary to dilute the sample?
  d. Why is it sometimes necessary to seed the sample?
  e. Why isn't ultimate BOD measured?

4.2. Incoming wastewater, with BOD_5 equal to about 200 mg/L, is treated in a well-run secondary treatment plant that removes 90 percent of the BOD. You are to run a 5-day BOD test with a standard 300-mL bottle, using a mixture of treated sewage and dilution water (no seed). Assuming that the initial DO is 9.2 mg/L:
  a. Roughly what maximum volume of treated wastewater should you put in the bottle if you want to have at least 2.0 mg/L of DO at the end of the test (filling the rest of the bottle with water)?
  b. If you make the mixture half water and half treated wastewater, what DO would you expect after 5 days?

4.3. A standard 5-day BOD test is run using a mix consisting of 4 parts distilled water and 1 part wastewater (no seed). The initial DO of the mix is 9.0 mg/L and the DO after 5 days is determined to be 1.0 mg/L. What is BOD_5?

4.4. If the BOD_5 for some waste is 200 mg/L and the ultimate BOD is 300 mg/L, find the reaction rate constant k.

4.5. A BOD test is run using 100 mL of treated wastewater mixed with 200 mL of pure water. The initial DO of the mix is 9.0 mg/L. After 5 days, the DO is 4.0 mg/L. After a long period of time, the DO is 2.0 mg/L and it no longer seems to be dropping. Assuming that nitrification has been inhibited so that the only BOD being measured is carbonaceous:
  a. What is the 5-day BOD of the wastewater?
  b. Assuming no nitrification effects, estimate the ultimate carbonaceous BOD.
  c. What would be the remaining BOD after 5 days have elapsed?
  d. Estimate the reaction rate constant k (day^{-1}).

4.6. Suppose you are to measure the BOD removal rate for a primary wastewater treatment plant. You take two samples of raw sewage on its way into the plant and two samples of the effluent leaving the plant. Standard 5-day BOD tests are run on the four samples, with no seeding, producing the following data:

Sample	Source	Dilution	DO_i (mg/L)	DO_f (mg/L)
1	Raw	1 : 30	9.2	2.2
2	Raw	1 : 15	9.2	?
3	Treated	1 : 20	9.0	2.0
4	Treated	?	9.0	>0

  a. Find BOD_5 for the raw and treated sewage, and the percent removal of BOD in the treatment plant.

   b. Find the DO that would be expected in sample 2 at the end of the test.

   c. What would be the maximum volume of treated sewage for sample 4 that could be put into the 300-mL BOD bottle and still have the DO after 5 days remain above zero?

4.7. A standard BOD test is run using seeded dilution water. In one bottle, the waste sample is mixed with seeded dilution water giving a dilution of 1 : 30. Another bottle contains just seeded dilution water. Both bottles begin the test with DO at the saturation value of 9.2 mg/L. After 5 days, the bottle containing waste has DO equal to 2.0 mg/L, while that containing just seeded dilution water has DO equal to 8.0 mg/L. Find the 5-day BOD of the waste.

4.8. A mixture consisting of 30 mL of waste and 270 mL of seeded dilution water has an initial DO of 8.55 mg/L; after 5 days, it has a final DO of 2.40 mg/L. Another bottle containing just the seeded dilution water has an initial DO of 8.75 mg/L and a final DO of 8.53 mg/L. Find the 5-day BOD of the waste.

4.9. Some wastewater has a BOD_5 of 150 mg/L at 20 °C. The reaction rate k at that temperature has been determined to be 0.23/day.

   a. Find the ultimate carbonaceous BOD.

   b. Find the reaction rate coefficient at 15 °C.

   c. Find BOD_5 at 15 °C.

4.10. Some waste has a 5-day BOD at 20 °C equal to 210 mg/L and an ultimate BOD of 350 mg/L. Find the 5-day BOD at 25 °C.

4.11. Suppose some wastewater has a BOD_5 equal to 180 mg/L and a reaction rate k equal to 0.22/day. It also has a total Kjeldahl nitrogen content (TKN) of 30 mg/L.

   a. Find the ultimate carbonaceous oxygen demand (CBOD).

   b. Find the ultimate nitrogenous oxygen demand (NBOD).

   c. Find the remaining BOD (nitrogenous plus carbonaceous) after 5 days have elapsed.

4.12. Suppose some pond water contains 10.0 mg/L of some algae which can be represented by the chemical formula $C_6H_{15}O_6N$. Using the following reactions

$$C_6H_{15}O_6N + 6O_2 \longrightarrow 6CO_2 + 6H_2O + NH_3$$

$$NH_3 + 2O_2 \longrightarrow NO_3^- + H^+ + H_2O$$

   a. Find the theoretical carbonaceous oxygen demand (see Section 2.2).

   b. Find the total theoretical (carbonaceous plus nitrogenous) oxygen demand.

4.13. For a solution containing 200 mg/L of glycine [$CH_2(NH_2)COOH$], whose oxidation can be represented as

$$2CH_2(NH_2)COOH + 3O_2 \longrightarrow 4CO_2 + 2H_2O + 2NH_3$$

$$NH_3 + 2O_2 \longrightarrow NO_3^- + H^+ + H_2O$$

   a. Find the theoretical CBOD.

   b. Find the ultimate NBOD.

   c. Find the total theoretical BOD.

4.14. A wastewater treatment plant discharges 1.0 m³/s of effluent having an ultimate BOD of 40.0 mg/L into a stream flowing at 10.0 m³/s. Just upstream from the discharge point, the stream has an ultimate BOD of 3.0 mg/L. The deoxygenation constant k_d is estimated at 0.22/day.

a. Assuming complete and instantaneous mixing, find the ultimate BOD of the mixture of waste and river just downstream from the outfall.

b. Assuming a constant cross-sectional area for the stream equal to 55 m², what ultimate BOD would you expect to find at a point 10 000 m downstream?

4.15. The wastewater in Problem 4.14 has DO equal to 4.0 mg/L when it is discharged. The river has its own DO, just upstream from the outfall, equal to 8.0 mg/L. Find the initial oxygen deficit of the mixture just downstream from the discharge point. The temperatures of sewage and river are both 15 °C.

4.16. A single source of BOD causes an oxygen-sag curve with a minimum downstream DO equal to 6.0 mg/L. If the BOD of the waste is doubled (without increasing the waste flow rate), what would be the new minimum downstream DO? In both cases assume that the initial oxygen deficit just below the source is zero and the saturated value of DO is 10.0 mg/L. (Note that when the initial deficit is zero, the deficit at any point is proportional to the initial BOD.)

4.17. The oxygen sag caused by a cannery reaches a minimum DO equal to 3.0 mg/L. Upstream from the cannery, the river DO is saturated at 10.0 mg/L and it has no BOD of its own. Just downstream from the discharge point, the DO is still essentially saturated (i.e., consider the initial oxygen deficit to be zero). By what percentage should the BOD of the cannery waste be reduced to assure a healthy stream with at least 5.0 mg/L DO everywhere?

4.18. Sketch an oxygen-sag curve and then directly below it (using the same x axis scale) sketch the shape of curves representing

a. The rate of reaeration.

b. The rate of deoxygenation.

c. At what point are these two rates equal to each other?

4.19. Sketch an oxygen-sag curve and indicate on it

a. The stretch of river for which the rate of deoxygenation exceeds the rate of reaeration.

b. The stretch of river for which the rate of reaeration exceeds the rate of deoxygenation.

c. The point where the two rates are equal to each other.

4.20. The ultimate BOD of a river just below a sewage outfall is 50.0 mg/L and the DO is at the saturation value of 10.0 mg/L. The deoxygenation rate coefficient k_d is 0.30/day and the reaeration rate coefficient k_r is 0.90/day. The river is flowing at the speed of 48.0 miles per day. The only source of BOD on this river is this single outfall.

a. Find the critical distance downstream at which DO is a minimum.

b. Find the minimum DO.

c. If a wastewater treatment plant is to be built, what fraction of the BOD would have to be removed from the sewage to assure a minimum of 5.0 mg/L everywhere downstream?

4.21. If the river in Problem 4.20 has an initial oxygen deficit, just below the outfall, of 2.0 mg/L, find the critical distance downstream at which DO is a minimum and find that minimum DO.

4.22. A city of 200 000 people deposits 37.0 cubic feet per second (cfs) of sewage having a BOD of 28.0 mg/L, and 1.8 mg/L of DO, into a river that has a flow rate of 250.0 cfs and a flow speed of 1.2 ft/s. Just upstream of the release point, the river has a BOD of 3.6 mg/L and a DO of 7.6 mg/L. The saturation value of DO is 8.5 mg/L. The deoxygenation coefficient k_d is 0.61/day and the reaeration coefficient k_r is 0.76/day. Assuming complete and instantaneous mixing of the sewage and river, find:

a. The initial oxygen deficit and ultimate BOD just downstream of the outfall.

b. The time and distance to reach the minimum DO.

c. The minimum DO.

d. The DO that could be expected 10 miles downstream.

4.23. For the following waste and river characteristics, find the minimum downstream DO that could be expected:

Parameter	Wastewater	River
Flow (m³/s)	0.30	0.90
Ultimate BOD (mg/L)	6.40	7.00
DO (mg/L)	1.00	6.00
k_d (day^{-1})	—	0.20
k_r (day^{-1})	—	0.37
Speed (m/s)	—	0.65
DO$_{sat}$ (mg/L)	8.0	8.0

4.24. Redo Example 4.8 at 30 °C. Assume initial DO is 6.3 mg/L.

4.25. Consider a lake with 100×10^6 m² of surface area for which the only source of phosphorus is the effluent from a wastewater treatment plant. The effluent flow rate is 0.4 m³/s and its phosphorus concentration is 10.0 mg/L (= 10.0 g/m³). The lake is also fed by a stream having 20 m³/s of flow with no phosphorus. If the phosphorus settling rate is estimated to be 10 m/year, estimate the average phosphorus concentration in the lake. What level of phosphorus removal at the treatment plant would be required to keep the average lake concentration below 0.010 mg/L?

4.26. A 50-cm³ sample of dry soil from an aquifer weighs 100 g. When it is poured into a graduated cylinder, it displaces 35 cm³ of water.

a. What is the porosity of the soil?

b. What is the average density of the actual solids contained in the soil?

4.27. Using data from Table 4.9, what volume of water would be removed from an unconfined gravel aquifer 10 000 m² in area if the water table drops by 100 cm?

4.28. Consider three monitoring wells, each located at the vertex of an equilateral triangle. The distance between any pair of wells is 300 m. The head at each well, referenced to some common datum, is as follows: well 1, 100 m; well 2, 100.3 m; well 3, 100.3 m. Sketch the well field and find the magnitude and direction of the hydraulic gradient.

4.29. Three wells are located in an x, y plane at the following coordinates: well 1, (0,0); well 2, (100 m, 0); and well 3, (100 m, 100 m). The ground surface is level and the distance from the surface to the water table for each well is as follows: well 1, 10 m; well 2, 10.2 m; well 3, 10.1 m. Sketch the well field and find the hydraulic gradient.

4.30. Suppose the aquifer described in Problem 4.28 has a hydraulic conductivity of 1000 m/day and a porosity of 0.23.

    a. What is the Darcy velocity?

    b. What is the average linear velocity of the groundwater?

    c. If a plume front is perfectly straight, how long would it take to travel to well 1 after first arriving (simultaneously) at wells 2 and 3, assuming a retardation factor of 2?

4.31. A 750-m section of river runs parallel to a channel 1000 m away (see Figure P4.31). An aquifer connecting the two has hydraulic conductivity equal to 7.0 m/day and an average thickness of 10.0 m. The surface of the river is 5.0 m higher than the surface of the channel. Estimate the rate of seepage from the river to the channel.

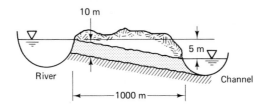

Figure P4.31

4.32. Based on observations of the rate of travel between two observation wells of a tracer element having retardation factor equal to 1.0, it is determined that the average linear flow velocity in an aquifer is 1.0 m/day when the hydraulic gradient is 0.0005. A sample of the aquifer is tested and found to have a porosity of 0.20. Estimate the hydraulic conductivity.

4.33. Derive the expression for the cone of depression in a confined aquifer as given by Eq. 4.52 in the text.

4.34. Drawdown at a 0.20-m-diameter, fully penetrating well, which has been pumping at the rate of 1000 m^3/day for a long enough time that steady-state conditions have been reached, is determined to be 0.70 m. The aquifer is unconfined and 10.0 m thick. An observation well 20.0 m away has been drawn down by 0.20 m. Determine the hydraulic conductivity of the aquifer.

4.35. A confined aquifer 30.0 m thick has been pumped from a fully penetrating well at a steady rate of 5000 m^3/d for a long time. Drawdown at an observation well 15.0 m from the pumped well is 3.0 m and drawdown at a second observation well 150.0 m away is 0.30 m. Find the hydraulic conductivity of the aquifer.

4.36. Derive the following expression for the length of time required for groundwater to flow from an observation well to a pumped well in a confined aquifer:

$$t = \frac{\pi B \eta}{Q} (R^2 - r_w^2)$$

where B = thickness of the confined aquifer

      η = aquifer porosity

      R = radial distance from the observation well to the pumped well

      r_w = radius of the pumped well

      Q = pumping rate

Hint: Combine Eq. 4.51, $Q = 2\pi r B K (dh/dr)$, with the average linear velocity from Eq. 4.45:

$$v' = -\frac{K}{\eta}\frac{dh}{dr} = \frac{dr}{dt}$$

4.37. For the aquifer described in Problem 4.35, and using the equation given in Problem 4.36, determine the time required for groundwater to travel from the observation well 15.0 m away to the pumped well with diameter 0.40 m. The porosity is 0.30.

4.38. A *stagnation point* in a capture-zone-type curve is a spot where groundwater would have no movement. For the case of a single extraction well, the stagnation point is located where the capture-zone curve crosses the x axis. Use the fact that for small angles $\tan \theta \approx \theta$ to show that the x-axis intercept of the capture-zone curve for a single well is $x = -Q/2Bv\pi$.

4.39. Suppose a spill of 0.10 m³ of TCE distributes itself evenly throughout an aquifer 10.0 m thick, forming a rectangular plume 2000 m long and 250 m wide (see Figure P4.39). The aquifer has porosity 0.40, hydraulic gradient 0.001, and hydraulic conductivity 0.001 m/s.

   a. What would be the concentration of TCE in this idealized groundwater plume?

   b. Using capture-zone-type curves, design an extraction field to pump out the plume under the assumption that the wells are all lined up along the leading edge of the plume, with each well to be pumped at the same rate, not to exceed 0.003 m³/s per well. What is the smallest number of wells that could be used to capture the whole plume?

   c. What minimum pumping rate would be required for each well?

   d. What would the optimal spacing be between the wells (at that minimum pumping rate)?

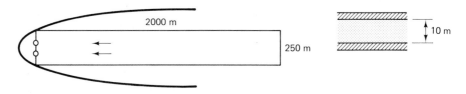

Figure P4.39

4.40. Starting with Eq. 4.56, show that the width of the capture zone along the y axis for n optimally spaced wells is equal to $nQ/2Bv$.

4.41. A single well is to be used to remove a symmetrical oblong plume of contaminated groundwater in an aquifer 20.0 m thick with porosity 0.30, hydraulic conductivity 1.0 × 10⁻⁴ m/s, and hydraulic gradient 0.0015. With the plume and capture-zone curve superimposed as shown in Figure P4.41, the angle from the well to the point where the two just touch is 45°, and the width of the plume is 100.0 m. What pumping rate would create these conditions?

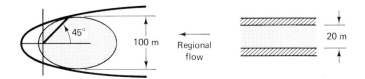

Figure P4.41

REFERENCES

American Public Health Association, 1960, *Standard Methods for the Examination of Water and Wastewater,* 11th ed., New York.

BERNER, E. K., and R. A. BERNER, 1987, *The Global Water Cycle, Geochemistry and Environment,* Prentice-Hall, Englewood Cliffs, NJ.

BUDYKO, M. I., 1974, *Climate and Life,* Academic, New York.

DAVIS, M. L., and D. A. CORNWELL, 1985, *Introduction to Environmental Engineering,* PWS Engineering, Boston.

FREDERICK, K. D., and J. C. HANSON, 1982, *Water for Western Agriculture,* Resources for the Future, Washington, DC.

FREEZE, R. A., and J. A. CHERRY, 1979, *Groundwater,* Prentice-Hall, Englewood Cliffs, NJ.

GALLOWAY, J. N., and E. B. COWLING, 1978, The effects of precipitation on aquatic and terrestrial ecosystems, A proposed precipitation chemistry network, *Journal of the Air Pollution Control Association* 28(3).

GUPTA, R. S., 1989, *Hydrology and Hydraulic Systems,* Prentice-Hall, Englewood Cliffs, NJ.

HARTE, J., 1985, *Consider a Spherical Cow, A Course in Environmental Problem Solving,* William Kaufmann, Inc., Los Altos, CA.

HEATH, R. C., 1983, *Basic Ground-Water Hydrology,* U.S. Geological Survey Water-Supply Paper 2220, U.S. Government Printing Office.

HENRIKSEN, A., 1980, *Proceedings of the International Conference on the Ecological Impact of Acid Precipitation,* D. Drabløs and A. Tollan (eds.), p. 68.

JAVANDEL, I., and C. TSANG, 1986, Capture-zone type curves: A tool for aquifer cleanup, *Ground Water,* 24(5):616–625.

LINSLEY, R. K., M. A. KOHLER, and J. L. H. PAULHUS, 1975, *Hydrology for Engineers,* 2d ed., McGraw-Hill, New York.

MORRISON, A., 1983, In third world villages, A simple handpump saves lives, *Civil Engineering/ASCE* October:68–72.

O'CONNOR, D. J., and W. E. DOBBINS, 1958, Mechanism of reaeration in natural streams, *Transactions of the American Society of Civil Engineers* 153:641.

REISNER, M., 1986, *Cadillac Desert, The American West and Its Disappearing Water,* Viking, New York.

ROBERTS, P. V., M. N. GOLTZ, and D. M. MACKAY, 1986, A natural gradient experiment on solute transport in a sand aquifer, 3, retardation estimates and mass balances for organic solutes, *Water Resources Research* 22(13):2047–2058.

SAWYER, C. N., 1947, Fertilization of lakes by agricultural and urban drainage, *Journal of the New England Water Works Association* 41(2).

SAWYER, C. N., AND P. L. MCCARTY, 1978, *Chemistry for Environmental Engineering,* 3d ed., McGraw-Hill, New York.

SHEPARD, M., 1986, Restoring life to acidified lakes, *EPRI Journal,* April/May.

SKOGERBOE, G. V., and J. P. LAW, 1971, *Research Needs for Irrigation Return Flow Quality Control,* Project No. 13030, U.S. Environmental Protection Agency, Washington, DC.

STREETER, N. W., and E. B. PHELPS, 1925, U.S. Public Health Service Bulletin No. 146.

STUMM, W., and J. J. MORGAN, 1981, *Aquatic Chemistry,* 2d ed., Wiley, New York.

TCHOBANOGLOUS, G., and E. D. SCHROEDER, 1985, *Water Quality,* Addison-Wesley, Reading, MA.

THOMANN, R. V., and J. A. MUELLER, 1987, *Principles of Surface Water Quality Modeling and Control,* Harper & Row, New York.

USEPA, 1984, *Acidic Deposition Phenomenon and Its Effects, Critical Assessment Review Papers,* A. P. Atshuller and R. A. Linthurst, Eds., EPA-600/8-83-016bF.

USEPA, 1986, Drinking water in America: An overview, *EPA Journal,* September.

USEPA, 1990, *Basics of Pump-and-Treat Ground-Water Remediation Technology,* J. W. Mercer, D. C. Skipp, and D. Giffin, EPA-600/8-90/003.

U.S. Geological Survey, 1984, *Estimated Use of Water in the United States, 1980,* Department of the Interior, Washington DC.

U.S. Geological Survey, 1985, *Study and Interpretation of the Chemical Characteristics of Natural Water,* Water-Supply Paper 2254, Department of the Interior, Washington, DC.

U.S. Water Resources Council, 1978, *The Nation's Water Resources, 1975–2000,* Second National Water Assessment, Washington, DC, December.

VIESSMAN, W., JR., and M. J. HAMMER, 1985, *Water Supply and Pollution Control,* 4th ed., Harper & Row, New York.

VOLLENWEIDER, R. A., 1975, Input–output models with special reference to the phosphorus loading concept in limnology, *Schweiz. Z. Hydrol.* 37:53–83.

WELCH, E. B., 1980, *Ecological Effects of Waste Water,* Cambridge University Press, Cambridge.

WILLIS, R., and W. W-G. YEH, 1987, *Groundwater Systems Planning & Management,* Prentice-Hall, Englewood Cliffs, NJ.

WRIGHT, R. F., 1984, Norwegian models for surface water chemistry: An overview, in *Modeling of Total Acid Precipitation Impacts,* J. L. Schnoor, Ed., Butterworth, Boston.

CHAPTER 5

Hazardous Substances and Risk Analysis

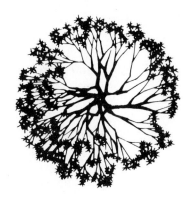

All substances are poisons; there is none which is not a poison. The right dose differentiates a poison and a remedy.
Paracelsus (1493–1541)

5.1 INTRODUCTION

In the past decade or so, there has been a marked shift in emphasis among those involved with the environment and public health, away from infectious diseases such as cholera and typhoid, which are caused by microorganisms, toward noninfectious human health problems such as cancer and birth defects that are induced by toxic substances. Dramatic incidents in the late 1970s such as occurred at Love Canal, New York, where hazardous substances from an abandoned dump site oozed into backyards and basements, and Times Beach, Missouri, where tens of thousands of gallons of dioxin-laced oil were carelessly sprayed onto the dusty streets, helped motivate the coming decade of hazardous waste controls.

In response to growing public pressure to do something about hazardous wastes, Congress passed the *Comprehensive Environmental Response, Compensation, and Liability Act* (CERCLA) in 1980 to deal with *already contaminated* sites, and in 1980 and 1984, it strengthened the *Resource Conservation and Recovery Act* (RCRA) that controls *new sources* of hazardous waste. By the end of the 1980s, the Environmental Protection Agency (EPA), which administers these programs, had over 1200 sites formally listed on its *National Priorities List* (NPL) of locations determined to be sufficiently threatening to human health and the environment to warrant federal action. The potential exists for thousands of additional sites to be added. Cleanup efforts that have already cost billions of dollars could ultimately require the expenditure of several hundreds of billions more.

One of the major issues raised in cleaning up hazardous waste is the question of how clean is clean? To completely eliminate human exposure to toxics in the environment would require enormous expenditures of money. At some point in the cleanup process, the remaining health and environmental risks may not justify the continued costs, and, from a risk perspective, society might be better off spending the money elsewhere. Identifying and quantifying risks has become a central element in the hazardous substance regulatory process and is emphasized in this chapter.

5.2 HAZARDOUS SUBSTANCES DEFINED

What is a hazardous substance? Unfortunately, the answer is largely a matter of definition, with various pieces of environmental legislation defining it somewhat differently. In general, the federal government defines hazardous waste as "anything which, because of its quantity, concentration, or physical, chemical, or infectious characteristics may cause, or significantly contribute to, an increase in mortality; or cause an increase in serious irreversible, or incapacitating reversible, illness; or pose a substantial present or potential hazard to human health and the environment when improperly treated, stored, transported, or disposed of, or otherwise managed."

More specifically, RCRA defines a substance as hazardous if it possesses any of the following four characteristic attributes: *reactivity, ignitability, corrosivity,* or *toxicity.* Briefly,

Ignitable substances are easily ignited and burn vigorously and persistently. Examples include volatile liquids, such as solvents, whose vapors ignite at relatively low temperatures (defined as 60 °C or less).

Corrosive substances include liquids with pH less than 2 or greater than 12.5, and those that are capable of corroding metal containers.

Reactive substances are unstable under normal conditions. They can cause explosions and/or liberate toxic fumes, gases, and vapors when mixed with water.

Toxic substances are harmful or fatal when ingested or absorbed. Toxicity is determined using a standardized laboratory test, called the *extraction proce-*

TABLE 5.1 MAXIMUM CONCENTRATION OF CONTAMINANTS FOR EP TOXICITY TEST

EPA hazardous waste number	Contaminant	Maximum concentration (mg/L)
D004	Arsenic	5.0
D005	Barium	100
D006	Cadmium	1.0
D007	Chromium	5.0
D008	Lead	5.0
D009	Mercury	0.2
D010	Selenium	1.0
D011	Silver	5.0
D012	Endrin	0.02
D013	Lindane	0.4
D014	Methoxychlor	10.0
D015	Toxaphene	0.5
D016	2,4-D (2,4-dichlorophenoxyacetic acid)	10.0
D017	2,4,5-TP Silvex (2,4,5-trichlorophenoxypropionic acid)	1.0

dure. A substance is designated as being *EP Toxic* if an extract contains any of the eight toxic elements or six pesticides listed in Table 5.1 in concentrations greater than the values given.

Note that the term *hazardous substance* is more all-encompassing than *toxic substance,* though the two are often used interchangeably.

While the four characteristic attributes given above provide guidelines for considering a substance to be hazardous, the EPA maintains an actual list of specific hazardous wastes. EPA *listed wastes* are organized into three categories: *source-specific wastes, generic wastes,* and *commercial chemical products.* Source-specific wastes include sludges and wastewaters from treatment and production processes in *specific industries,* such as petroleum refining and wood preserving. The list of generic wastes includes wastes from common manufacturing and industrial *processes,* such as solvents used in degreasing operations. The third list contains specific chemical products, such as benzene, creosote, mercury, and various pesticides. All listed wastes are presumed to be hazardous regardless of their concentrations.

As of 1987 there were over 400 listed hazardous wastes. All told, over 275 million tons of RCRA-defined hazardous wastes are generated each year in the United States (USEPA, 1987). That is more than one ton of hazardous waste per person per year. Examples of these hazardous wastes are given in Table 5.2.

TABLE 5.2 EXAMPLES OF HAZARDOUS WASTES GENERATED
BY BUSINESS AND INDUSTRIES

Waste generators	Waste types
Chemical manufacturers	Strong acids and bases Spent solvents Reactive wastes
Vehicle maintenance shops	Heavy metal paint wastes Ignitable wastes Used lead acid batteries Spent solvents
Printing industry	Heavy metal solutions Waste inks Spent solvents Spent electroplating wastes Ink sludges containing heavy metals
Leather products manufacturing	Waste toluene and benzene
Paper industry	Paint wastes containing heavy metals Ignitable solvents Strong acids and bases
Construction industry	Ignitable paint wastes Spent solvents Strong acids and bases
Cleaning agents and cosmetics manufacturing	Heavy metal dusts Ignitable wastes Flammable solvents Strong acids and bases
Furniture and wood manufacturing and refinishing	Ignitable wastes Spent solvents
Metal manufacturing	Paint wastes containing heavy metals Strong acids and bases Cyanide wastes Sludges containing heavy metals

5.3 HAZARDOUS SUBSTANCE LEGISLATION

There are many federal environmental laws that regulate hazardous substances in
the United States. Table 5.3 gives a very brief description of the most important
of these, along with the agency in charge of enforcement. In this section, we will
focus on the two Acts that are most crucial to the current management programs
for hazardous wastes. The first is the Resource Conservation and Recovery Act
(RCRA), which provides guidelines for prudent management of new and future
hazardous substances; the second is the Comprehensive Environmental Response, Compensation, and Liability Act (CERCLA), which deals primarily with
mistakes of the past: inactive and abandoned hazardous waste sites.

TABLE 5.3   ENVIRONMENTAL LAWS CONTROLLING HAZARDOUS SUBSTANCES

Legislation	Description
Atomic Energy Act (Nuclear Regulatory Commission)	Regulates nuclear energy production and nuclear waste disposal.
Clean Air Act (EPA)	Regulates the emission of hazardous air pollutants.
Clean Water Act (EPA)	Regulates the discharge of hazardous pollutants into the nation's surface water.
Comprehensive Environmental Response, Compensation, and Liability Act (*Superfund*) (EPA)	Provides for the cleanup of inactive and abandoned hazardous waste sites.
Federal Insecticide, Fungicide, and Rodenticide Act (EPA)	Regulates the manufacture, distribution, and use of pesticides and the conduct of research into their health and environmental effects.
Hazardous Materials Transportation Act (U.S. Department of Transportation)	Regulates the transportation of hazardous materials.
Marine Protection, Research, and Sanctuaries Act (EPA)	Regulates waste disposal at sea.
Occupational Safety and Health Act (U.S. Occupational Safety and Health Administration)	Regulates hazards in the workplace, including worker exposure to hazardous substances.
Resource Conservation and Recovery Act (EPA)	Regulates hazardous waste generation, storage, transportation, treatment, and disposal.
Safe Drinking Water Act (EPA)	Regulates contaminant levels in drinking water and the disposal of wastes into injection wells.
Surface Mining Control and Reclamation Act (U.S. Department of the Interior)	Regulates the environmental aspects of mining (particularly coal) and reclamation.
Toxic Substances Control Act (EPA)	Regulates the manufacture, use, and disposal of chemical substances.

Resource Conservation and Recovery Act

The Resource Conservation and Recovery Act regulates the generation, storage, transportation, treatment, and disposal of hazardous substances. It is our single most important law dealing with the management of hazardous waste and it is perhaps the most comprehensive piece of legislation that EPA has ever promulgated. Its origins are in the 1965 Solid Waste Disposal Act, which was the first federal law to address the enormous problem of how to safely dispose of household, municipal, commercial, and industrial refuse. Congress amended that law in 1970 with the passage of the Resource Recovery Act, and, finally, RCRA itself was passed in 1976. Revisions to RCRA were made in 1980 and again in 1984.

The 1984 amendments, referred to as the *Hazardous and Solid Waste Amendments* (HSWA), significantly expanded the scope of RCRA, particularly in the area of land disposal.

Transportation, Storage, and Disposal

The key concept in RCRA is that hazardous substances must be properly managed from the moment they are generated until their ultimate disposal. This step-by-step management is often referred to as the *cradle-to-grave* approach, and it has three key elements:

1. A *tracking* system, in which a *manifest* document accompanies any waste that is transported from one location to another.
2. A *permitting* system that helps assure safe operation of facilities that treat, store, or dispose of hazardous wastes.
3. A system of controls and restrictions governing the *disposal* of hazardous wastes onto, or into, the land.

About 96 percent of U.S. hazardous wastes are treated or disposed of at the site where they were originally generated (USEPA, 1987c). The remaining 4 percent still represents a substantial volume of material that is transported from the source to *treatment, storage, or disposal facilities* (TSDFs). To help eliminate improper handling of such transported wastes, EPA requires generators to prepare a *hazardous waste manifest* that must accompany the waste. The manifest identifies the type and quantity of waste, the generator, the transporter, and the TSD facility to which the waste is being shipped. One copy of the manifest is sent to the EPA when the waste leaves the generator, and another when the waste arrives at a TSD facility. The generator, who is ultimately responsible for the waste tracking system, also receives a copy of the manifest after the waste arrives at the TSDF. Figure 5.1 illustrates the manifest system.

Under RCRA, the treatment, storage, and disposal facilities that accept hazardous waste must first obtain permits from the EPA. The permitting process has been established to give the EPA power to enforce a number of standards and requirements, including the power to inspect facilities and to bring civil actions against violators of any of RCRA's provisions. To help assure proper management of hazardous wastes in the longer term, RCRA regulations require that TSDFs acquire sufficient financial assurance mechanisms to ensure that the facility can be properly maintained after it ultimately closes, including a provision for 30 years of facility maintenance, security measures, and groundwater monitoring.

Disposal of hazardous wastes by TSDFs is becoming increasingly difficult. In the past, about 80 percent of U.S. hazardous wastes were disposed of on land, either by pumping them underground using deep injection wells, burying them in landfills, or containing them in surface impoundments. The 1984 Hazardous and Solid Waste Amendments to RCRA, however, actively discourage disposal of

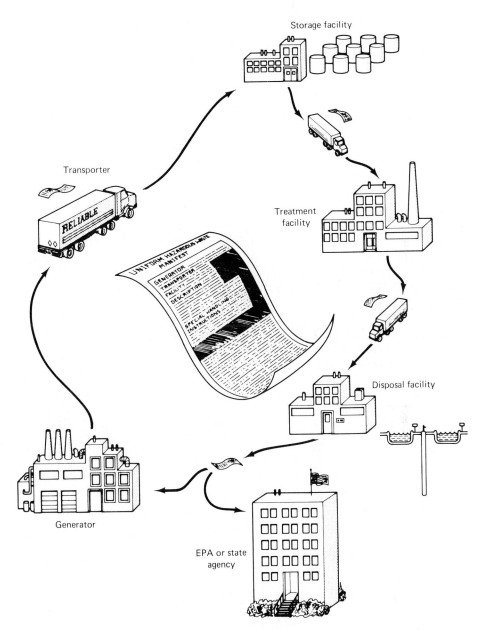

Figure 5.1   A one-page manifest must accompany every waste shipment. The resulting paper trail documents the waste's progress through treatment, storage, and disposal, providing a mechanism for alerting the generator and the EPA if irregularities occur (USEPA 1986c).

wastes on land. In fact, land disposal is to be banned unless EPA determines that for a particular site, and type of waste, there will be no migration of hazardous constituents for as long as the wastes remain hazardous. To the extent that land disposal will be continued to be used, new landfills and surface impoundments are required to have double liners, leachate collection systems, and groundwater monitoring facilities to help assure long-term containment; and, given the ongoing problem of groundwater contamination from landfills, all liquid hazardous wastes are to be banned. In view of the land ban, alternatives are being actively pursued and treatment technologies are improving rapidly. Some of these approaches will be described in Chapter 6.

Waste Reduction

It is clear that rising economic costs of hazardous waste management are dictating a fresh look at the very origins of the problem itself. Is it really necessary for us to generate over one ton of hazardous waste per person per year to provide the goods and services that we have come to expect? It is almost always cheaper and simpler to control the amount of pollution generated in the first place than to attempt to find engineering fixes that deal with it after it has been created. The 1984 amendments to RCRA recognize this important role of waste reduction, by stating:

> The Congress hereby declares it to be the national policy of the United States that, wherever feasible, the generation of hazardous waste is to be reduced or eliminated as expeditiously as possible.

Figure 5.2 suggests a priority system for ranking various approaches to hazardous waste reduction. The first priority is to find ways to *eliminate* uses of hazardous substances altogether. Elimination might be achieved by changing manufacturing processes or by substituting products that can satisfy the same need without creating hazardous wastes (an example would be the substitution of concrete posts for toxic, creosote-treated wooden posts). The next priority is to

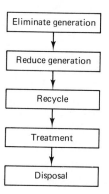

Figure 5.2 Heirarchy of priorities in hazardous waste management.

reduce the amounts generated. Again, manufacturing process changes can be important. The third strategy is to *recycle* hazardous substances such as solvents and acids to maximize their use before treatment and disposal becomes necessary. Finally, hazardous substances can be *treated* to reduce their volume and toxicity.

Eliminating, reducing, and recyling hazardous substances reduces treatment and disposal costs, and oftentimes generators can find it in their own economic best interest to actively seek out these source reduction opportunities as a way to reduce the overall cost of doing business. RCRA, in fact, requires generators to certify on their hazardous waste manifests that they have taken steps to reduce the volume and toxicity of the hazardous waste that they are creating.

Superfund

The Comprehensive Environmental Response, Compensation, and Liabilities Act was enacted in 1980 to deal with abandoned hazardous waste sites. It is more often referred to as the Superfund law as a result of one of its key provisions in which a $1.6 billion trust fund was created from special crude oil and chemical company taxes. In 1986, CERCLA was reauthorized and the trust fund was increased to $8.5 billion, with half of the money coming from petroleum and chemical companies, and most of the rest coming from a special corporate environmental tax augmented by general revenues. The revised law is designated as the *Superfund Amendments and Reauthorization Act* of 1986 (SARA).

Under Superfund, the EPA can deal with both short-term, emergency situations triggered by the actual or potential release of hazardous substances into the environment, as well as with long-term problems involving abandoned or uncontrolled hazardous waste sites for which more permanent solutions are required.

Short-term *removal actions* can be taken at any site where there is an imminent threat to human health or the environment, such as might occur during a spill or fire, or when wastes that have been illegally disposed of are discovered ("midnight dumping"). In addition to removing and disposing of hazardous substances and securing the endangered area, EPA may take more extensive actions as necessary, such as providing alternative drinking water supplies to local residents if their drinking water has been contaminated, or even temporarily relocating residents.

More complex and extensive problems that are not immediately life-threatening are handled under the EPA's *remedial* programs, which are outlined in Figure 5.3. Central to the Superfund process is the creation of a National Priorities List (NPL) of sites that are eligible for federally financed remedial activities. The NPL identifies the worst sites in the nation based on such factors as the quantities and toxicity of wastes involved, the exposure pathways, the number of people potentially exposed, and the importance and vulnerability of the underlying groundwater. In other words, a *risk assessment* of the sort that will be described in this chapter is used to rank potential NPL sites.

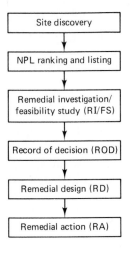

Figure 5.3 The Superfund remedial process.

Once a site is listed, the EPA begins a *Remedial Investigation/Feasibility Study* (RI/FS) to determine appropriate remedial actions for the site. The remedial investigation phase involves gathering information needed to fully characterize the contamination problem at the site, including environmental and public health risks. The feasibility study that follows uses the remedial investigation information to identify, evaluate, and select cleanup alternatives. These alternative cleanup approaches are analyzed based on their relative effectiveness and cost. The RI/FS process culminates with the signing of a *Record of Decision* (ROD) in which the remedial action that has been selected is set forth. After the ROD has been signed, the detailed *Remedial Design* of the selected alternative takes place. For complex sites, the entire process from listing to the beginning of the actual cleanup itself has often taken five years or more.

The final *Remedial Action* may consist of a number of short-term steps, such as installation of surface water runoff controls, excavating soil, building below grade containment walls, and capping the site. Often, as when contaminated groundwater must be pumped out and cleaned, the long-term Remedial Action measures can take decades. The cleanup of a typical NPL site is a slow, complex, expensive process and because of these characteristics, Superfund has been subject to considerable criticism.

One of the most important and guiding policies of CERCLA is that those parties who are responsible for hazardous waste problems will be forced to pay the entire cost of cleanup. *Responsible parties* (RPs) may be the historic owners or operators of the site, any generators who disposed of wastes at the site (whether legally or not), or even the transporters who brought wastes to the site. The courts have recognized the concept of *joint and several* liability, which means, in essence, that if damages cannot be individually apportioned by the responsible parties themselves, then each and every party is subject to liability for the entire cleanup cost. In other words, no matter how little an individual RP may have contributed to the overall problem, he can, theoretically, be liable for the entire cost of the cleanup. If responsible parties do not voluntarily perform the appropriate response actions, the EPA can use Superfund monies to clean up the site. If they do so, then the EPA is empowered to collect three times the cleanup cost from the RPs.

5.4 RISK ASSESSMENT

One of the most important shifts in environmental policy in the 1980s was the acceptance of the role of risk assessment and risk management in environmental decision making. In early environmental legislation, such as the Clean Air and Clean Water Acts, the concept of risk is hardly mentioned; instead, these acts required that pollution standards be set that would allow adequate margins of safety to protect public health. Intrinsic to these standards was the assumption that pollutants have thresholds, and that exposure to concentrations below these thresholds would produce no harm. All of that changed when the problems of toxic waste were finally recognized and addressed. Now, it is necessary to speak in terms of risks. For a given exposure, what is the risk of cancer or birth defects, and what risk is acceptable? How can standards be set when there is a paucity of data for chemicals already in use, and when testing programs cannot begin to keep up with the rate at which new chemicals are introduced? Policymakers have had to grapple with these issues in the creation of legislation and in the administration of the resulting regulations. And, rather than wait for definitive answers to emerge from the scientific community, they have had to act.

The result has been the emergence of the quite controversial field of environmental risk assessment. Hardly anyone is comfortable with it. Scientists often deplore the notion of condensing masses of frequently conflicting, highly uncertain, often ambiguous data, which has been extrapolated well beyond anything actually measured, down to a single number. Officials in government agencies often find themselves facing a hostile public while defending the idea that, for example, ''one cancer in one million people over a 70-year exposure period'' should be acceptable. Those in the regulated community may look at the conservative assumptions built into standard risk assessment techniques and feel risks are overstated, and that regulations should be postponed until more data have been collected. Many concerned citizens think assessments based on exposure to one chemical at a time grossly underestimate risk, and that synergistic effects caused by exposure to many different toxic substances together are likely to amplify risks. The general public seems to demand that there be no toxic risk at all in their water, air, or food while simultaneously it condones cigarette smoking.

Some of the above conflicts can best be dealt with if we make the distinction between *risk assessment* and *risk management*. Risk assessment is the scientific side of the story. It is the gathering of data that are used to relate response to dose. Such dose–response data can then be combined with estimates of likely human exposure to produce overall assessments of risk. Risk management, on the other hand, is the process of deciding what to do. Given the estimates of risk already established, political and social judgment is required to decide whether a one-in-one-million risk is acceptable and, if it is, how to go about trying to achieve it.

The usual starting point for an explanation of risk is to point out that there is some risk in everything we do and that we ought to place the risks we face from exposure to various toxicants into that perspective. We will all die at some point so, as a trivial example, we could say the lifetime risk of death from all causes is 1.0, or 100 percent. Of the total 2.1 million deaths per year in the United States in the late 1980s, there were roughly 460 000 cancer deaths. Neglecting age structure complications, we could say that, on the average, the risk of dying of cancer is therefore about 22 percent (460 000/2 100 000 = 0.22). Smokers of a pack of cigarettes per day face a risk of dying of cancer or heart disease of about 25 percent. They also shorten their life expectancy by about 5 min for each cigarette smoked (roughly the time required to smoke that cigarette) (Wilson and Crouch, 1987). As another example, living in an average home with 1.5 pCi/L of radon is thought to produce a lifetime risk of dying of cancer of about 0.003, or 0.3 percent. These and other risks are listed in Table 5.4.

For comparison, the EPA attempts to control our exposure to toxics to levels that will pose lifetime risks of on the order of one in 10^{-7} to 10^{-4} (0.00001 percent to 0.01 percent).

The National Academy of Sciences (1983) suggests that risk assessment be divided into the following four steps (Figure 5.4):

- *Hazard identification* is the process of determining whether a particular chemical is causally linked to particular health effects such as cancer or birth

TABLE 5.4 EXAMPLES OF SOME COMMONPLACE RISKS IN THE UNITED STATES

Action	Lifetime risk
Cigarette smoking, one pack per day	0.25
All cancers	0.22
Death in a motor vehicle accident	0.02
Homicide	0.01
Home accident deaths	0.01
Radon in homes, cancer deaths	0.003
Alcohol, light drinker, cancers	0.001
Sea level background radiation, cancers	0.001
4 tablespoons peanut butter per day (aflatoxin)	0.0006
Typical EPA maximum contaminant level	0.0000001–0.0001

Source: Based on data given in Wilson and Crouch (1987) and The National Center for Health Statistics.

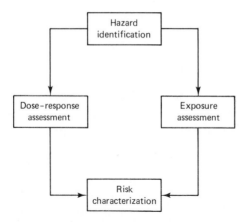

Figure 5.4 Risk assessment is usually considered to be a four-step process.

defects. Since human data are so often difficult to obtain, this step usually focuses on whether a chemical is toxic in animals or other test organisms.

- *Dose–response assessment* is the process of characterizing the relation between the dose of an agent administered or received and the incidence of an adverse health effect. Many different dose–response relationships are possible for any given agent depending on such conditions as whether the response is carcinogenic (cancer causing) or noncarcinogenic and whether the experiment is a one-time acute test or a long-term chronic test. Since most tests are performed with high doses, the dose–response assessment must include a consideration for the proper method of extrapolating data to low exposure rates that humans are likely to experience. Part of the assessment must also include a method of extrapolating animal data to humans.

- *Exposure assessment* involves a determination of the size and nature of the population that has been exposed to the toxicant under consideration, and the length of time and toxicant concentration to which they have been exposed. Consideration must be given to such factors as the age and health of the exposed population, smoking history, the likelihood that members of the population might be pregnant, and whether or not synergistic effects might occur due to exposure to multiple toxicants.

- *Risk characterization* is the integration of the above three steps that results in an estimate of the magnitude of the public-health problem.

5.5 HAZARD IDENTIFICATION

The first step in a risk analysis is to determine whether the chemicals that a population has been exposed to are likely to have any adverse health effects. This is the work of toxicologists, who study both the nature of the adverse effects caused by toxic agents as well as the probability of their occurrence. We start our description of this hazard identification process by summarizing the pathways that a chemical may take as it passes through a human body and the kinds of damage that may result.

A toxicant can enter the body using any of three pathways: by ingestion with food or drink, through inhalation, or by contact with the skin (dermal) or other exterior surfaces, such as the eyes. Once in the body it can be absorbed by the blood and distributed to various organs and systems. The toxicant may then be stored, for example, in fat as in the case of DDT, or it may be eliminated from the body by transformation to something else and/or by excretion. The biotransformation process usually yields metabolites that can be more readily eliminated from the body than the original chemicals; however, metabolism can also convert chemicals to more toxic forms. Figure 5.5 presents the most important movements of chemical toxicants in the body, showing absorption, distribution, storage, and excretion. Although these are shown as separate operations, they all occur simultaneously.

There are several organs that are especially vulnerable to toxicants. The liver, for example, which filters the blood before it is pumped through the lungs, is often the target. Since toxics are transported by the bloodstream, and since the liver is exposed to so much of the blood supply, it can be directly damaged by toxics. Moreover, since a major function of the liver is to metabolize substances, converting them into forms that can more easily be excreted from the body, it is also susceptible to chemical attack by toxic chemicals formed during the biotransformation process itself. Chemicals that can cause liver damage are called *hepatoxins*. Examples of hepatoxic agents include a number of synthetic organic compounds, such as carbon tetrachloride, chloroform, and trichloroethylene; pesticides such as DDT and paraquat; heavy metals such as arsenic, iron, and manganese; and drugs, such as acetaminophen and anabolic steroids. The kidneys are also responsible for filtering the blood, and they too are frequently susceptible to damage.

Toxic chemicals often injure other organs and organ systems as well. *Hematotoxicity* is the term used to describe the toxic effects of substances on the blood. Some hematotoxins, such as carbon monoxide in polluted air, and nitrates in groundwater, affect the ability of blood to transport oxygen to the tissues. Other toxicants, such as benzene, affect the formation of platelets which are necessary for blood clotting. The lungs and skin, due to their proximity to pollutants, are also often affected by chemical toxicants. Lung function can be impaired by such substances as cigarette smoke, ozone, asbestos, and quartz rock dust. The skin reacts in a variety of ways to chemical toxicants but the most common, and serious, environmentally related skin problem is cancer induced by excessive ultraviolet radiation, as will be described in Chapter 8.

Mutagenesis

Deoxyribonucleic acid (DNA) is an essential component of all living things and a basic material in the chromosomes of the cell nucleus. It contains the genetic code that determines the overall character and appearance of every organism.

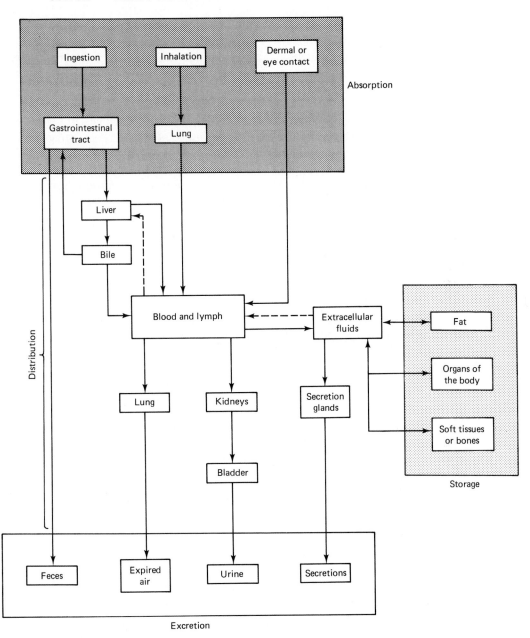

Figure 5.5 Fate of chemical toxicants in the body. (*Source:* Environ, 1988).

Each molecule of DNA has the ability to replicate itself exactly, transmitting that genetic information to new cells. Our interest here in DNA results from the fact that certain chemical agents, as well as ionizing radiation, are genotoxic; that is, they are capable of altering DNA. Such changes, or *mutations,* in the genetic material of an organism can cause cells to misfunction, leading in some cases to cell death, cancer, reproductive failure, or abnormal offspring. Chemicals that are capable of causing cancer are called *carcinogens;* chemicals that can cause birth defects are *teratogens.*

Mutations may affect somatic cells, which are the cells that make up the tissues and organs of the body itself, or they may cause changes in germ cells (sperm or ovum) that may be transmitted to future offspring. As is suggested in Figure 5.6, one possible outcome of a mutagenic event in a somatic cell is the death of the cell itself. If it survives, however, the change may be such that the cell no longer responds to signals that normally control cell reproduction. If that occurs, the cell may undergo rapid and uncontrolled cellular division, forming a tumor.

While mutations in somatic cells may end the life of the affected individual, germ cell mutations may become established in the gene pool and affect future generations.

Carcinogenesis

Chemically induced carcinogenesis is thought to involve two distinct stages, referred to as *initiation* and *promotion.* In the initiation stage, a mutation alters a

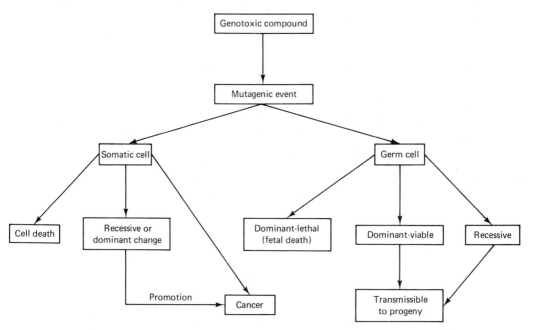

Figure 5.6 Possible consequences of a mutagenic event in somatic and germinal cells. (*Source: Industrial Toxicology* by Williams and Burson © 1985 by Van Nostrand Reinhold. Reprinted by permission of the publisher. All Rights Reserved.)

cell's genetic material in a way that may or may not result in the uncontrolled growth of cells that characterizes cancer. In the second, or promotion, stage of development, affected cells no longer recognize growth constraints that normally apply and a tumor develops. Promoters can increase the incidence rate of tumors among cells that have already undergone initiation, or they can shorten the latency period between initiation and the full carcinogenic response. The model of initiation followed by promotion suggests that some carcinogens may be initiators, others may be promoters, and some may be complete carcinogens capable of causing both stages to occur. Current regulations do not make this distinction, however, and any substance capable of increasing the incidence of tumors is considered a carcinogen, subject to the same risk assessment techniques.

Tumors, in turn, may be *benign* or *malignant* depending on whether or not the tumor is contained within its own boundaries. If a tumor undergoes *metastasis,* that is, it breaks apart and portions of it enter other areas of the body, it is said to be malignant. Once a tumor has metastasized it is obviously much harder to treat or remove.

The theoretical possibility that a single genotoxic event can lead to a tumor is referred to as the *one-hit hypothesis.* Based on this hypothesis, exposure to even the smallest amount of a carcinogen leads to some nonzero probability that a malignancy will result. That is, in a conservative, worst-case risk assessment for carcinogens, it is assumed that there is no threshold dose below which the risk is zero.

A brief glossary of carcinogenesis terminology is presented in Table 5.5.

Toxicity Testing in Animals

With several thousand new chemicals coming onto the market each year, and a backlog of tens of thousands of chemicals already being manufactured, it is not possible to fully test each and every chemical for its toxicity. As a result, a hierarchy of testing procedures has been developed that can be used to select those chemicals that are most likely to pose serious risks.

The prevailing carcinogenesis theory, that human cancers are initiated by gene mutations, has led to the development of short-term, *in vitro* (outside the whole animal) screening procedures that are the usual first step in determining whether a chemical is carcinogenic. It is thought that if a chemical can be shown to be mutagenic, then it *may* be carcinogenic, and further testing may be called for. The most widely used short-term test, called the *Ames mutagenicity assay,* subjects special tester strains of bacteria to the chemical in question. These tester strains have previously been rendered incapable of normal bacterial division, so, unless they mutate back to a form that is capable of division, they will die. Bacteria that survive and form colonies do so through mutation; therefore, the greater the survival rate of these special bacteria, the more mutagenic is the chemical.

TABLE 5.5   GLOSSARY OF CARCINOGENESIS TERMINOLOGY

Acute toxicity	Adverse effects caused by a toxic agent occurring within a short period of time following exposure.
Benign tumor	A new tumor composed of cells that, though proliferating in an abnormal manner, do not spread to surrounding, normal tissue.
Cancer	An abnormal process in which cells begin a phase of uncontrolled growth and spread.
Carcinogen	Any cancer-producing substance.
Carcinoma	A malignant tumor in the tissue that covers internal or external surfaces of the body such as the stomach, liver, or skin.
Chronic toxicity	Adverse effects caused by a toxic agent after a long period of exposure.
Initiator	A chemical that initiates the change in a cell that irreversibly converts the cell into a cancerous or precancerous state.
Malignant tumor	Relatively autonomous growth of cells or tissue that invade surrounding tissue and have the ability to metastasize.
Mutagenesis	Alteration of DNA in either somatic or germinal cells not associated with the normal process of recombination.
Mutation	A permanent, transmissible change in DNA that changes the function or behavior of the cell.
Neoplasm	Literally, new growth, usually of an abnormally fast-growing tissue.
Oncogenic	Giving rise to tumors or causing tumor formation.
Pharmacokinetics	The study of how a chemical is absorbed, distributed, metabolized, and excreted.
Promoter	A chemical that can increase the incidence of response to a carcinogen previously administered.
Sarcoma	A cancer that arises from mesodermal tissue, e.g., fat, muscle, bone.
Teratogen	Any substance capable of causing malformation during development of the fetus.
Toxicity	A relative term generally used in comparing the harmful effect of one chemical on some biologic mechanism with the effect of another chemical.

Source: Based on Williams and Burson (1985).

Intermediate testing procedures involve relatively short-term (several months duration) carcinogenesis bioassays in which specific organs in mice and rats are subjected to known mutagens to determine whether tumors develop.

Finally, the most costly, complex, and long-lasting test, called a *chronic carcinogenesis bioassay,* involves hundreds or thousands of animals over a time period of several years. To assure comparable test results and verifiable data, the National Toxicology Program in the United States has established minimum test requirements for an acceptable chronic bioassay, which include:

- *Two species of rodents must be tested.* Mice and rats, using specially inbred strains for consistency, are most often used. They have relatively short lifetimes and their small size makes them easier to test in large numbers.

- *At least 50 males and 50 females of each species for each dose must be tested.* Many more animals are required if the test is to be sensitive enough to detect risks of less than a few percent.
- *At least two doses must be administered (plus a no-dose control).* One dose is traditionally set at the maximum tolerated dose (MTD), a level that can be administered for a major portion of an animal's lifetime without significantly impairing growth or shortening the lifetime. The second dose is usually one-half or one-fourth the MTD.

Exposure begins at 6 weeks of age, and ends when the animal reaches 24 months of age. At the end of the test, all animals must be subjected to detailed pathological examinations.

Notice, following this protocol, the minimum number of animals required for a bioassay is 600 (2 species × 100 animals × 3 doses), and at that number it is still only relatively high risks that can be detected. For example, suppose at a particular dose the probability of causing a tumor is 5 percent (that is, the "risk" is 5 percent). That means that, *on the average,* out of every 100 animals exposed we would expect to see 5 excess tumors (above those seen in the controls). Suppose we gave the same dose to 10 groups of 100 animals each, and for simplicity, suppose no tumors are ever found in the controls. In some exposed groups we might find the 5 tumors expected; in some there might be 4 or 6, in others perhaps 1 or 2 would appear, but the average would be 5 tumors per group. Statistically, with the actual risk being 5 percent, we might consider it very unlikely that there would ever be a group of 100 without at least one excess tumor. Suppose, then, we run the test on only a *single* group of 100 and find one tumor. The risk might in actuality be 1 percent as a first pass at the data might imply; however, as the above example suggests, the risk might be perhaps as high as 5 percent. In other words, a test using only 100 animals cannot, in a statistically significant way, ensure that the risk is less than about 5 percent. Bioassays designed to detect lower risks require many thousands of animals and, in fact, the largest experiment ever performed involved over 24 000 mice and yet it was still insufficiently sensitive to measure less than a one percent increase in tumor incidence (Environ, 1988).

The inability of a bioassay to detect small risks presents one of the greatest difficulties in applying the data so obtained to human risk assessment. Regulators try to restrict human risks due to exposure to carcinogens to levels of about one-in-one-million, yet animal studies are only capable of detecting risks of perhaps one-in-one-hundred. It is necessary, therefore, to find some way to extrapolate the data taken for animals exposed to high doses, to humans who will be exposed to doses that are at least several orders of magnitude lower.

Human Studies

Another shortcoming in the animal testing methods just described, besides the necessity to extrapolate the data toward zero risk, is the obvious difficulty in interpreting the data for humans. How does the fact that some substance causes tumors in mice relate to the likelihood that it will cause cancer in humans as well? Animal testing can always be criticized in this way, but since we are not inclined to perform the same sorts of tests directly on humans, other methods must be used to gather evidence of human toxicity.

While some data can be obtained from clinical studies of humans who have inadvertently been exposed to a suspected toxicant, the most important source of information relating exposure to risk is obtained with *epidemiologic* studies. Epidemiology is the study of the incidence rate of diseases in real populations. By attempting to find correlations between elevated rates of incidence of a particular disease in certain groups of people, and some measure of their exposure to various environmental factors, an epidemiologist attempts to show in a quantitative way the relationship between exposure and risk. Such data can be used to complement animal data, clinical data, and scientific analyses of the characteristics of the substances in question.

Caution must be exercised in interpreting every epidemiologic study, since any number of confounding variables may lead to invalid conclusions. For example, the study may be biased because workers are compared with nonworkers (workers are generally healthier), or because relative rates of smoking have not been accounted for, or there may be other variables that are not even hypothesized in the study that may be the actual causal agent. As an example of the latter, consider an attempt to compare lung cancer rates in a city having high ambient air pollution levels with rates in a city having less pollution. Suppose the rates are higher in the more polluted city, even after accounting for smoking history, age distribution, and working background. To conclude that ambient air pollution is causing those differences may be totally invalid. Instead, it might well be different levels of radon in homes, for example, or differences in other indoor air pollutants associated with the type of fuel used for cooking and heating that are causing the cancer variations.

Weight-of-Evidence Categories for Potential Carcinogens

Based on the accumulated evidence from case studies, epidemiologic studies, and animal data, the EPA uses the following categories to describe the likelihood that a chemical substance is carcinogenic (USEPA, 1986a). Using both human and animal data, five categories, A–E, have been established as follows:

> *Group A: Human Carcinogen.* A substance is put into this category only when there is *sufficient* epidemiologic evidence to support a causal association between exposure to the agent and cancer.
>
> *Group B: Probable Human Carcinogen.* This group is actually made up of two subgroups. An agent is categorized as **B1** if there is *limited* epidemiologic evidence; and it is put into **B2** if there is inadequate human data but sufficient evidence of carcinogenicity in animals.

TABLE 5.6 CATEGORIZATION OF EVIDENCE FOR HUMAN CARCINOGENICITY

Human evidence	Animal evidence				
	Sufficient	Limited	Inadequate	No data	No evidence
Sufficient	A	A	A	A	A
Limited	B1	B1	B1	B1	B1
Inadequate	B2	C	D	D	D
No data	B2	C	D	D	E
No evidence	B2	C	D	D	E

Source: USEPA (1986a).

Group C: Possible Human Carcinogen. This group is used for agents with limited evidence of carcinogenicity in animals and an absence of human data.

Group D: Not Classified. For agents with inadequate human and animal evidence or for which no data are available.

Group E: Evidence of Noncarcinogenicity. Used for agents that show no evidence for carcinogenicity in at least two adequate animal tests in different species or in both adequate epidemiologic and animal studies.

Table 5.6 summarizes this categorization scheme.

5.6 DOSE–RESPONSE ASSESSMENT

As the name suggests, the fundamental goal of a dose–response assessment is to obtain a mathematical relationship between the amount of a toxicant to which a human is exposed and the risk that there will be an unhealthy response to that dose. The abscissa is dose, usually expressed as milligrams of substance per kilogram of body weight per day (mg/kg/day). The dose is an exposure averaged over an entire lifetime (for humans, assumed to be 70 years). The ordinate is the response, which is the risk that there will be some adverse health effect. Note that response (risk) has no units; it is a probability that there will be some adverse health effect. For example, if a given exposure would be expected to produce 100 cancers in a population of one million during the next 70 years, the response could be expressed as 0.0001, 1×10^{-4}, or 0.01 percent.

For substances that induce a carcinogenic response, it is always conservatively assumed that exposure to any amount of the carcinogen will create some likelihood of cancer. That is, a plot of response versus dose is required to go through the origin. For noncarcinogenic responses, it is usually assumed that there is some *threshold dose,* below which there will be no response. As a result of these two assumptions, the dose–response curves and the methods used to apply them are quite different for carcinogenic and noncarcinogenic effects, as suggested in Figure 5.7. Realize that the same chemical may be capable of causing both kinds of response.

To apply dose–response data obtained from animal bioassays to humans, a *scaling factor* must be introduced. Sometimes the scaling factor is based on the assumption that doses are equivalent if the dose per unit of body weight in the animal and human is the same. Sometimes (as is done by the EPA) equivalent doses are based on equal dose per unit of body surface area. In either case, the resulting human dose–response curve is specified with the standard mg/kg/day units for dose. Adjustments may also have to be made to account for differences in the rates of chemical absorption. If enough is known about the differences between the absorption rates in test animals and in humans for the particular substance in question, it is possible to account for those differences later in the risk assessment. Usually though, there is insufficient data and it is simply assumed that the absorption rates are the same.

Dose–Response Curves for Carcinogens

The most controversial aspect of dose–response curves for carcinogens is the method chosen to extrapolate from the high doses actually administered to test animals to the low doses to which humans are likely to be exposed. Recall that even with extremely large numbers of animals in a bioassay, the lowest risks that can be measured are usually a few percent. Since regulators attempt to control human risk to several orders of magnitude less than that, there will be no actual animal data anywhere near the range of most interest.

Many mathematical models have been proposed for the extrapolation to low doses. Unfortunately, no model can be proved or disproved from the data, so there is no way to know which model is the most accurate. That means the choice of models is strictly a policy decision. One commonly used model is called the *one-hit model* in which the relationship between dose d and lifetime risk (probabil-

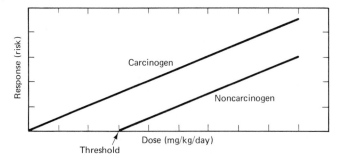

Figure 5.7 Dose–response curves for carcinogens are assumed to have no threshold; that is, any exposure produces some chance of causing cancer.

ity) of cancer, $P(d)$, is given in the following form (Crump, 1984):

$$P(d) = 1 - e^{-(q_0 + q_1 d)} \tag{5.1}$$

where q_0 and q_1 are parameters picked to fit the data. The one-hit model corresponds to the simplest mechanistic model of carcinogenesis in which it is assumed that a single chemical hit is capable of inducing a tumor.

If we substitute $d = 0$ into (5.1), the result will be an expression for the background rate of cancer incidence, $P(0)$. Using the mathematical expansion for an exponential

$$e^x = 1 + x + \frac{x^2}{2!} + \cdots + \frac{x^n}{n!} \cong 1 + x \quad \text{(for small } x) \tag{5.2}$$

and assuming that the background cancer rate is small allows us to write

$$P(0) = 1 - e^{-q_0} \cong 1 - [1 + (-q_0)] = q_0 \tag{5.3}$$

That is, the background rate for cancer incidence corresponds to the parameter q_0. Using the exponential expansion again, the one-hit model suggests the lifetime probability of cancer for small dose rates can be expressed as

$$P(d) \cong 1 - [1 - (q_0 + q_1 d)] = q_0 + q_1 d = P(0) + q_1 d \tag{5.4}$$

For low doses, the additional risk of cancer above the background rate would be

$$\text{Additional risk} = A(d) = P(d) - P(0) \tag{5.5}$$

or, using (5.4),

$$A(d) \cong q_1 d \tag{5.6}$$

That is, the excess lifetime probability of cancer is linearly related to dose.

The one-hit model relating risk to dose is not the only one possible. Another mathematical model that has been proposed has its roots in the multistage model of tumor formation; that is, that tumors are the result of a sequence of biological events (Crump, 1984). The *multistage* model expresses the relationship between risk and dose as

$$P(d) = 1 - \exp[-(q_0 + q_1 d + q_2 d^2 + \cdots q_n d^n)] \tag{5.7}$$

where the individual parameters q_i are positive constants picked to best fit the dose–response data. Again, it is easy to show that for small values of d, the multistage model also has the desirable feature of producing a linear relationship between additional risk and dose. Figure 5.8 illustrates the use of a one-hit model and a multistage model to fit experimental data. The multistage model will always fit the data better since it includes the one-hit model as a special case.

Since the choice of an appropriate low-dose model is not based on experimental data, there is no "correct" one. To protect public health, the EPA chooses to err on the side of safety and overemphasize risk. Their model of choice is a modified multistage model, called the *linear multistage model*. It is

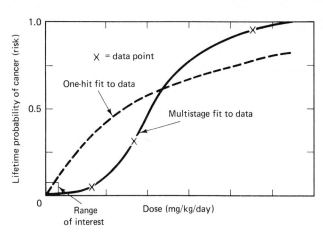

Figure 5.8 Dose–response curves showing two methods of fitting an equation to the data. The range of interest is well below the point where any data actually exist (based on Crump, 1984).

linear at low doses with the constant of proportionality picked in a way that statistically will produce less than a 5% chance of underestimating risk.

At low doses, the slope of the dose–response curve produced by the linear multistage model is called the *potency factor* (PF), or *slope factor*. It has units of $(mg/kg/day)^{-1}$. As suggested in Figure 5.9, another way to interpret the potency factor is that it is the risk produced by a lifetime average daily dose of 1 mg/kg/day. The key dose–response equation for a carcinogen is thus

$$\text{Lifetime risk} = \text{Average daily dose (mg/kg/day)} \times \text{Potency factor (mg/kg/day)}^{-1}$$
$$(5.8)$$

In (5.8) the risk is the probability of getting cancer (not the probability of dying of cancer) and the dose is an average taken over an assumed 70-year lifetime. The dose is called the *lifetime average daily dose* (LADD) or the *chronic daily intake* (CDI).

The EPA maintains a database of information on toxic substances called the *Integrated Risk Information System* (IRIS). Included in the rather extensive background information on each substance is the potency factor and the weight-of-evidence category as described above. Table 5.7 gives a short list of some of the chemicals, potency factors, and categories for which assessments have been made. Note that the exposure route may be either oral or by inhalation. The EPA

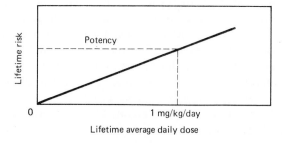

Figure 5.9 The potency factor is the slope of the dose–response curve at low doses.

TABLE 5.7 TOXICITY DATA FOR SELECTED POTENTIAL CARCINOGENS

Chemical	Category	Potency factor oral route $(mg/kg/day)^{-1}$	Potency factor inhalation route $(mg/kg/day)^{-1}$
Arsenic	A	1.75	50
Benzene	A	2.9×10^{-2}	2.9×10^{-2}
Benzo(a)pyrene	B2	11.5	6.11
Cadmium	B1	—	6.1
Carbon tetrachloride	B2	0.13	—
Chloroform	B2	6.1×10^{-3}	8.1×10^{-2}
Chromium VI	A	—	41
DDT	B2	0.34	—
1,1-Dichloroethylene	C	0.58	1.16
Dieldrin	B2	30	—
Heptachlor	B2	3.4	—
Hexachloroethane	C	1.4×10^{-2}	—
Methylene chloride	B2	7.5×10^{-3}	1.4×10^{-2}
Nickel and compounds	A	—	1.19
Polychlorinated biphenyls (PCBs)	B2	7.7	—
2,3,7,8-TCDD (dioxin)	B2	1.56×10^{5}	—
Tetrachloroethylene	B2	5.1×10^{-2}	$1.0 - 3.3 \times 10^{-3}$
1,1,1-Trichloroethane (1,1,1-TCA)	D	—	—
Trichloroethylene (TCE)	B2	1.1×10^{-2}	1.3×10^{-2}
Vinyl chloride	A	2.3	0.295

Source: USEPA IRIS database (1989).

emphasizes that the potency factors given are *upper-bound* estimates of human risk.

Consider the example of estimating risk caused by a lifetime of drinking water that contains some carcinogen. To determine the chronic daily intake, we would need to know the amount of water consumed each day, the concentration of carcinogen in that water, and, to put the dose into proper units, the body weight of the individual. Similarly, to estimate the dose produced by inhaling contaminants in air, we would need to know the amount of air we breathe, the pollutant concentration, and the individual's body weight. For risk assessments applied to large groups of individuals, it is convenient to make certain assumptions about average body weight, amount of water ingested daily, and the amount of air breathed. The EPA recommends the standard values for such *exposure constants* given in Table 5.8. The following example shows how they can be used.

Example 5.1   Risk Assessment for Chloroform in Drinking Water

When drinking water is chlorinated, the trihalomethane, chloroform, is routinely, and inadvertently, created in concentrations of approximately 30–70 $\mu g/L$. Suppose a municipal drinking water supply has an average chloroform concentration of 70 $\mu g/L$.

TABLE 5.8 EPA RECOMMENDED STANDARD VALUES
FOR DAILY INTAKE CALCULATIONS

Parameter	Standard value
Average body weight, adult	70 kg
Average body weight, child	10 kg
Amount of water ingested daily, adult	2 liters
Amount of water ingested daily, child	1 liter
Amount of air breathed daily, adult	20 m^3
Amount of air breathed daily, child	5 m^3
Amount of fish consumed daily, adult	6.5 g
If exposure is for an entire lifetime, use	70 years

Source: USEPA (1986b).

a. Estimate the maximum lifetime risk of cancer, for an adult, associated with the chloroform in that drinking water.

b. If a city of 500 000 is served by this supply, what number of extra cancers per year would be expected?

c. Compare the extra cancers per year caused by chloroform in the drinking water with the expected number of cancer deaths from all causes (the cancer death rate in the United States is 193 per 100 000 per year).

Solution

a. From Table 5.8, an adult is assumed to weigh 70 kg and drink 2 L of water per day, so the chronic daily intake (CDI) would be

$$\text{CDI} = \frac{70 \times 10^{-6}\ \text{g/L} \times 10^3\ \text{mg/g} \times 2\text{L/day}}{70\ \text{kg}} = 0.002\ \text{mg/kg/day}$$

From Table 5.7, the oral carcinogenic potency factor is 6.1×10^{-3} (mg/kg/day)$^{-1}$. Using (5.8), the lifetime risk of getting cancer would be

$$\text{Risk} = \text{CDI} \times \text{Potency factor}$$

$$= 0.002\ \text{mg/kg/day} \times \frac{6.1 \times 10^{-3}}{\text{mg/kg/day}} = 12.2 \times 10^{-6}$$

This calculation suggests that over a 70-year period, the individual probability of cancer is 12.2 per million. Since potency factors supposedly represent probabilities at the 95-percent confidence level, we could say that 12.2 per million is an overestimate of the actual probability of cancer.

b. If there are 12.2 cancers per million people over a 70-year period, then in any given year, in a population of one-half million, the number of cancers caused by chloroform would be

$$\text{Cancer/yr} = 500\ 000\ \text{people} \times \frac{12.2\ \text{cancer}}{10^6\ \text{people}} \times \frac{1}{70\ \text{yr}}$$

$$= 0.09 \approx 0.1\ \text{cancer/yr}$$

c. The total number of annual cancer deaths that would normally be expected in this city of 500 000 at the given rate of 193 deaths/year per 100 000, would be

$$\text{Total cancer deaths/yr} = \frac{193 \text{ deaths/yr}}{100\ 000 \text{ people}} \times 500\ 000 \text{ people}$$

$$= 965 \text{ deaths/yr}$$

It would seem that an additional 0.1 new (treatable) cancers per year in a population that will already have 965 deaths would be nondetectable.

Once again it is necessary to emphasize that the science behind a risk assessment calculation of the sort demonstrated in Example 5.1 is extremely primitive and that there are enormous uncertainties associated with any particular answer so computed. There is still great value, however, to this sort of procedure since it does organize a mass of data into a format that can be communicated to a much wider audience, and it can greatly help that audience find legitimate perspectives based on that data. For example, it matters little whether the annual extra cancers associated with chloroform in the above example are found to be 0.09 or 0.9—the conclusion that the extra cancers would be nondetectable would not change.

The dose rate used in the above example is based on an assumed 70-year lifetime of exposure, standardized values for ingestion rates, and 100 percent absorption of the toxicant. Any of these assumptions may be modified. In more general terms, the chronic daily intake can be obtained as follows:

$$\text{CDI (mg/kg/day)} = \frac{\text{Total dose (mg)}}{\text{Body weight (kg)} \times \text{Lifetime (days)}} \tag{5.9}$$

where

$$\begin{pmatrix} \text{Total} \\ \text{dose} \end{pmatrix} = \begin{pmatrix} \text{Contaminant} \\ \text{concentration} \end{pmatrix} \times \begin{pmatrix} \text{Intake} \\ \text{rate} \end{pmatrix} \times \begin{pmatrix} \text{Exposure} \\ \text{duration} \end{pmatrix} \times \begin{pmatrix} \text{Absorption} \\ \text{fraction} \end{pmatrix} \tag{5.10}$$

Example 5.2   An Occupational Exposure

Estimate the cancer risk for a 60-kg worker exposed to a particular carcinogen under the following circumstances. Exposure time is 5 days per week, 50 weeks per year, over a 20-year period of time. The worker is assumed to breathe heavily for 2 hr per workday at the rate of 1.5 m³/hr and 6 hr per workday at a moderate breathing rate of 1 m³/hr. The carcinogen has a potency factor of 0.02 (mg/kg/day)$^{-1}$, the absorption factor is estimated at 80 percent, and its average concentration in the air is thought to be about 0.05 mg/m³.

Solution   We can begin by calculating the total amount of air breathed per workday:

$$\text{Daily intake rate} = 1.5 \text{ m}^3\text{/hr} \times 2 \text{ hr} + 1 \text{ m}^3\text{/hr} \times 6 \text{ hr} = 9 \text{ m}^3\text{/day}$$

The total dose is then

$$\text{Total dose} = 9 \text{ m}^3\text{/day} \times 5 \text{ days/week} \times 50 \text{ weeks/yr} \times 20 \text{ yrs} \times 0.05 \text{ mg/m}^3 \times 0.8$$
$$= 1800 \text{ mg}$$

Using a standard estimate of 70 years for a lifetime, the CDI is

$$\text{CDI} = \frac{1800 \text{ mg}}{60 \text{ kg} \times 70 \text{ yr} \times 365 \text{ days/yr}} = 0.00117 \text{ mg/kg/day}$$

The cancer risk is thus

$$\text{Risk} = 0.00117 \text{ mg/kg/day} \times 0.02 \text{ (mg/kg/day)}^{-1} = 2.3 \times 10^{-5}$$

or about 23 chances in one million.

Dose–Response Curves for Noncarcinogens

The key assumption for noncarcinogens is that there exists an exposure threshold; that is, any exposure less than the threshold would be expected to show no increase in adverse effects above natural background rates. One of the principal goals of toxicant testing is therefore to identify and quantify such thresholds. Unfortunately, for the usual case, inadequate data are available to establish such thresholds with any degree of certainty and, as a result, it has been necessary to introduce a number of special assumptions and definitions.

Suppose there exists a precise threshold for some particular toxicant for some particular animal species. To experimentally determine the threshold, we might imagine a testing program in which animals would be exposed to a range of doses. Doses below the threshold would elicit no response; doses above the threshold would produce responses. The lowest dose administered that results in a response is given a special name: the *lowest-observed-effect level* (LOEL). Conversely, the highest dose administered that does not create a response is called the *no-observed-effect level* (NOEL). NOELs and LOELs are often further refined by noting a distinction between effects that are *adverse* to health and effects that are not. Thus, there are also *no-observed-adverse-effect levels* (NOAELs) and *lowest-observed-adverse-effect levels* (LOAELs).

Figure 5.10 illustrates these levels and introduces another exposure called the *reference dose,* or RfD. The RfD used to be called the *acceptable daily intake* (ADI) and, as that name implies, it is intended to give an indication of a level of human exposure that is likely to be without appreciable risk. The units of RfD are mg/kg/day averaged over a lifetime. The RfD is obtained by dividing the NOEL by an appropriate *uncertainty factor* (sometimes called a safety factor). A 10-fold uncertainty factor is used to account for differences in sensitivity between the most sensitive individuals in an exposed human population, such as pregnant women, babies, and the elderly, and "normal, healthy" people. Another factor of 10 is introduced when the NOEL is based on animal data that are to be extrapolated to humans. And finally, another factor of 10 is sometimes applied when there are no good human data and the animal data available are limited. Thus, depending on the strength of the available data, human RfD levels are established at doses that are anywhere from one-tenth to one-thousandth of the NOEL, which is itself somewhat below the actual threshold.

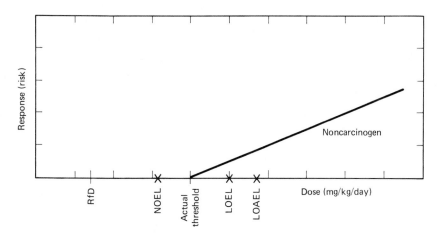

Figure 5.10 Illustrating the various estimates of threshold dose for noncarcino-genic effects. The reference dose (RfD) is the NOEL divided by an uncertainty factor typically between 10 and 1000.

While the RfD is a useful reference point, there is no way to guarantee that it has been established at a level that is without risk. Conversely, there may be levels of exposure considerably higher than the RfD that are completely safe. Table 5.9 gives a short list of some commonly encountered toxicants and their RfDs.

The reference dose can be used in quantitative risk assessments following a procedure similar to that used for carcinogens. The main difference is that the dose is shifted by an amount equal to the threshold as the following expression

TABLE 5.9 RfDs FOR CHRONIC NONCARCINOGENIC EFFECTS OF SELECTED CHEMICALS

Chemical	RfD (mg/kg/day)
Acetone	0.1
Cadmium	5×10^{-4}
Chloroform	0.01
1,1-Dichloroethylene	0.009
cis-1,2-Dichloroethylene	0.01
Methylene chloride	0.06
Phenol	0.04
PCB	0.0001
Tetrachloroethylene	0.01
Toluene	0.3
1,1,1-Trichloroethane	0.09
1,1,2-Trichloro-1,2,2-trifluoroethane (CFC-113)	30
Xylene	2.0

Source: USEPA, IRIS database (1989).

suggests:

$$\text{Risk} = (\text{CDI} - \text{RfD}) \times \text{Potency factor} \tag{5.11}$$

where the potency factor is the slope of the dose–response curve. This type of risk calculation is not often done, however, and the RfD is mostly used as a simple indicator of potential risk. When used in this way, the chronic daily intake is compared with the RfD and if it is below the RfD, negligible risk is assumed to exist for almost all members of the exposed population. If the CDI exceeds the RfD, then there might be cause for concern, though there is no clear demarcation between safe and unsafe.

5.7 HUMAN EXPOSURE ASSESSMENT

One of the most elementary concepts of risk assessment is one that is all too often overlooked in public discussions. And that is that risk has two components: the toxicity of the substance involved, *and* the amount of exposure to that substance. Unless individuals are exposed to the toxicants, there is no human risk.

A human exposure assessment is itself a two-part process. First, pathways that allow toxic agents to be transported from the source to the point of contact with people must be evaluated. And second, an estimate must be made of the amount of contact that is likely to occur between people and those contaminants. Figure 5.11 suggests some of the transport mechanisms that are common at a toxic waste site. Substances that are exposed to the atmosphere may volatilize and be transported with the prevailing winds (in which case, plume models such as the ones introduced in Chapter 7 are often used). Substances in contact with soil may leach into groundwater and eventually be transported to local drinking water wells (groundwater flows were analyzed in Chapter 4). As pollutants are transported from one place to another, they may undergo various transformations that can change their toxicity and/or concentration. A careful study of the fate and transport of pollutants is beyond the scope of this text. A useful summary of exposure pathway models that the EPA uses is given in the *Superfund Exposure Assessment Manual* (USEPA, 1988).

Once the exposure pathways have been analyzed, an estimate of the concentrations of toxicants in the air, water, soil, and food at a particular exposure point can be made. With the concentrations of various toxic agents established, the second half of an exposure assessment begins. Human contact with those contaminants must be estimated. Necessary information includes numbers of people exposed, duration of exposure, and amounts of contaminated air, water, food, and soil that find their way into each exposed person's body. Often, the human intake estimates are based on a lifetime of exposure, assuming standard, recommended, daily values of amounts of air breathed, water consumed, and body weight, such as are given in Table 5.8. In some circumstances, the exposure may be intermittent and adjustments might need to be made for various body weights, rates of absorption, and exposure periods, as was illustrated in Example 5.2.

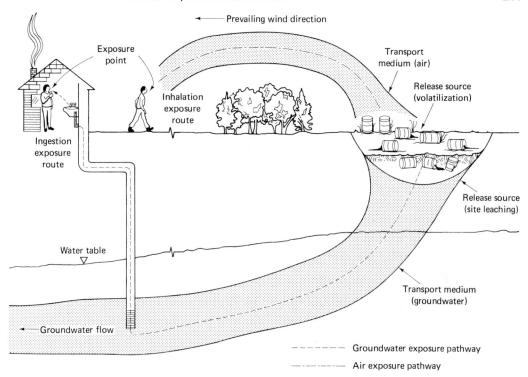

Figure 5.11   Illustration of exposure pathways (USEPA, 1986b).

Bioconcentration

One potentially important exposure route corresponds to human consumption of contaminated fish. It is relatively straightforward to estimate concentrations of contaminants in water, and it also reasonable to make estimates of consumption of fish that individuals may consume (for example, the standard intake values given in Table 5.8). What is more difficult is the task of estimating the concentration of a contaminant in fish, given only the chemical concentration in water. The *bioconcentration factor* (BCF) provides the key link. It is a measure of the tendency for a substance to accumulate in fish tissue. The equilibrium concentration of a chemical in fish can be estimated by multiplying its concentration in water by the bioconcentration factor:

Concentration in fish = (Concentration in water) × (Bioconcentration factor)

$$(5.12)$$

The units of BCF (L/kg) are picked to allow the concentration of substance in water to be the usual milligram per liter and the concentration in fish to be milligram of substance per kilogram of fish. Some example values of BCF are given in Table 5.10. Note the high bioconcentration factors for chlorinated hydro-

TABLE 5.10 BIOCONCENTRATION FACTORS (BCFs) FOR A
SELECTED LIST OF CHEMICALS

Chemical	Bioconcentration factor (L/ kg)
Aldrin	28
Arsenic and compounds	44
Benzene	5.2
Cadmium and compounds	81
Carbon tetrachloride	19
Chlordane	14 000
Chloroform	3.75
Chromium III, VI and compounds	16
Copper	200
DDE	51 000
DDT	54 000
1,1-Dichloroethylene	5.6
Dieldrin	4 760
Formaldehyde	0
Heptachlor	15 700
Hexachloroethane	87
Nickel and compounds	47
Polychlorinated biphenyls (PCBs)	100 000
2,3,7,8-TCDD (dioxin)	5 000
Tetrachloroethylene	31
1,1,1-Trichloroethane	5.6
Trichloroethylene (TCE)	10.6
Vinyl chloride	1.17

Source: USEPA (1986b).

carbon pesticides such as chlordane, DDT, and heptachlor, and the especially high concentration factor for polychlorinated biphenyls (PCBs). These high bioconcentration factors have been an important part of the decision to reduce or eliminate use of these chemicals in the United States.

The following example illustrates the use of bioconcentration factors in a carcinogenic risk assessment.

Example 5.3   Bioconcentration of TCE

Suppose a 70-kg individual consumes an average of 6.5 g of fish per day taken from water with a concentration of trichloroethylene (TCE) equal to 100 ppb (0.1 mg/L). Estimate the (maximum) lifetime cancer risk that this exposure would cause.

Solution: From Table 5.7 the cancer potency factor for an oral dose of TCE is 1.1×10^{-2} (mg/kg/day)$^{-1}$. In Table 5.10 the biocentration factor for TCE is given as 10.6 L/kg. The expected concentration of TCE in fish is therefore (5.12):

Concentration = 0.1 mg/L × 10.6 L/kg = 1.06 mg TCE/kg fish

If a 70-kg individual consumes 6.5 g per day (0.0065 kg/day) of this fish, the chronic daily intake would be

$$CDI = 0.0065 \text{ kg fish/day} \times 1.06 \text{ mg TCE/kg fish/70 kg}$$

$$= 9.8 \times 10^{-5} \text{ mg/kg/day}$$

Using (5.8), the lifetime risk of cancer is

$$Risk = CDI \times \text{Potency factor}$$

$$= 9.8 \times 10^{-5} \text{ mg/kg/day} \times 1.1 \times 10^{-2} \text{ (mg/kg/day)}^{-1} = 1.08 \times 10^{-6}$$

or about 1 in one million.

Contaminant Degradation

Many toxic chemicals of concern are nonconservative; that is, they degrade with time. Degradation may be the result of a number of processes that remove it from the medium in which it resides. There may be phase-transfer as a chemical volatilizes, chemical transformation if it reacts with other substances, or biological transformation if it is consumed by microorganisms. The persistence of a chemical as it moves through various environmental media may be affected by some combination of these mechanisms. A convenient way to deal with such complexity is to simply combine the degradation processes into a single, overall *half-life*. The half-life of a given substance will depend on whether it appears in soil, air, surface water, or groundwater. Some representative half-lives are given in Table 5.11.

TABLE 5.11   RANGE OF HALF-LIVES (DAYS) OF VARIOUS CONTAMINANTS IN AIR AND SURFACE WATER

	Air		Surface water	
Chemical name	Low	High	Low	High
Benzene	6	—	1	6
Benzo(a)pyrene	1	6	0.4	—
Carbon tetrachloride	8030	—	0.3	300
Chlordane	40	—	420	500
Chloroform	80	—	0.3	30
DDT	—	—	56	110
1,1-Dichloroethane	45	—	1	5
Formaldehyde	0.8	—	0.9	3.5
Heptachlor	40	—	0.96	—
Hexachloroethane	7900	—	1.1	9.5
Polychlorinated biphenyls (PCBs)	58	—	2	12.9
2,3,7,8-TCDD (dioxin)	—	—	365	730
1,1,1-Trichloroethane	803	1752	0.14	7
Trichloroethylene	3.7	—	1	90
Vinyl chloride	1.2	—	1	5

Source: USEPA (1986b).

The relationship between reaction rate coefficient K and half-life $T_{1/2}$ was derived in Chapter 3. Recall that if the concentration of a substance is modeled with a simple exponential decay relationship

$$C(t) = C(0)e^{-Kt} \qquad (5.13)$$

then the time required for the concentration to be decreased by 50 percent is the half-life, given by

$$T_{1/2} = \ln 2/K = 0.693/K \qquad (5.14)$$

An example of how to use half-lives is given in the following example.

Example 5.4   A Leaking Underground Storage Tank Exposure Assessment

Suppose a leaking underground storage tank (LUST) is contaminating the groundwater, causing a concentration directly beneath the site of 300 μg/L. The groundwater is flowing at the rate of 1.0 ft per day toward a public drinking water well one mile away. Preliminary measurements indicate that the toxicant is subject to degradation in the aquifer with a half-life of 10 years. It also appears that there is a *retardation factor* in effect, whereby the rate at which the toxicant flows is only half the rate of groundwater flow. To be conservative, assume the pollutant does not disperse within the aquifer, but merely flows along with it. The public well pumps 75 000 gallons of water per day that is mixed with uncontaminated water from other sources, for a total of 1 million gallons per day delivered to 50 000 customers.

   a. Estimate the steady-state pollutant concentration expected at the well.
   b. Estimate the daily dose for a 70-kg adult ingesting 2 L/day of this water.
   c. If the carcinogenic potency of the contaminant is 0.02 (mg/kg/day)$^{-1}$, estimate the excess cancers per year in this population if they were to drink this water for 70 years.

Solution

   a. The contaminant is flowing at half the rate of the groundwater flow, or 0.5 ft per day. The time required to travel to the well one mile away would be

$$\text{Time} = \frac{5280 \text{ ft}}{0.5 \text{ ft/day}} = 10\,560 \text{ days}$$

The pollutant is assumed to degrade exponentially, so the reaction rate coefficient K can be found using (5.14):

$$K = \frac{0.693}{T_{1/2}} = \frac{0.693}{10 \text{ yr} \times 365 \text{ days/yr}} = 1.9 \times 10^{-4}/\text{day}$$

In the 10 560 days required to travel to the drinking water well, (5.13) suggests the initial 300 μg/L would be reduced to

$$C(t) = C(0)e^{-Kt} = 300e^{-(1.9 \times 10^{-4}/\text{day} \times 10\,560 \text{ day})} = 40.4 \ \mu\text{g/L}$$

   b. When 75 000 gallons of this contaminated water is mixed with unpolluted water to give 1 million gallons of mixture, the concentration in drinking water would be

$$C_m = 40.4 \ \mu g/L \times \frac{75\ 000 \text{ gal}}{1 \times 10^6 \text{ gal}} = 3.0 \ \mu g/L$$

A 70-kg individual drinking 2 L per day of this water would receive a daily dose of

$$\text{CDI} = \frac{3.0 \times 10^{-3} \text{ mg/L} \times 2 \text{ L/day}}{70 \text{ kg}} = 8.6 \times 10^{-5} \text{ mg/kg/day}$$

   c. It would be expected that 50 000 people drinking this water throughout their lives would experience an annual cancer rate increase of no more than

$$\text{Annual increase} = \text{Dose} \times \text{Potency} \times \text{Population/Lifetime}$$

$$= \frac{8.6 \times 10^{-5} \text{ mg/kg/day} \times 0.02 \ (\text{mg/kg/day})^{-1} \times 50\ 000 \text{ people}}{70 \text{ yr}}$$

$$= 1.2 \times 10^{-3} \text{ cancer/yr}$$

5.8 RISK CHARACTERIZATION

The final step in a risk assessment is to bring the various studies together into an overall risk characterization. In its most primitive sense, this step could be interpreted to mean simply multiplying the exposure (dose) by the potency to get individual risk, and then multiplying that by the number of people exposed to get an estimate of overall risk to some specific population.

    While there are obvious advantages to presenting a simple, single number for extra cancers, or some other risk measure, a proper characterization of risk should be much more comprehensive. The final expressions of risk derived in this step will be used by regulatory decision makers in the process of weighing health risks against other societal costs and benefits, and the public will use them to help them decide on the adequacy of proposed measures to manage the risks. Both groups need to appreciate the extraordinary ''leaps of faith'' that, by necessity, have had to be used to determine these simple quantitative estimates. It must always be emphasized that these estimates are preliminary, subject to change, and extremely uncertain.

    The National Academy of Sciences (1983) suggests a number of questions that should be addressed in a final characterization of risk, including:

- What are the statistical uncertainties in estimating the extent of health effects? How are these uncertainties to be computed and presented?
- What are the biological uncertainties? What is their origin? How will they be estimated? What effect do they have on quantitative estimates? How will the uncertainties be described to agency decision-makers?
- Which dose–response assessments and exposure assessments should be used?

- Which population groups should be the primary targets for protection, and which provide the most meaningful expression of the health risk?

5.9 AN APPLICATION OF RISK ASSESSMENT

In 1987, the EPA released a report entitled *Unfinished Business: A Comparative Assessment of Environmental Problems* (USEPA, 1987b), in which the concepts of risk assessment were applied to a variety of pressing environmental problems. The goal of the study was to attempt to use risk as a policy tool for ranking major environmental problems to help the agency establish broad, long-term priorities.

At the outset it was realized that direct comparisons of different environmental problems would be next to impossible. Not only are the data usually insufficient to quantify risks, but the kinds of risk associated with some problems, such as global warming, are virtually incomparable with risks of others, such as hazardous waste. In most cases, considerable professional judgment, rather than hard data, was required to finalize the rankings. In spite of difficulties such as these, the report is noteworthy in terms of both its methodology and its conclusions.

The study was organized around a list of 31 environmental problems including topics as diverse as conventional (criteria) air pollutants, indoor radon, stratospheric ozone depletion, global warming, active (RCRA) and inactive (Superfund) hazardous waste sites, damage to wetlands, mining wastes, and pesticide residues on foods. Each of the 31 problems was analyzed in terms of four different types of risk: cancer risks, noncancer health risks, ecological effects, and welfare effects (visibility impairment, materials damage, etc.). In each assessment, it was assumed that existing environmental control programs continue so that the results represent risks as they exist now, rather than what they would have been had abatement programs not already been in place.

The ranking of cancer risks was perhaps the most straightforward part of the study since the EPA already has established risk assessment procedures and there are considerable data already available from which to work. Rankings were based primarily on overall cancer risk to the entire U.S. population, though high risks to specific groups of individuals, such as farm workers, were noted. A number of caveats were emphasized in the final rankings on such issues as lack of complete data, uneven quality of data, and the usual uncertainties in any risk assessment that arise from such factors as interspecies comparisons, adequacy the of low-dose extrapolation model, and estimations of exposures. Ordinal rankings were given, but it was emphasized that these should not be interpreted as being precise, especially when similarly ranked problems are being compared.

Given all of the uncertainties, in the cancer working group's final judgment two problems were tied at the top of the list: (1) worker exposure to chemicals, which does not involve a large number of individuals, but does result in high individual risks to those exposed; and (2) indoor radon exposure, which is causing considerable risk to a large number of people. Inactive (Superfund) hazardous waste sites ranked 8th and active (RCRA) hazardous waste sites were 13th. Interestingly, it was noted that with the exception of pesticide residues on food, the major route of exposure for carcinogens is inhalation. Their ranking of carcinogenic risks is reproduced in Table 5.12.

    The other working groups had considerably greater difficulty ranking the 31 environmental problem areas since there are no accepted guidelines for quantitatively assessing relative risks. As noted in *Unfinished Business,* a perusal of the rankings of the 31 problem areas for each of the four types of risk (cancer,

TABLE 5.12 CONSENSUS RANKING OF ENVIRONMENTAL PROBLEM AREAS ON THE BASIS OF POPULATION CANCER RISK

Rank	Problem area[a]	Selected comments
1 (tied)	Worker exposure to chemicals	About 250 cancer cases per year estimated based on exposure to 4 chemicals; but workers face potential exposures to over 20 000 substances. Very high individual risk possible.
1 (tied)	Indoor radon	Estimated 5000–20 000 lung cancers annually from exposure in homes.
3	Pesticide residues on foods	Estimated 6000 cancers annually, based on exposure to 200 potential oncogens.
4 (tied)	Indoor air pollutants (nonradon)	Estimated 3500–6500 cancers annually, mostly due to tobacco smoke.
4 (tied)	Consumer exposure to chemicals	Risk from 4 chemicals investigated is about 100–135 cancers annually; an estimated 10 000 chemicals in consumer products. Cleaning fluids, pesticides, particleboard, and asbestos-containing products especially noted.
6	Hazardous/toxic air pollutants	Estimated 2000 cancers annually based on an assessment of 20 substances.
7	Depletion of stratospheric ozone	Ozone depletion projected to result in 10 000 additional annual deaths in the year 2100. Not ranked higher because of the uncertainties in future risk.
8	Hazardous waste sites, inactive	Cancer incidence of 1000 annually from 6 chemicals assessed. Considerable uncertainty since risk based on extrapolation from 35 sites to about 25 000 sites.
9	Drinking water	Estimated 400–1000 annual cancers, mostly from radon and trihalomethanes.
10	Application of pesticides	Approximately 100 cancers annually; small population exposed but high individual risks.

(continued)

TABLE 5.12 (*continued*)

Rank	Problem area[a]	Selected comments
11	Radiation other than radon	Estimated 360 cancers per year. Mostly from building materials. Medical exposure and natural background levels not included.
12	Other pesticide risks	Consumer and professional exterminator uses estimated cancers of 150 annually. Poor data.
13	Hazardous waste sites, active	Probably fewer than 100 cancers annually; estimates sensitive to assumptions regarding proximity of future wells to waste sites.
14	Nonhazardous waste sites, industrial	No real analysis done, ranking based on consensus of professional opinion.
15	New toxic chemicals	Difficult to assess; done by consensus.
16	Nonhazardous waste sites, municipal	Estimated 40 cancers annually not including municipal surface impoundments.
17	Contaminated sludge	Preliminary results estimate 40 cancers annually, mostly from incineration and landfilling.
18	Mining waste	Estimated 10–20 cancers annually, largely due to arsenic. Remote locations and small population exposure reduce overall risk though individual risk may be high.
19	Releases from storage tanks	Preliminary analysis, based on benzene, indicates low cancer incidence (<1).
20	Non-point-source discharges to surface water	No quantitative analysis available; judgment.
21	Other groundwater contamination	Lack of information; individual risks considered less than 10^{-6}, with rough estimate of total population risk at < 1.
22	Criteria air pollutants	Excluding carcinogenic particles and VOCs (included under Hazardous/toxic air pollutants); ranked low because remaining criteria pollutants have not been shown to be carcinogens.
23	Direct point-source discharges to surface water	No quantitative assessment available. Only ingestion of contaminated seafood was considered.
24	Indirect point-source discharges to surface water	Same as above.
25	Accidental releases, toxics	Short duration exposure yields low cancer risk; noncancer health effects of much greater concern.
26	Accidental releases, oil spills	See above. Greater concern for welfare and ecological effects.

[a] Not ranked: Biotechnology; global warming; other air pollutants; discharges to estuaries, coastal waters and oceans; discharges to wetlands.

Source: Based on data from USEPA (1987b).

noncancer health effects, ecological, and welfare effects) produced the following general results:

- No problems rank relatively high in all four types of risk, or relatively low in all four.
- Problems that rank relatively high in three of the four risk types, or at least medium in all four, include criteria air pollutants (see Chapter 7), stratospheric ozone depletion, pesticide residues on food, and other pesticide risks (runoff and air deposition of pesticides).
- Problems that rank relatively high in cancer and noncancer health risks but low in ecological and welfare risks include hazardous air pollutants; indoor radon; indoor air pollution other than radon; pesticide application; exposure to consumer products, and worker exposures to chemicals.
- Problems that rank relatively high in ecological and welfare risks, but low in both health risks, include global warming, point and nonpoint sources of surface water pollution, physical alteration of aquatic habitats (including estuaries and wetlands), and mining wastes.
- Areas related to groundwater consistently rank medium or low.

In spite of the great uncertainties involved in making their assessments, the divergence between EPA effort and relative risks is noteworthy. As concluded in the study, areas of relatively high risk but low EPA effort include indoor radon, indoor air pollution, stratospheric ozone depletion, global warming, nonpoint sources, discharges to estuaries, coastal waters, and oceans, other pesticide risks, accidental releases of toxics, consumer products, and worker exposures. Areas of high EPA effort but relatively medium or low risks include RCRA sites, Superfund sites, underground storage tanks, and municipal nonhazardous waste sites.

PROBLEMS

5.1. The cancer risk caused by exposure to radiation is thought to be approximately 1 lethal cancer per 8000 person-rems of exposure. Exposure to cosmic radiation increases with increasing altitude; at sea level it is about 40 mrem/year and at the elevation of Denver (about one mile high) it is about 65 mrem/year.

    a. Compare the lifetime probability of dying of cancer induced by cosmic radiation for a person living at sea level with that of a person living in Denver.

    b. Estimate the number of cancer deaths per year in Denver, population 0.5 million, caused by cosmic radiation. Compare it to total cancer deaths in Denver, assuming the typical cancer death rate of 193 per 100 000 per year.

    c. If all 240 million Americans lived at sea level, estimate the total cancer deaths per year caused by cosmic radiation.

5.2. Living in an average U.S. home with 1.5 pCi/L of radon is thought to cause a cancer risk equivalent to that caused by approximately 400 mrem/year of radiation. Using the radiation potency factor and population given in Problem 5.1:

   a. Estimate the annual cancers in the United States caused by radon in homes.

   b. Estimate the lifetime risk to an individual living in a home with that amount of radon.

5.3. The dose to residents living near Chernobyl when it exploded has been estimated at about 5 rems. Using data from Problem 5.1, estimate the individual cancer risk faced by someone exposed to that amount of radiation.

5.4. It has been estimated that about 75 million people in the Ukraine and Byelorussia were exposed to an average of 0.4 rem of radiation as a result of the Chernobyl nuclear accident.

   a. How many extra cancer deaths might eventually be expected from this exposure (see Problem 5.1)?

   b. If the normal probability of dying of cancer from all causes is 0.22, how many cancer deaths would you normally expect among those 75 million people? What is the percentage increase in cancer deaths that would be expected due to Chernobyl?

5.5. The average radiation exposure to someone living within 20 miles of a nuclear power plant has been estimated at about 0.02 mrem/year. How many extra cancer deaths would be expected per year in the United States (population 240 million) due to nuclear power if that were the average exposure? (Use potency data given in Problem 5.1.)

5.6. One way to estimate maximum acceptable concentrations of toxicants in drinking water or air is to pick an acceptable lifetime risk and calculate the concentration that would give that risk assuming standard daily intakes given in Table 5.8. Find the "acceptable concentrations" and the chronic daily intake (CDI) of the following substances:

   a. Benzene in drinking water (mg/L), at a lifetime acceptable risk of 1×10^{-5}.

   b. Trichloroethylene in air (mg/m^3), at a lifetime acceptable risk of 1×10^{-6}.

5.7. Suppose an individual eats fish from a river contaminated by benzene. Also suppose that the bioconcentration factor (BCF) is 5.2 L of water per kilogram of fish. What concentration of benzene (mg/L) in the water would produce a lifetime risk of 1×10^{-6} to an individual who eats 6.5 g of fish per day ?

5.8. The reference dose (RfD) for methylene chloride (set to allow a large margin of safety for noncarcinogenic effects) has been set at 0.06 mg/kg/day. At that dose, what is the lifetime (oral) carcinogenic risk? Which would be more stringent, a methylene chloride oral concentration standard based on a carcinogenic risk of 1×10^{-6} or a standard based on RfD?

5.9. Estimate the chronic daily intake (CDI) of DDT for a 70-kg individual consuming 2 g of fish per day for a lifetime, from a stream with 20 ppb of DDT. What would be the maximum lifetime cancer risk to this individual due to DDT?

5.10. Suppose 100 g of the pesticide heptachlor is accidently leaked into a 30 000-m^3 pond. What would be the concentration in the pond if it immediately mixes uniformly throughout the pond? If the half-life is 2 days, what would be the concentration after 1 week?

5.11. Suppose 1.0 g/day of heptachlor leaks into a 30 000-m³ pond. If heptachlor has a half-life of 2 days, and complete mixing occurs:
   a. What would be the steady-state concentration in the pond?
   b. Suppose a 70-kg individual drank 2 L/day of that water for 5 years. Estimate the maximum risk of cancer due to that exposure to heptachlor.

5.12. Mainstream smoke inhaled by a 70-kg smoker contains roughly 30 μg/cigarette of the class B2 carcinogen, benzo(a)pyrene. For an individual who smokes 20 cigarettes per day for 40 years, estimate:
   a. The chronic daily intake (CDI) in mg/kg/day averaged over a 70-year period.
   b. The lifetime risk of cancer caused by that benzo(a)pyrene (there are other carcinogens in cigarettes as well, by the way).

5.13. Consider the problem of indoor air pollution caused by sidestream smoke (unfiltered smoke from an idling cigarette) emitting roughly 100 μg/cigarette of benzo(a)pyrene. What average concentration of benzo(a)pyrene in air would produce a one-in-one-million lifetime cancer risk using standard values for inhalation? (See also Problem 7.37.)

5.14. The maximum contaminant level (MCL) for trichloroethylene (TCE) has been set at 5 ppb (=5 μg/L) in drinking water.
   a. At that concentration of TCE, what would be the lifetime individual cancer risk, using standard values for daily intake?
   b. If all 240 million Americans drank water with that concentration of TCE, what maximum number of extra cancers per year would be expected?

5.15. One way to express cancer potency for substances that are inhaled is in terms of risk caused by a lifetime of breathing air with a concentration of 1.0 μg/m³ of carcinogen. The potency for formaldehyde in these terms is 1.3×10^{-5} cancer/μg/m³. What is the cancer risk caused by a lifetime of breathing formaldehyde at the not unusual (in smoggy cities) concentration of 50 μg/m³ (the threshold of eye irritation).

5.16. Trichloroethylene (TCE) is a common groundwater contaminant. In terms of cancer risk, which would be better: (1) to drink unchlorinated groundwater with 10 ppb of TCE; or (2) to switch to a surface water supply that, as a result of chlorination, has a chloroform concentration of 50 ppb?

5.17. Suppose a 70-kg man is exposed to 0.1 mg/m³ of tetrachloroethylene in the air at his workplace. If he inhales 1 m³/hr, 8 hr per day, 5 days a week, 50 weeks a year, for 30 years, and if tetrachloroethylene has an absorption factor of 90 percent, an oral potency of 2×10^{-3} (mg/kg/day)$^{-1}$, what would be his lifetime cancer risk?

5.18. Suppose a factory releases a continuous flow of wastewater into a local stream resulting in an in-stream carcinogen concentration of 100 μg/L just below the outfall. Suppose this carcinogen has an oral potency factor of 0.30 (mg/kg/day)$^{-1}$ and that it is degradable with a reaction rate coefficient K of 0.10/day. To keep the problem simple, assume the stream is uniform in cross section, flowing at the rate of 1 mph, and there are no other sources or sinks for this carcinogen. At a distance of 100 miles downstream, a town uses this stream as its only source of water. Estimate the individual lifetime cancer risk caused by drinking this water.

5.19. The following tumor data were collected for rats exposed to ethylene thiourea (ETU) (data from Crump, 1984):

Dietary concentration	Animals with tumors
125 ppm	3%
250 ppm	23%
500 ppm	88%

A one-hit model fitted to the data has coefficients: $q_0 = 0.01209$ and $q_1 = 0.001852/$ ppm. A multistage model has the coefficients $q_0 = 0.02077$, $q_1 = q_2 = 0.0$, $q_3 = 1.101 \times 10^{-8}/(ppm)^3$, and $q_4 = 1.276 \times 10^{-11}/(ppm)^4$.

a. For each of the three concentrations given above, compare the measured data with the values given by each of these two models.

b. For a concentration of 1 ppm, compare the values that each of the two models would predict for percent tumors.

REFERENCES

ANDERSON, E. L., et al., 1983, Quantitative approaches in use to assess cancer risk, *Risk Analysis*, 3(4):277–295.

CRUMP, K. S., 1984, An improved procedure for low-dose carcinogenic risk assessment from animal data, *Journal of Environmental Pathology, Toxicology, and Oncology* 5-4/5:339–349.

Environ, 1988, *Elements of Toxicology and Chemical Risk Assessment*, Environ Corporation, Washington, DC.

FREEMAN, H. M. (ed.), 1989, *Standard Handbook of Hazardous Waste Treatment and Disposal*, McGraw Hill, New York.

IRVINE, R. L., and P. A. WILDERER, 1989, Aerobic processes, in *Standard Handbook of Hazardous Waste Treatment and Disposal*, H. M. Freeman (ed.), McGraw Hill, New York.

National Academy of Sciences, 1983, *Risk Assessment in the Federal Government: Managing the Process*, National Academy Press, Washington, DC.

TEAF, C. M., 1985, Mutagenis, in *Industrial Toxicology, Safety and Health Applications in the Workplace*, P. L. Williams and J. L. Burson, Eds., Van Nostrand Reinhold, New York.

USEPA, 1986a, *Guidelines for Carcinogen Risk Assessment*, Federal Register, Vol. 51, No. 185, pp. 33 992–34 003, September 24, 1986.

USEPA, 1986b, *Superfund Public Health Evaluation Manual*, Office of Emergency and Remedial Response, Washington, DC.

USEPA, 1986c, *Solving the Hazardous Waste Problem: EPA's RCRA Program*, Office of Solid Waste, Washington, DC.

USEPA, 1987a, *The Hazardous Waste System*, Environmental Protection Agency, Office of Solid Waste and Emergency Response, Washington, DC.

USEPA, 1987b, *Unfinished Business: A Comparative Assessment of Environmental Problems,* Office of Policy, Planning and Evaluation, EPA/230/2-87/025, Washington, DC.

USEPA, 1987c, *The Hazardous Waste System,* Office of Solid Waste and Emergency Response, Washington, DC.

USEPA, 1988, *Superfund Exposure Assessment Manual,* Environmental Protection Agency, Office of Remedial Response, EPA/540/1-88/001, Washington, DC.

WENTZ, C. A., 1989, *Hazardous Waste Management,* McGraw Hill, New York.

WILLIAMS, P. L., and J. L. BURSON, 1985, *Industrial Toxicology, Safety and Health Applications in the Workplace,* Van Nostrand Reinhold, New York.

WILSON, R., and E. A. C. CROUCH, 1987, Risk assessment and comparisons: An introduction, *Science,* 17 April 1987, pp. 267–270.

CHAPTER 6

Treatment of Water and Wastes

Our planet is shrouded in water, and yet 8 million children under the age of five will die this year from lack of safe water.
United Nations Environmental Program

6.1 DRINKING WATER QUALITY

One of the first things that a world traveler learns to ask is whether or not the water is safe to drink. The unfortunate answer in most places in the world is no. It has been estimated that 80 percent of all sickness in the world is attributable to inadequate water or sanitation. Furthermore, an estimated three-fourths of the population in Asia, Africa, and Latin America lack a safe supply of water for drinking, washing, and sanitation (Morrison, 1983).

Developing and maintaining an adequate supply of safe drinking water requires the coordinated efforts of scientists, engineers, water plant operators, and regulatory officials.

The Safe Drinking Water Act

Legislation to protect drinking water quality in the United States began with the Public Health Service Act of 1912. In this Act, the nation's first water quality standards were created. While these standards slowly evolved over the years, it was not until the passage of the *Safe Drinking Water Act* (SDWA) of 1974 that federal responsibility was extended beyond interstate carriers to include all community water systems serving 15 or more outlets, or 25 or more customers. This original Safe Drinking Water Act had two basic thrusts: (1) It required the EPA to establish national standards for drinking water quality and (2) it required the operators of some 60 000 public water systems in the country to monitor the quality of water being delivered to customers and to treat that water, if necessary, to assure compliance with the standards.

Twelve years later, in 1986, strengthening amendments were added to the SDWA. These required the EPA to quicken the pace of standard setting and required the equivalent of filtration for all surface water supplies and required disinfection for all water systems. The regulations also recognized the potential for lead contamination in water distribution systems, including the plumbing systems in homes and nonresidential facilities. To help prevent lead poisoning, the 1986 amendments require the use of ''lead-free'' pipe, solder, and flux for all such systems (in the past, solder normally contained about 50 percent lead; now it may contain no more than 0.2 percent lead).

Drinking water standards fall into two categories: *primary standards,* which specify *maximum contaminant levels* (MCLs) based on health related criteria, and *secondary standards,* which are unenforceable guidelines based on both aesthetics such as taste, odor, and color of drinking water, as well as nonaesthetic characteristics such as corrosivity and hardness. In setting MCLs, the EPA is required to balance the public health benefits of the standard against what is technologically and economically feasible. In this way, MCLs are quite different from National Ambient Air Quality Standards, which must be set at levels that protect public health regardless of cost or feasibility.

In setting standards, the EPA creates unenforceable maximum contaminant level goals (MCLGs; formerly known as recommended maximum contaminant levels, or RMCLs) set at levels that present no known or anticipated health effects, including a margin of safety, regardless of technological feasibility or cost. The Safe Drinking Water Act requires the EPA to periodically review the actual MCLs to determine whether they can be brought closer to the desired MCLGs.

Primary Standards

Contaminants for which MCLs and MCLGs are established are classified as being either inorganic chemicals, organic chemicals, radionuclides, or microbiological contaminants. Under the impetus of the SDWA, the list of substances in each category is continuously growing. A list of the standards for inorganic chemicals is given in Table 6.1. Additional inorganic chemicals for which MCLs are being proposed include asbestos and copper.

Organic chemical contaminants for which MCLs are being promulgated are conveniently classified using the following three groupings:

1. *Synthetic organic chemicals* (SOCs), are compounds used in the manufacture of a wide variety of agricultural and industrial products. This includes primarily pesticides and herbicides.

TABLE 6.1   MAXIMUM CONTAMINANT LEVELS (MCLs) FOR CERTAIN INORGANIC
CHEMICALS

Contaminant	Principal health effects	Maximum contaminant levels (mg/L)
Arsenic	Dermal and nervous system toxicity effects	0.05
Barium	Circulatory system effects	1.0
Cadmium	Kidney effects	0.010
Chromium	Liver/kidney effects	0.05
Fluoride	Skeletal damage	1.8 (at 20 °C)
Lead	Central and peripheral nervous system damage; kidney effects; highly toxic to infants and pregnant women	0.05
Mercury	Central nervous system disorders; kidney effects	0.002
Nitrate and nitrite	Methemoglobinemia (blue-baby syndrome)	10.0 (as N)
Selenium	Gastrointestinal effects	0.01
Silver	Skin discoloration (argyria)	0.05

Source: Based on USEPA (1986a).

2. *Trihalomethanes* (THMs) are the by-products of water chlorination. They include chloroform ($CHCl_3$), bromodichloromethane ($CHBrCl_2$), dibromochloromethane ($CHBr_2Cl$), and bromoform ($CHBr_3$).

3. *Volatile organic chemicals* (VOCs) are synthetic chemicals that readily vaporize at room temperature. These include degreasing agents, paint thinners, glues, dyes, and some pesticides. Representative chemicals include benzene, carbon tetrachloride, 1,1,1-trichloroethane (TCA), trichlorethylene (TCE), and vinyl chloride.

A list of organic chemicals for which MCLs had been set as of 1986 is given in Table 6.2.

Radioactivity in public drinking water supplies is the third category of contaminants regulated by the Safe Drinking Water Act. Some radioactive compounds, or *radionuclides,* are naturally occurring substances such as radon and radium-226 (which are often found in groundwater), while others, such as strontium-90 and tritium, are surface water contaminants resulting from atmospheric nuclear weapons testing fallout. The MCL for the combination of radium-226 and radium-228 is 5 pCi/L, while the gross α-particle activity MCL (including radium-226 but excluding radon and uranium) is 15 pCi/L. The MCL for β-particle and photon radioactivity is an annual dose either to the whole body or to any particular organ, of 4 mrem/year (USEPA, 1976).

The most significant radionuclide associated with drinking water is dissolved radon gas. It is a colorless, odorless, and tasteless gas that occurs naturally in

TABLE 6.2 MAXIMUM CONTAMINANT LEVELS (MCLs) FOR CERTAIN ORGANIC CHEMICALS

Contaminant	Principal health effects	MCL (mg/L)
Endrin	Nervous system/kidney effects	0.0002
Lindane	Nervous system/kidney effects	0.004
Methoxychlor	Nervous system/kidney effects	0.1
2,4-D	Liver/kidney effects	0.1
2,4,5-TP Silvex	Liver/kidney effects	0.01
Toxaphene	Cancer risk	0.005
Trihalomethanes (THMs)	Cancer risk	0.10

Source: Based on USEPA (1976, 1986a).

some groundwater. It is an unusual contaminant because the danger arises not from drinking radon contaminated water, but from breathing the gas after it has been released into the air. When radon-laden water is heated or agitated, such as occurs in showers or washing machines, the dissolved radon gas is released. As will be discussed in Chapter 7, inhaled radon gas is thought to be a major cause of lung cancer.

The fourth category of primary MCLs is *microbiological contaminants.* While it would be desirable to evaluate the safety of a given water supply by individually testing for specific pathogenic microorganisms, such tests are too difficult to perform on a routine basis or to be used as a standard. Instead, a much simpler technique is used, based on testing water for evidence of any fecal contamination. In this test, coliform bacteria (typically *Escherichia coli*) are used as indicator organisms whose presence suggests that the water is contaminated. Since the number of coliform bacteria excreted in feces is on the order of 50 million per gram, and the concentration of coliforms in untreated domestic wastewater is usually several million per 100 mL, it would be highly unlikely that water contaminated with human wastes would have no coliforms. That conclusion is the basis for the drinking water standard for microbiological contaminants which specifies, in essence, that on the average, water should contain no more than one coliform per 100 mL.

The assumption that the absence of coliforms implies an absence of pathogens is based primarily on the following two observations. First, in our society it is the excreta from relatively few individuals that adds pathogens to a wastestream, while the entire population contributes coliforms. Thus, the number of coliforms should far exceed the number of pathogens. Second, for many of the waterborne diseases that have plagued humankind, the survival rate of pathogens outside the host is much lower than the survival rate of coliforms. The combination of these factors suggests that, statistically speaking, the ratio of pathogens to coliforms should be sufficiently small that we can conclude that it is extremely unlikely that a sample of water would contain a pathogen without also containing numerous coliforms.

TABLE 6.3   CALCULATED VIRUS–COLIFORM RATIOS FOR SEWAGE AND
POLLUTED SURFACE WATER

	Virus	Coliform	Virus/coliform ratio
Sewage	500/100 mL	46×10^6/100 mL	1 : 92 000
Polluted surface water	1/100 mL	5×10^4/100 mL	1 : 50 000

Source: Robeck et al. (1962).

This approach to testing for microbiological purity has been quite effective, but it is not an absolutely certain measure. There is considerable evidence, for example, that some nonbacterial pathogens, notably viruses and *Giardia* cysts, survive considerably longer outside of their hosts than coliform bacteria. This increases the probability of encountering pathogens without accompanying coliforms. However, as suggested by the approximations given in Table 6.3 for viruses, the risk of such an encounter is still considered to be acceptably small under normal circumstances.

The coliform test is also used to assess the safety of water-contact recreational activities, with many states recommending a limit of 1000 coliforms per 100 mL. However, proper interpretation of a coliform test made on surface water is complicated by the fact that fecal coliforms are discharged by animals as well as humans. Thus, a high fecal coliform count is not necessarily an indication of human contamination. When it is important to distinguish between human and animal contamination, more sophisticated testing can be performed. Such testing is based on the fact that the ratio of fecal coliform to fecal streptococci is different in human and animal discharges.

TABLE 6.4   SECONDARY STANDARD FOR
DRINKING WATER

Contaminant	Level
Chloride	250 mg/L
Color	15 color units
Copper	1 mg/L
Corrosivity	Noncorrosive
Foaming agents	0.5 mg/L
Iron	0.3 mg/L
Manganese	0.05 mg/L
Odor	3 threshold odor number
pH	6.5–8.5
Sulfate	250 mg/L
Total dissolved solids	500 mg/L
Zinc	5 mg/L

Secondary Standards

Secondary standards are nonenforceable, maximum contaminant levels based on aesthetic factors such as taste, color, and odor, rather than on health effects. The limits suggested in Table 6.4 for chloride, copper, total dissolved solids, and zinc are in large part based on taste. Excessive sulfate is undesirable because of its laxative effect; iron and manganese are objectionable because of taste and their ability to stain laundry and fixtures; foaming and color are visually upsetting; and odor from various dissolved gases may make water unacceptable to the drinker.

6.2 WATER TREATMENT SYSTEMS

The purpose of a water treatment system is to bring raw water up to drinking water quality. The particular type of treatment equipment required to meet these standards will depend to a large extent on the quality of the source water. Some water requires only simple disinfection. Surface water will usually need to be filtered and disinfected, while groundwater will often need to have hardness (calcium and magnesium) removed before disinfection.

As suggested in Figure 6.1, a typical treatment plant for surface water might include the following sequence of steps:

Screening to remove relatively large floating and suspended debris.

Mixing the water with chemicals that encourage suspended solids to coagulate into larger particles that will more easily settle.

Flocculation, which is the process of gently mixing the water and coagulant, allowing the formation of large particles of floc.

Sedimentation in which the flow is slowed enough so that gravity will cause the floc to settle.

Sludge processing where the mixture of solids and liquids collected from the settling tank are dewatered and disposed of.

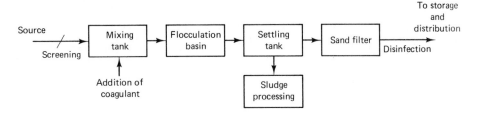

Figure 6.1 Schematic of a typical water treatment plant for surface water. Softening may be required as an additional step for groundwater.

Disinfection of the liquid effluent to ensure that the water is free of harmful pathogens.

Hardness removal can be added to this generalized flow diagram if needed.

Coagulation and Flocculation

Raw water may contain suspended particles of color, turbidity, and bacteria that are too small to settle in a reasonable time period, and cannot be removed by simple filtration. The object of coagulation is to alter these particles in such a way as to allow them to adhere to each other. Thus, they can grow to a size that will allow removal by sedimentation and filtration. Coagulation is considered to be a *chemical* treatment process that destabilizes colloidal particles (particles in the size range of about 0.001–1 μm), as opposed to the *physical* treatment operations of flocculation, sedimentation, and filtration that follow.

Most colloids of interest in water treatment remain suspended in solution because they have a net negative surface charge that causes the particles to repel each other. The intended action of the coagulant is to neutralize that charge, allowing the particles to come together to form larger particles that can be more easily removed from the raw water. The exact way that coagulants work is still not well understood, so past experience with similar treatment systems is invaluable to the design of new systems.

The usual coagulant is alum [$Al_2(SO_4)_3 \cdot 18H_2O$], though $FeCl_3$, $FeSO_4$, and other coagulants, such as polyelectrolytes, can be used. Since the intention here is simply to introduce the concepts of water treatment, leaving the complexities for more specialized books, let us just look at the reactions involving alum. Alum ionizes in water, producing Al^{3+} ions, some of which neutralize the negative charges on the colloids. Most of the aluminum ions, however, react with alkalinity in the water (bicarbonate) to form insoluble aluminum hydroxide [$Al(OH)_3$]. The aluminum hydroxide adsorbs ions from solution and forms a precipitate of $Al(OH)_3$ and adsorbed sulfates. The overall reaction is

$$Al_2(SO_4)_3 \cdot 18H_2O + 6HCO_3^- \rightleftharpoons 2Al(OH)_3\downarrow + 6CO_2 + 18H_2O + 3SO_4^{2-}$$

$$(6.1)$$

If insufficient bicarbonate is available for this reaction to occur, the pH must be raised, usually by adding lime [$Ca(OH)_2$] or sodium carbonate (Na_2CO_3).

Coagulants are added to the raw water in a chamber that has rapidly rotating paddles to mix the chemicals. Detention times in the rapid mix tank are typically on the order of one minute. Flocculation follows in a tank that provides gentle agitation for approximately one-half hour. During this time, the precipitating aluminum hydroxide attracts colloidal particles, forming a plainly visible floc. The mixing in the flocculation tank must be done very carefully. It must be sufficient to encourage particles to make contact with each other, enabling the floc to grow in size, but it cannot be so vigorous that the fragile floc particles will break apart. Mixing also helps keep the floc from settling in this tank, rather than in the sedimentation tank that follows.

While Figure 6.1 suggests separate tanks for rapid mixing, flocculation, and sedimentation, the three physical processes can be combined into a single unit, as shown in Figure 6.2.

Sedimentation and Filtration

After flocculation, the water flows through a sedimentation basin, or clarifier. A sedimentation basin is a large circular, or rectangular, concrete tank designed to hold the water for a long enough time to allow most of the suspended solids to settle out. Typical detention times range from 1 to 10 hr. The longer the detention time the bigger and more expensive the tank must be, but, correspondingly, the better will be the tank's performance. Solids that collect on the bottom of the tank may be removed manually by periodically shutting down the tank and washing out the collected sludge, or the tank may be continuously and mechanically cleaned using a bottom scraper. The effluent from the tank is then filtered.

One of the most widely used filtration units is called a *rapid-sand filter,* which consists of a layer of carefully sieved sand on top of a bed of graded gravels. The pore openings between grains of sand are often greater than the size of the floc particles that are to be removed, so much of the filtration is accomplished by means other than simple straining. Adsorption, continued flocculation, and sedimentation in the pore spaces are also important removal mechanisms. When the filter becomes clogged with particles, which occurs roughly once a day, the filter is shut down for a short period of time and cleaned by forcing water backwards through the sand for 10–15 min. After cleaning, the sand settles back in place and operation resumes.

Disinfection

During coagulation, settling, and filtration, practically all of the suspended solids, most of the color, and all but a few percent of the bacteria, are removed. The final step is disinfection to kill any remaining pathogenic organisms.

Chlorination using chlorine gas (Cl_2), sodium hypochlorite (NaOCl), or calcium hypochlorite [$Ca(OCl)_2$], is the most commonly used method of disinfection in this country. Chlorine is a powerful oxidizing agent that is easy to use, inexpensive, and reliable. The precise mechanism by which chlorine kills microorganisms is not well understood. It is thought that its strong oxidizing power destroys the enzymatic processes necessary for cell life. Though chlorination is completely effective against bacteria, its effectiveness is less certain with protozoal cysts, most notably those of *Giardia lamblia,* or with viruses. Complete treatment, including coagulation, sedimentation, filtration, as well as disinfection, is especially recommended where *Giardia lamblia* is problematic.

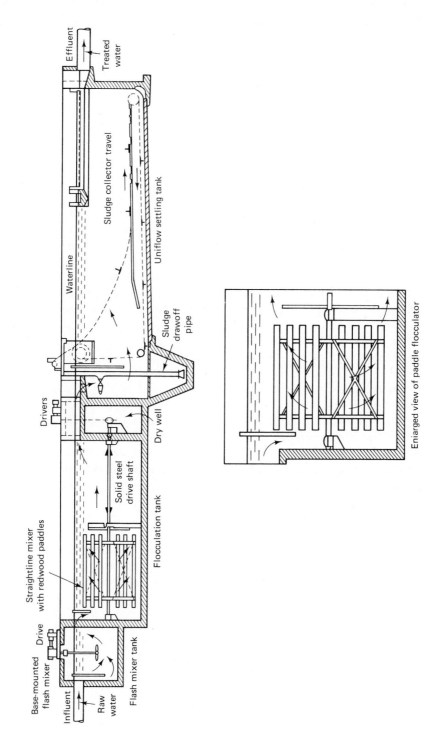

Figure 6.2 Cross section of a rapid mixing, flocculation, and sedimentation tank. (*Source:* Courtesy Materials Handling Systems Div., FMC Corp.).

Using chlorine gas to illustrate the chemical reactions occurring during chlorination, the key reaction is

$$Cl_2 + H_2O \rightleftharpoons HOCl + H^+ + Cl^- \tag{6.2}$$

The hypochlorous acid formed (HOCl) is the prime disinfecting agent. Its dissociation is pH dependent, yielding less effective hypochlorite (OCl^-) ions at higher pH values:

$$HOCl \rightleftharpoons H^+ + OCl^- \tag{6.3}$$

Together, HOCl and OCl^- are called the *free available chlorine*.

A principal advantage of chlorination over other forms of disinfection is that a chlorine residual is created that can protect the treated water after leaving the treatment plant. This guards against possible contamination that might occur in the water distribution system. To increase the lifetime of this residual, some systems add ammonia to the treated water, forming *chloramines* (NH_2Cl, $NHCl_2$, NCl_3). Chloramines, although they are less effective as oxidants than HOCl, are more persistent. Residual chlorine that exists as chloramine is referred to as *combined* available residual chlorine.

A disadvantage of chlorination is the potential formation of trihalomethanes (THMs) such as the carcinogen chloroform ($CHCl_3$). THMs are created when chlorine combines with natural organic substances, such as decaying vegetation, that may be present in the water itself. One approach to reducing THMs is to remove more of the organics before any chlorination takes place. It was common practice in the past, for example, to chlorinate the incoming raw water before coagulation and filtration. This contributed to formation of THMs. By eliminating this step, and by increasing the organic removal during treatment before chlorination, some degree of control is achieved. In the future, the actual removal of THMs from the treated water, perhaps by aeration or adsorption on activated carbon, may become necessary.

The problem of THMs is helping spur interest in alternatives to chlorination as the preferred method of disinfection. Alternative disinfectants include chloramines, chlorine dioxide, and ozone. Each has the advantage of not creating THMs, but there are uncertainties and known disadvantages to each that have restricted their more widespread use. Chloramines, as mentioned above, are less certain disinfectants than free chlorine, so chloramination is usually used in combination with some other more effective method of disinfection. Chlorine dioxide (ClO_2), on the other hand, is a potent bactericide and viricide, and it does form a residual capable of protecting water in the distribution system. However, there is concern for certain toxic chlorate and chlorite substances that it may create, and it is a very costly method of disinfection. Ozonation involves the passage of ozone (O_3) through the water. Ozone is a very power disinfectant that is even more effective against cysts and viruses than chlorine, and it has the added advantage of leaving no taste or odor problems. Though ozonation is widely used in European water treatment facilities, it has the disadvantages of not forming a protective residual in the treated water, and it is more expensive than chlorination.

Hardness and Alkalinity

The presence of multivalent cations, most notably calcium and magnesium ions, is referred to as water *hardness*. Groundwater is especially prone to excessive concentrations of these ions, leading to two distinct problems. First, the reaction between hardness and soap produces a sticky, gummy deposit called "soap curd" (the ring around the bathtub). Essentially all home cleaning activities, from bathing and grooming to dishwashing and laundering, are made more difficult with hard water.

While the introduction of synthetic detergents has decreased, but not eliminated, the impact of hardness on cleaning, the second problem, that of scaling, remains significant. When hard water is heated, calcium carbonate ($CaCO_3$) and magnesium hydroxide [$Mg(OH)_2$] readily precipitate out of solution, forming a rock-like scale that clogs hot water pipes and reduces the efficiency of water heaters, boilers, and heat exchangers. Pipes filled with scale must ultimately be replaced, usually at great expense. Heating equipment that has scaled up not only transmits heat less readily, thus increasing fuel costs, but also is prone to failure at a much earlier time. For both of these reasons, if hardness is not controlled at the water treatment plant itself, many individuals and industrial facilities find it worth the expense to provide their own water softening.

Hardness is defined as the concentration of all multivalent metallic cations in solution. The principal ions causing hardness in natural water are calcium (Ca^{2+}) and magnesium (Mg^{2+}). Others, including iron (Fe^{2+}), manganese (Mn^{2+}), strontium (Sr^{2+}), and aluminum (Al^{3+}), may be present, though in much smaller quantities.

It is conventional practice in reporting hardness (as well as alkalinity) to measure the concentrations of these individual ions not in terms of mg/L, but in terms of *equivalents*. By noting, for example, that 8 g of oxygen combine with 1 g of hydrogen (forming H_2O), while 3 g of carbon combine with 1 g of hydrogen (forming CH_4), the idea of equivalents suggests that 8 g of oxygen ought to combine with 3 g of carbon (which it does in CO_2). In this example, the *equivalent weight* of oxygen would be 8 g, while that of carbon would be 3 g, and one equivalent weight of oxygen would combine with one equivalent weight of carbon (to form CO_2). This idea of equivalents, when first proposed in the 18th century, was not taken very seriously because it is complicated by the fact that individual elements can have more than one equivalent weight, depending on the reaction in question (e.g., carbon forms CO as well as CO_2). With the understanding of the structure of the atom that came later, however, it regained favor as a convenient method of handling certain chemical computations.

The *equivalent weight* of a substance is its atomic or molecular weight divided by a number *n*. For the reactions of interest in hardness and alkalinity calculations, n is simply the ionic charge. For compounds, it is the number of hydrogen ions that would be required to replace the cation.

$$\text{Equivalent weight (EW)} = \frac{\text{Atomic (molecular) weight}}{n} \qquad (6.4)$$

Thus, for example, $CaCO_3$ has $n = 2$ since it would take two hydrogen ions to replace the cation (Ca^{2+}). Its equivalent weight is therefore

$$\text{Equivalent weight of } CaCO_3 = \frac{(40 + 12 + 3 \times 16)}{2} = \frac{100}{2}$$

$$= 50 \text{ g/eq} = 50 \text{ mg/meq}$$

where (mg/meq) are the units of milligrams per milliequivalent. For the calcium ion itself (Ca^{2+}), which has an atomic weight of 40.1 and a charge of 2, the equivalent weight is

$$\text{Equivalent weight of } Ca^{2+} = \frac{40.1}{2} = 20.05 \text{ mg/meq}$$

and for magnesium (Mg^{2+}) with atomic weight 24.3,

$$\text{Equivalent weight of } Mg^{2+} = \frac{24.3}{2} = 12.15 \text{ mg/meq}$$

In measuring hardness, the concentrations of the multivalent cations are converted to mg/L as $CaCO_3$ using the following expression

$$\text{mg/L of } X \text{ as } CaCO_3 = \frac{\text{Concentration of } X \text{ (mg/L)} \times 50 \text{ mg } CaCO_3/\text{meq}}{\text{Equivalent weight of } X \text{ (mg/meq)}]}$$

$$(6.5)$$

The *total hardness* (TH) as $CaCO_3$ is the sum of each individual hardness:

$$\text{Total hardness} = Ca^{2+} + Mg^{2+} \qquad (6.6)$$

where it has been assumed in (6.6) that calcium and magnesium are the only two multivalent cations with appreciable concentrations. The following example shows how to work with these units.

Example 6.1   Total Hardness as $CaCO_3$

A sample of groundwater has 100 mg/L of Ca^{2+} and 10 mg/L of Mg^{2+}. Express its hardness in units of meq/L and mg/L as $CaCO_3$.

Solution   The contribution of calcium in meq/L is

$$\frac{100 \text{ mg/L}}{20.05 \text{ mg/meq}} = 4.99 \text{ meq/L}$$

and in mg/L as $CaCO_3$ is

$$4.99 \text{ meq/L} \times 50 \text{ mg } CaCO_3/\text{meq} = 249.4 \text{ mg/L as } CaCO_3$$

For magnesium, the hardness is

$$\frac{10 \text{ mg/L}}{12.15 \text{ mg/meq}} = 0.82 \text{ meq/L}$$

or

$$0.82 \text{ meq/L} \times 50 \text{ mg CaCO}_3/\text{meq} = 41.0 \text{ mg/L as CaCO}_3$$

The total hardness is

$$4.99 \text{ meq/L} + 0.82 \text{ meq/L} = 5.8 \text{ meq/L}$$

or

$$(249.4 + 41.0) \text{ mg/L} = 290.4 \text{ mg/L as CaCO}_3$$

While public acceptance of hardness is quite dependent on the past experiences of individual consumers, hardness above about 150 mg/L as $CaCO_3$ is noticed by most consumers. Though there are no absolute distinctions, the qualitative classification of hardness given in Table 6.5 is often used (Tchobanoglous and Schroeder, 1985).

It is useful at times to separate total hardness, which is almost entirely the combination of calcium and magnesium cations, into two components: *carbonate hardness* (CH), associated with the anions HCO_3^- and CO_3^{2-}, and *noncarbonate hardness* (NCH) associated with other anions. If carbonate hardness exceeds the total hardness (TH), then CH is given the same value as TH. Carbonate hardness is especially important since it leads to scaling, as the following reaction suggests:

$$Ca^{2+} + 2HCO_3^- \longrightarrow CaCO_3 + CO_2 + H_2O \qquad (6.7)$$

Carbonate hardness is sometimes referred to as temporary hardness because it can be removed by simply heating the water.

Another important characteristic of water is its *alkalinity,* which is a measure of the water's ability to absorb hydrogen ions without significant pH change. That is, alkalinity is a measure of the buffering capacity of water. In most natural water, the total amount of H^+ that can be neutralized is dominated by the carbon-

TABLE 6.5 QUALITATIVE CLASSIFICATION
OF WATERS ACCORDING TO HARDNESS

	Hardness	
Description	meq/L	mg/L as CaCO$_3$
Soft	<1	<50
Moderately hard	1–3	50–150
Hard	3–6	150–300
Very hard	>6	>300

Source: Tchobanoglous and Schroeder (1985).

ate buffering system. Thus,

$$\text{Alkalinity (mol/L)} = [HCO_3^-] + 2[CO_3^{2-}] + [OH^-] - [H^+] \quad (6.8)$$

Notice that the concentration of carbonate $[CO_3^{2-}]$ is multiplied by 2 since each ion can neutralize two H^+ ions. This assumes that the concentrations are being measured in molarity units (mol/L). More often, the concentrations are measured in terms of equivalents, or in mg/L as $CaCO_3$, in which case the 2 is already accounted for in the conversions and concentrations are added directly:

$$\text{Alkalinity (meq/L)} = (HCO_3^-) + (CO_3^{2-}) + (OH^-) - (H^+) \quad (6.9)$$

where the quantities in parentheses are concentrations in meq/L or mg/L as $CaCO_3$.

The following example demonstrates these alkalinity calculations.

Example 6.2   Calculating Alkalinity

A sample of water at pH 10.0 has 32.0 mg/L of CO_3^{2-} and 56.0 mg/L of HCO_3^-. Find the alkalinity as $CaCO_3$.

Solution   The carbonate has a molecular weight of $12 + 3 \times 16 = 60$ and $n = 2$, so the equivalent weight is $60/2 = 30$ mg/meq. Converting the given concentration into mg/L as $CaCO_3$ gives

$$CO_3^{2-} = 32.0 \text{ mg/L} \times \frac{1}{30 \text{ mg/meq}} \times 50 \text{ mg } CaCO_3/\text{meq} = 53.3 \text{ mg/L as } CaCO_3$$

The bicarbonate has molecular weight 61 with $n = 1$, so its equivalent weight is 61 mg/meq:

$$HCO_3^- = 56.0 \text{ mg/L} \times \frac{1}{61 \text{ mg/meq}} \times 50 \text{ mg } CaCO_3/\text{meq} = 45.9 \text{ mg/L as } CaCO_3$$

The pH is 10, so $[H^+] = 10^{-10}$ mol/L, and its EW is 1 mg/meq; expressed as $CaCO_3$ it becomes

$$H^+ = 10^{-10} \text{ mol/L} \times 1 \text{ g/mol} \times \frac{10^3 \text{ mg/g}}{1 \text{ mg/meq}} \times 50 \text{ mg } CaCO_3/\text{meq}$$

$$H^+ = 5.0 \times 10^{-6} \text{ mg/L as } CaCO_3$$

Since $[H^+][OH^-] = 10^{-14}$, then $[OH^-] = 10^{-4}$ mol/L; its EW is 17 mg/meq.

$$OH^- = 10^{-4} \text{ mol/L} \times 17 \text{ g/mol} \times \frac{10^3 \text{ mg/g}}{17 \text{ mg/meq}} \times 50 \text{ mg } CaCO_3/\text{meq}$$

$$OH^- = 5.0 \text{ mg/L as } CaCO_3$$

Total alkalinity, then, is just $(CO_3^{2-}) + (HCO_3^-) + (OH^-) - (H^+)$

$$\text{Alkalinity} = 53.3 + 45.9 + 5.0 - 5.0 \times 10^{-6} = 104.2 \text{ mg/L as } CaCO_3$$

For nearly neutral water (pH around 6–8) the concentrations of H^+ and OH^- are insignificant and alkalinity is determined entirely by the carbonates:

$$\text{Alkalinity (meq/L)} = (HCO_3^-) + (CO_3^{2-}) \quad (6.10)$$

Example 6.3   A Chemical Analysis of a Sample of Water

An analysis of a sample of water with pH 7.5 has produced the following concentrations (mg/L):

	Cations		Anions
Ca^{2+}	80	Cl^-	100
Mg^{2+}	30	SO_4^{2-}	201
Na^+	72	HCO_3^-	165
K^+	6		

Find the total hardness, the carbonate hardness, the noncarbonate hardness, and the alkalinity, all expressed as $CaCO_3$. Find the total dissolved solids (TDS) in mg/L.

Solution   It is helpful to set up a table in which each of the concentrations can be expressed in terms of $CaCO_3$. Equation 6.5 and the approach used in Example 6.1 will be followed. For example, for Ca^{2+}

$$Ca^{2+} = 80 \text{ mg/L} \times \frac{1}{20.05 \text{ mg/meq}} \times 50 \text{ mg } CaCO_3/\text{meq}$$

$$= 199.5 \text{ mg/L as } CaCO_3$$

Ion	mg/L	MW	n	mg/meq	meq/L	mg/L as $CaCO_3$
Ca^{2+}	80	40.1	2	20.05	3.99	199.5
Mg^{2+}	30	24.3	2	12.15	2.47	123.5
Na^+	72	23.0	1	23.0	3.13	156.5
K^+	6	39.1	1	39.1	0.15	7.7
Cl^-	100	35.5	1	35.5	2.82	140.8
SO_4^{2-}	201	96.1	2	48.05	4.18	209.2
HCO_3^-	165	61.0	1	61.0	2.70	135.2

As a first check on the chemical analysis, we can compare the sum of the concentrations of cations and anions as $CaCO_3$ or as meq/L to see if they are nearly equal.

$$\Sigma \text{ cations} = 199.5 + 123.5 + 156.5 + 7.7 = 487.2 \text{ mg/L as } CaCO_3$$

$$\Sigma \text{ anions} = 140.8 + 209.2 + 135.2 = 485.2 \text{ mg/L as } CaCO_3$$

which is quite close. The difference would probably be associated with small concentrations of other ions as well as measurement error.

   a. The total hardness is the sum of the multivalent cations, Ca^{2+} and Mg^{2+}:

$$TH = 199.5 + 123.5 = 323.0 \text{ mg/L as } CaCO_3$$

   b. The carbonate hardness is that portion of total hardness associated with carbonates, which in this case is just bicarbonate HCO_3^-:

$$CH = 135.2 \text{ mg/L as } CaCO_3$$

c. The noncarbonate hardness is the difference between the total hardness and the carbonate hardness:

$$NCH = TH - CH = 323.0 - 135.2 = 187.8 \text{ mg/L as } CaCO_3$$

d. Since the pH is nearly neutral, the concentrations of H^+ and OH^- are negligible, so the alkalinity is given by just the bicarbonate:

$$\text{Alkalinity} = (HCO_3^-) = 135.2 \text{ mg/L as } CaCO_3$$

e. The total dissolved solids is simply the sum of the cation and anion concentrations expressed in mg/L:

$$TDS = 80 + 30 + 72 + 6 + 100 + 201 + 165 = 654 \text{ mg/L}$$

It is sometimes helpful to display the ionic constituents of water using a bar graph such as that shown in Figure 6.3.

Softening

Surface waters seldom have hardness levels above 200 mg/L as $CaCO_3$, so softening is not usually part of the water treatment process. For groundwater, however, where hardness levels are often over 1000 mg/L, it is quite common. There are two common approaches to softening water: the *lime–soda* process and the *ion-exchange* process. Either may be used in a central treatment plant prior to distribution, but individual home units use the ion-exchange process.

In the lime–soda process, either quick lime (CaO) or hydrated lime $[Ca(OH)_2]$ is added to the water, raising the pH to about 10.3 and converting soluble bicarbonate ions (HCO_3^-) into insoluble carbonate (CO_3^{2-}). The carbonate

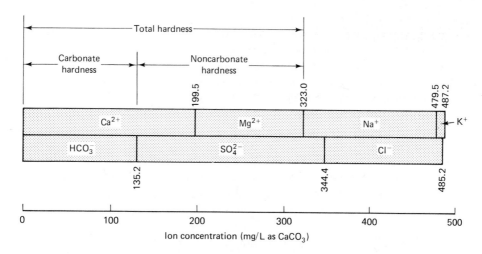

Figure 6.3 Bar graph illustrating the ionic constituents of Example 6.3.

then precipitates out as $CaCO_3$, as is suggested in the following reaction:

$$Ca(HCO_3)_2 + Ca(OH)_2 \longrightarrow 2\ CaCO_3\downarrow + 2H_2O \qquad (6.11)$$

Similarly, magnesium ions can be removed by forming the precipitate $Mg(OH)_2$:

$$Mg(HCO_3)_2 + 2Ca(OH)_2 \longrightarrow 2CaCO_3\downarrow + Mg(OH)_2\downarrow + H_2O \qquad (6.12)$$

If insufficient natural bicarbonate alkalinity (HCO_3^-) is available to cause reaction (6.11) or (6.12), it may be necessary to add carbonate in the form of soda ash (Na_2CO_3). The reaction for magnesium would then be

$$Mg(SO_4) + Ca(OH)_2 + Na_2CO_3 \longrightarrow Mg(OH)_2\downarrow + CaCO_3\downarrow + Na_2SO_4$$
$$(6.13)$$

The additional carbonate required makes the process of noncarbonate hardness removal more expensive than the removal of carbonate hardness.

Most of the precipitates formed are removed in a sedimentation basin. Particles too finely divided to settle can later contribute to the clogging of filters and distribution piping, so the water is often recarbonated with CO_2 to convert those carbonate particles into soluble bicarbonates.

The alternative softening process uses ion exchange whereby hard water is forced through a column containing solid resin beads made of naturally occurring clays called *zeolites,* or synthetic resins. In an ion-exchange unit, the resin removes Ca^{2+} and Mg^{2+} ions from the water and replaces them with sodium ions, which form soluble salts. Using $Ca(HCO_3)_2$ as an example, the reaction can be represented as follows:

$$Ca(HCO_3)_2 + Na_2R \longrightarrow CaR + 2NaHCO_3 \qquad (6.14)$$

where R represents the solid ion-exchange resin. The calcium reacts with the resin and is removed from the water as CaR. The alkalinity (HCO_3^-) remains unchanged. The sodium salts that are formed do not cause hardness, but the dissolved sodium ions remain in the treated water and may be harmful to individuals with heart problems.

The hardness removal is essentially 100 percent effective as long as the ion-exchange medium has sodium remaining. When the sodium is depleted, the ion-exchange bed must be regenerated by removing it from service and backwashing it with a solution of NaCl, forming new Na_2R. The wastewater produced during regeneration must be properly disposed of, since it contains a high concentration of chlorides. The regeneration reaction involving Ca can be represented as

$$CaR + 2NaCl \longrightarrow Na_2R + CaCl_2 \qquad (6.15)$$

The ion-exchange process can be used in waste treatment as well as water treatment. In such cases, the process can enable recovery of valuable chemicals for reuse, or harmful ones for disposal. It is, for example, often used to recover chromic acid from metal finishing waste, for reuse in chrome-plating baths. It is even used for the removal of radioactivity.

6.3 WASTEWATER TREATMENT

Municipal wastewater is typically over 99.9 percent water. The characteristics of the remaining portion vary somewhat from city to city, with the variation depending primarily on inputs from industrial facilities that mix with the somewhat predictable residential flows. Given the almost limitless combinations of chemicals found in wastewater, it is too difficult to list them individually. Instead, they are often described by a few general categories, as has been done in Table 6.6.

In Table 6.6, a distinction is made between total dissolved solids and *suspended solids* (SS). The sum of the two is *total solids* (TS). The suspended solids portion is, by definition, the portion of total solids that can be removed by a membrane filter (having a pore size of about 1.2 μm). The remainder (TDS) that cannot be filtered includes dissolved solids, colloidal solids, and very small suspended particles.

Wastewater treatment plants are usually designated as providing either *primary, secondary,* or *advanced* treatment, depending on the degree of purification. Primary treatment plants utilize physical processes, such as screening and sedimentation, to remove a portion of the pollutants that will settle, float, or that are too large to pass through simple screening devices. This is followed by disinfection. Primary treatment typically removes about 35 percent of the BOD and 60 percent of the suspended solids. In the early 1970s, the sewage of about 50 million people in the United States was receiving no better treatment than this. While the most visibly objectionable substances are removed in primary treatment, and some degree of safety is provided by the disinfection, the effluent still has enough BOD to cause oxygen depletion problems and enough nutrients, such as nitrogen and phosphorus, to accelerate eutrophication.

The *Clean Water Act* (CWA) of 1977, in essence, requires at least secondary treatment for all publicly owned treatment works (POTWs) by stipulating that such facilities provide at least 85 percent BOD removal (with possible case-by-case variances that allow lower percentages for marine discharges). This translates into an effluent requirement of 30 mg/L for both 5-day BOD and suspended

TABLE 6.6   TYPICAL RANGE OF COMPOSITION OF UNTREATED DOMESTIC WASTEWATER

Constituent	Abbreviation	Concentration (mg/L)
5-day biochemical oxygen demand	BOD$_5$	100–300
Chemical oxygen demand	COD	250–1000
Total dissolved solids	TDS	200–1000
Suspended solids	SS	100–350
Total Kjeldahl nitrogen	TKN	20–80
Total phosphorus (as P)	TP	5–20

Source: Adapted from Davis and Cornwell (1985).

solids (monthly average). In secondary treatment plants, the physical processes that make up primary treatment are augmented with processes that involve the microbial oxidation of wastes. Such biological treatment mimics nature by utilizing microorganisms to oxidize the organics, with the advantage being that the oxidation can be done under controlled conditions in the treatment plant itself, rather than in the receiving body of water. When properly designed and operated, secondary treatment plants remove about 90 percent of the BOD and 90 percent of the suspended solids.

While the main purpose of primary treatment (in addition to disinfecting the wastes) is to remove objectionable solids, and the principal goal of secondary treatment is to remove most of the BOD, neither is effective at removing nutrients, dissolved material, or biologically resistant (refractory) substances. For example, typically no more than half of the nitrogen and one-third of the phorphorus are removed during secondary treatment. This means the effluent can still be a major contributor to eutrophication problems. In circumstances where either the raw sewage has particular pollutants of concern or the receiving body of water is especially sensitive, so-called advanced treatment (previously called *tertiary* treatment) may be required. Advanced treatment processes are varied and specialized, depending on the nature of the pollutants that must be removed. In most circumstances, advanced treatment follows primary and secondary treatment, although in some cases, especially in the treatment of industrial waste, it may completely replace those conventional processes.

An example wastewater treatment plant that provides primary and secondary treatment is illustrated in Figure 6.4.

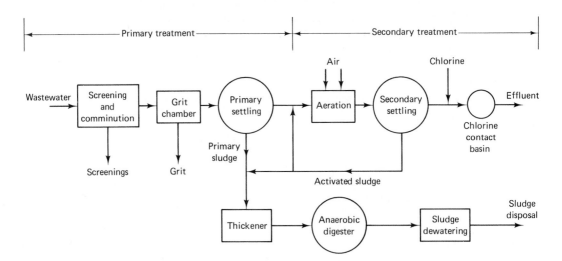

Figure 6.4 Schematic of an example wastewater treatment facility providing primary and secondary treatment using the activated sludge process.

Primary Treatment

As suggested in Figure 6.4, primary treatment begins with simple screening. Screening removes large floating objects such as rags, sticks, old shoes, and whatever else that might damage the pumps or clog small pipes. Screens vary, but typically consist of parallel steel bars spaced anywhere from 2 to 7 cm apart, perhaps followed by a wire mesh screen with smaller openings. One way to avoid the problem of disposal of materials collected on screens is to use a device called a comminuter, which grinds those coarse materials into small enough pieces that they can be left right in the wastewater flow.

After screening, the wastewater passes into a grit chamber where it is held for a few minutes. The *detention time* (the tank volume divided by the flow rate) is chosen to be long enough to allow sand, grit, and other heavy material to settle out but is too short to allow lighter, organic materials to settle. By collecting only these heavier materials, the disposal problem is simplified since those materials are usually nonoffensive and, after washing, can be easily disposed of in a landfill.

From the grit chamber, the sewage passes to a primary settling tank (also known as a "sedimentation basin" or a "clarifier") where the flow speed is reduced sufficiently to allow most of the suspended solids to settle out by gravity. Detention times of approximately 2–3 hr are typical, resulting in the removal of from 50 to 65 percent of the suspended solids and 25 to 40 percent of the BOD. Primary settling tanks are either round or rectangular and their behavior is similar to that of the clarifiers already described for water treatment facilities. The solids that settle, called *primary sludge* or raw sludge, are removed for further processing, as is the grease and scum that floats to the top of the tank.

If this is just a primary treatment plant, the effluent at this point is chlorinated to destroy bacteria and help control odors. Then it is released.

Secondary (Biological) Treatment

The main purpose of secondary treatment is to provide BOD removal beyond what is achievable by simple sedimentation. There are three commonly used approaches, all of which take advantage of the ability of microorganisms to convert organic wastes into stabilized, low-energy compounds. Two of these approaches, the *trickling filter* (and its variations) and the *activated sludge* process, sequentially follow normal primary treatment. The third, *oxidation ponds* (or lagoons), however, can provide equivalent results without preliminary treatment.

Trickling Filters

A trickling filter consists of a rotating distribution arm that sprays liquid wastewater over a circular bed of "fist size" rocks or other coarse materials (Figure 6.5). The spaces between the rocks allow air to circulate easily so that

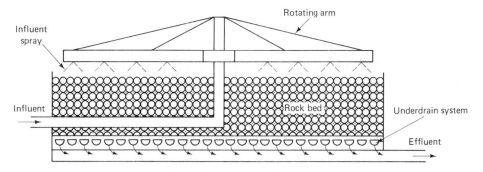

Figure 6.5 Sketch of the cross section of a trickling filter.

aerobic conditions can be maintained. Of course, the size of the openings is such that there is no actual filtering taking place, so the name trickling filter is somewhat of a misnomer. Instead, the individual rocks in the bed are covered by a layer of biological slime that adsorbs and consumes the wastes trickling through the bed. This slime consists mainly of bacteria, but it may also include fungi, algae, protozoa, worms, insect larvae, and snails. The accumulating slime periodically slides off individual rocks and is collected at the bottom of the filter, along with the treated wastewater, and passed on to the secondary settling tank where it is removed. Not shown is a provision for returning some of the effluent from the filter back into the incoming flow. Such recycling not only enables more effluent organic removal, it also provides a way to keep the biological slimes from drying out and dying during low flow conditions.

If ordinary rocks are used in the bed of a trickling filter, structural problems caused by their weight tend to restrict the bed depth to about 3 m. This makes it necessary for the beds to be quite large. Diameters as great as 60 m are not unusual. However, plastic media are becoming increasingly popular as a replacement for rocks, since in the same volume they can be designed to achieve greater surface areas for slime growth, and their lightness allows much deeper beds. The combination allows equivalent treatment to rock beds, but with much smaller land area requirements. They can also be designed to be less prone to plugging by the accumulating slime, and modestly higher rates of BOD removal are possible. These filters, made with plastic media, are sometimes referred to as *biological towers*.

Rotating Biological Contactor

Trickling filters (and biological towers) are examples of devices that rely on microorganisms that grow on the surface of rocks, plastic, or other media. A variation on this *attached growth* idea is provided by the *rotating biological contactor* (RBC). An RBC consists of a series of closely spaced, circular, plastic disks, that are typically 3.6 m in diameter and attached to a rotating horizontal shaft. The bottom 40 percent of each disk is submersed in a tank containing the wastewater to be treated. The biomass film that grows on the surface of the disks moves into and out of the wastewater as the RBC rotates. While the microorganisms are submerged in the wastewater, they adsorb organics; while they are rotated out of the wastewater, they are supplied with needed oxygen. By placing modular RBC units in series, treatment levels that exceed conventional secondary treatment can be achieved (Figure 6.6). These devices have been used in the United States only since 1969, and, although early units suffered from assorted mechanical problems, they are now generally accepted. They are easier to operate under varying load conditions than trickling filters, since it is easier to keep the solid medium wet at all times.

Activated Sludge

Trickling filters were first used in 1893 and they have been successfully used ever since, but they do cost more to build, are more temperature sensitive, and remove less BOD than the activated sludge plants that were later developed.

The example wastewater treatment plant flow diagram given in Figure 6.4 was drawn to illustrate the activated sludge process. As indicated there, the key biological unit in the process is the aeration tank, which receives effluent from the primary clarifier. It also receives a mass of recycled biological organisms from the

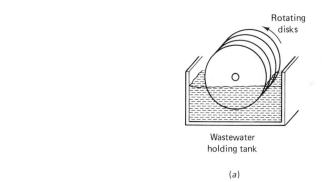

(a)

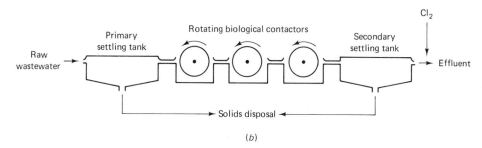

(b)

Figure 6.6 Rotating biological contactor cross section and treatment system: (a) RBC cross section; (b) RBC treatment system.

secondary settling tank, known as *activated sludge*. To maintain aerobic conditions, air or oxygen is pumped into the tank and the mixture is kept thoroughly agitated. After about 6–8 hr of agitation, the wastewater (now referred to as the "mixed liquor") flows into the secondary settling tank where the solids, mostly bacterial masses, are separated from the liquid by subsidence. A portion of those solids is returned to the aeration tank to maintain the proper bacterial population there, while the remainder must be processed and disposed of.

Figure 6.7 illustrates several variations of the basic activated sludge process (after primary settling). The earliest aeration tanks were designed so that the mixture of influent and recycled sludge was introduced at one end of a long, narrow tank, so that it tended to move uniformly from one end of the tank to the other (so-called "plug flow"). With such designs, the oxygen demand is exerted mostly at the entrance to the tank, with relatively little oxygen required at the

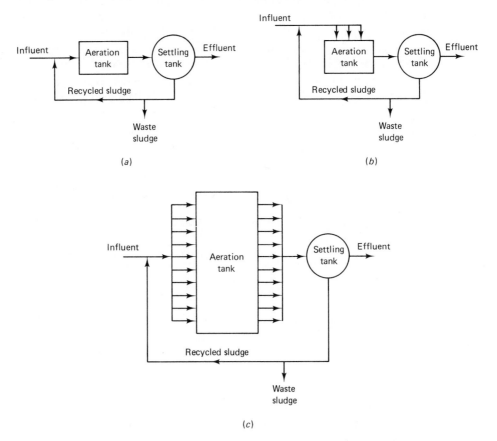

Figure 6.7 Modifications of the activated sludge process: (*a*) simple plug flow, (*b*) step aeration, and (*c*) complete mix. (*Source:* After Linsley, R. K., and Franzini, J. B. *Water-Resources Engineering,* 3d ed. Copyright © 1979 by McGraw-Hill, Inc. Used with permisison of the publisher.)

exit. This nonuniformity of demand requires careful control by the plant operators to maintain aerobic conditions throughout, especially under variations in
load. By introducing the mixture of wastes and recycled sludge at several points
along the tank, as is shown in Figure 6.7b, or continuously along the edge of the
tank, as is shown in Figure 6.7c, the stability of the process is increased, thus
reducing the likelihood of biological upsets that can render the plants useless.

By allowing greater contact between microorganisms and wastewater in a
given volume of space, activated sludge tanks can take up considerably less land
area than trickling filters with equivalent performance. They are also less expensive to construct than trickling filters, have fewer problems with flies and odors,
and can achieve higher rates of BOD removal. They do, however, require more
energy for pumps and blowers, and hence have higher operating costs.

Sludge Treatment

The processes described thus far have the purpose of removing solids and
BOD from the wastewater before the liquid effluent is released to a convenient,
nearby body of water. What remains to be disposed of is a mixture of solids and
water, called sludge. The collection, processing, and disposal of sludge can be the
most costly and complex aspect of wastewater treatment.

The quantity of sludge produced may be as high as 2 percent of the original
volume of wastewater, depending somewhat on the treatment process being
used. Since sludge can be as much as 97 percent water, and since the cost of
disposal will be related to the volume of sludge being processed, one of the
primary goals of sludge treatment is to separate as much of the water from the
solids as possible. The other goal is to stabilize the solids so that they are no
longer objectionable or environmentally damaging.

The traditional method of sludge processing utilizes anaerobic digestion.
That is, it involves bacteria that thrive in the absence of oxygen. Anaerobic
digestion is slower than aerobic digestion, but has the advantage that only a small
percentage of the wastes are converted into new bacterial cells. Instead, most of
the organics are converted into carbon dioxide and methane gas. The digestion
process is complex, but can be summarized by the two steps shown in Figure 6.8.
In the first phase, complex organics such as fats, proteins, and carbohydrates are
biologically converted into simpler organic materials, mostly organic fatty acids.
The bacteria that perform this conversion are commonly referred to as *acid form-*

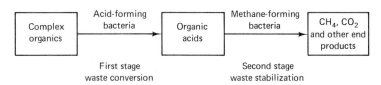

Figure 6.8 The two phases of anaerobic digestion.

ers. They are relatively tolerant to changes in temperature and pH, and they grow much faster than the *methane formers* that carry out the second stage of digestion.

Methane-forming bacteria slowly convert the organic acids into CO_2, CH_4, and other stable end products. These bacteria are very sensitive to temperature, pH, toxins, and oxygen. If their environmental conditions are not just right, the rate at which they convert organic acids to methane slows, and organic acids begin to accumulate, dropping the pH. A positive feedback loop can be established where the acid formers continue to produce acid, while the methane formers, experiencing lower and lower pH, become more and more inhibited. When this occurs, the digester is said to have gone sour, and massive doses of lime may be required to bring it back to operational status.

Most treatment plants utilizing anaerobic digestion for sludge stabilization use a two-stage digester as shown in Figure 6.9. Sludge in the first stage is thoroughly mixed and heated to increase the rate of digestion. Typical retention times are between 10 and 15 days. The second stage tank is neither heated nor mixed and is likely to have a floating cover to accommodate the varying amount of gas being stored. Stratification occurs in the second stage, which allows a certain amount of separation of liquids (called supernatant) and solids, as well as the accumulation of gas. The supernatant is returned to the main treatment plant for further BOD removal, and the settled sludge is removed, dewatered, and disposed of. The gas produced in the digester is about 60 percent methane, which is a valuable fuel with many potential uses within the treatment plant. The methane may be used to heat the first stage of the digester, and it can run in an engine/generator set to power pumps, compressors, and other electrical equipment.

Digested sludge removed from the second stage of the anaerobic digester is still mostly liquid. The solids have been well digested, so there is little odor. The most popular way of dewatering has been to pump the sludge onto large sludge drying beds where evaporation and seepage remove the water. Other methods include use of vacuum filters, filter presses, centrifuges, or incinerators. The

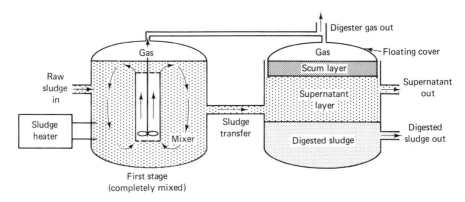

Figure 6.9 Schematic of a two-stage anaerobic digester.

digested and dewatered sludge is potentially useful as a soil conditioner, but most often it is simply trucked away and disposed of in a landfill.

Oxidation Ponds

Oxidation ponds are large, shallow ponds, typically 1–2 m deep, where raw or partially treated sewage is decomposed by microorganisms. The conditions are similar to those that prevail in a eutrophic lake. The ponds can be designed to maintain aerobic conditions throughout, but more often the decomposition taking place near the surface is aerobic, while that near the bottom is anaerobic. Such ponds, having a mix of aerobic and anaerobic conditions, are called *facultative ponds*. In ponds, the oxygen required for aerobic decomposition is derived from surface aeration and algal photosynthesis; deeper ponds, called *lagoons,* are mechanically aerated. A schematic diagram of the reactions taking place in a facultative pond is given in Figure 6.10.

Oxidation ponds can be designed to provide complete treatment to raw sewage, but they require a good deal of space. These ponds have been used extensively in small communities where land constraints are not so critical. The amount of pond surface area required is considerable, with 1 hectare per 240 people (1 acre per 100 people) often being recommended, although in areas with warm climates and mild winters, such as in the southwestern United States, about half that area is often used (Viessman and Hammer, 1985).

Ponds are easy to build and manage, they accommodate large fluctuations in flow, and they can provide treatment that approaches that of conventional biological systems but at a much lower cost. The effluent, however, may contain undesirable concentrations of algae and, especially in the winter when less oxygen is

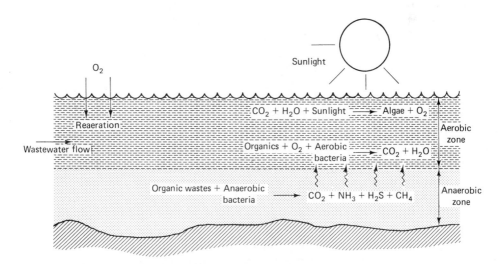

Figure 6.10   Schematic of an oxidation pond.

liberated by photosynthesis, they may produce unpleasant odors. Used alone, they also have the disadvantage that the effluent may not meet the EPA secondary treatment requirement of 30 mg/L BOD_5 and suspended solids. Their simplicity and effectiveness in destroying pathogenic organisms, however, make these ponds especially useful in developing countries.

Oxidation ponds are also used to augment secondary treatment, in which case they are often called *polishing ponds*.

Advanced Wastewater Treatment

Anything that follows conventional primary and biological treatment is considered to be advanced treatment. Many of the early advanced treatment facilities were designed with the primary purpose of removing nitrogen and phosphorus (the principal nutrients responsible for eutrophication), as well as to more completely reduce BOD. Very often now, advanced treatment is designed with the additional goal of removing various toxic substances, such as metals. Treatment technologies that can be used for hazardous wastes will be described in in the next section.

Nitrogen Removal

Nitrogen exists in a variety of forms in wastewater. As bacteria decompose waste, nitrogen that was bound up in complex organic molecules is released as ammonia nitrogen. Ammonia, in turn, exists in water in two forms: as ammonium ion (NH_4^+), which is highly soluble, and as ammonia gas (NH_3), which is not. As pH increases, the equilibrium relationship between these two forms is driven toward the less soluble ammonia gas (see Example 2.5):

$$NH_4^+ + OH^- \longrightarrow NH_3\uparrow + H_2O \qquad (6.16)$$

One method of nitrogen removal, *ammonia stripping,* is based on this reaction. In this process, the pH of treated wastewater is raised to at least 10, typically with quick lime (CaO), to form dissolved ammonia gas. The ammonia can then be liberated in a stripping tower of the sort illustrated in Figure 6.11. Unfortunately, these systems have been plagued by a number of problems that have limited the usefulness of this approach. For one, the lime reacts with CO_2 to form a calcium carbonate scale, which must be removed periodically from the stripping surfaces. This scaling can be so severe, as was the case at the tertiary treatment plant in South Lake Tahoe, that the tower may eventually cease to function. Ammonia stripping is also less effective in cold weather, in part because ammonia is more soluble in cold water, making it harder to strip, but also because towers can ice up. The process also has been criticized because it simply transfers the pollution problem from one medium to another, in this case from water to air, creating an additional burden on the atmosphere.

A second approach to nitrogen control utilizes aerobic bacteria to convert ammonia (NH_4^+) to (NO_3^-), which is nitrification, followed by an anaerobic stage

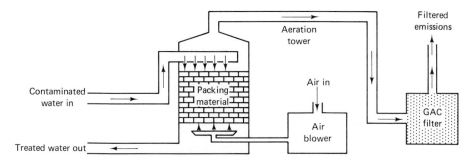

Figure 6.11   An air stripping tower followed by a granular activated carbon filter provides effective removal of VOCs. To remove nonvolatiles, the treated water coming out of the tower may be pumped through other carbon adsorbers.

where different bacteria convert nitrates to nitrogen gas (N_2), which is denitrification. The overall process then is referred to as *nitrification/denitrification*.

The nitrification step actually occurs in two stages. Ammonia is converted to nitrites (NO_2^-) by *Nitrosomonas,* while *Nitrobacter* oxidize nitrites to nitrates, as was described in Section 4.4. The combination of steps can be summarized by

$$NH_4^+ + 2O_2 + \xrightarrow{\text{bacteria}} NO_3^- + 2H^+ + H_2O \qquad (6.17)$$

Nitrification does not begin to be important until domestic wastewater is at least 5–8 days old. Thus, if this method of nitrogen control is to be used, the wastewater must be kept in the treatment plant for a much longer time than would normally be the case. Detention times of 15 days or more are typically required. If reaction (6.17) takes place in the treatment plant rather than in the receiving body of water, at least the oxygen demand for nitrification is satisfied. The nitrogen, however, remains in the effluent, and if the process were to stop here, that nitrogen could go on to contribute to unwanted algal growth. To avoid this, the denitrification step is required.

The second phase of the nitrification/denitrification process is anaerobic denitrification:

$$2NO_3^- + \text{organic matter} \xrightarrow{\text{bacteria}} N_2\uparrow + CO_2 + H_2O \qquad (6.18)$$

which releases harmless, elemental nitrogen gas. The energy to drive this reaction comes from the organic matter indicated in (6.18). Since this denitrification process occurs after waste treatment, there may not be enough organic material left in the waste stream to supply the necessary energy and an additional source, usually methanol (CH_3OH), must be provided.

Phosphorus Removal

Only about 30 percent of the phosphorus in municipal wastewater is removed during conventional primary and biological treatment. Since phosphorus is very often the limiting nutrient, its removal from the waste stream is especially important when eutrophication is a problem.

Phosphorus in wastewater exists in many forms, but all of it ends up as orthophosphate ($H_2PO_4^-$, HPO_4^{2-}, and PO_4^{3-}). Removing phosphates is most often accomplished by adding a coagulant, usually alum [$Al_2(SO_4)_3$] or lime [$Ca(OH)_2$]. The pertinent reaction involving alum is

$$Al_2(SO_4)_3 + 2PO_4^{3-} \longrightarrow 2AlPO_4\downarrow + 3SO_4^{2-} \qquad (6.19)$$

Alum is sometimes added to the aeration tank when the activated sludge process is being used, thus minimizing the need for additional equipment.

The reaction for precipitation with lime can be represented as

$$5Ca(OH)_2 + 3HPO_4^{2-} \longrightarrow Ca_5OH(PO_4)_3\downarrow + 3H_2O + 6OH^- \qquad (6.20)$$

where the precipitate formed is called calcium hydroxyphosphate, or, hydroxyl-apatite. When lime is used as the coagulant, it is often used after biological treatment, especially when ammonia stripping is also part of the treatment process. The lime not only causes the phosphate to precipitate out of solution, it also raises the pH of the waste stream so that soluble ammonium ions are converted to ammonia gas.

6.4 HAZARDOUS WASTE TREATMENT TECHNOLOGIES

Even with a much more vigorous hazardous waste reduction program, as RCRA requires, there will still be large quantities of hazardous wastes that will require treatment and disposal. In the past, there was little treatment, and disposal was most often on land. In both SARA (Superfund) and the 1984 Hazardous and Solid Waste Amendments of RCRA, emphasis is on the development and use of alternative and innovative treatment technologies that result in permanent destruction of wastes or a reduction in toxicity, mobility, and volume. Land disposal is greatly restricted under the 1984 RCRA amendments.

Treatment technologies are often categorized as being physical, chemical, biological, thermal, or stabilization/fixation. These categories are reasonably well defined, though there is room for confusion when technologies have overlapping characteristics.

Chemical, biological, and physical wastewater treatment processes are currently the most commonly used methods of treating aqueous hazardous waste. Chemical treatment transforms waste into less hazardous substances using such techniques as pH neutralization, oxidation or reduction, and precipitation. Biological treatment uses microorganisms to degrade organic compounds in the waste stream. Physical treatment processes include gravity separation, phase change systems, such as air and steam stripping of volatiles from liquid wastes, and various filtering operations, including carbon adsorption.

Thermal destruction processes include incineration, which is increasingly becoming a preferred option for the treatment of hazardous wastes, and pyrolosis, which is the chemical decomposition of waste brought about by heating the material in the absence of oxygen.

Fixation/stabilization techniques involve removal of excess water from a waste and solidifying the remainder either by mixing it with a stabilizing agent, such as Portland cement, or vitrifying it to create a glassy substance. Solidification is most often used on inorganic sludges.

Choosing an appropriate technology to use in any given situation is obviously beyond the scope of this text. Not only are there many different kinds of hazardous wastes, in terms of their chemical makeup, but the treatability of the wastes depends on their form. A technology suitable for treating PCBs in sludges, for example, may not be appropriate for treating the same contaminant in dry soil. Table 6.7 gives a partial listing of available treatment technologies appropriate for a variety of types of hazardous waste streams along with the applicable form of waste (liquid, gaseous, solids/sludges.) For a more complete list, as well as detailed descriptions of each technology, see for example Freeman (1989).

Physical Treatment

Sedimentation.   The simplest physical treatment systems that separate solids from liquids take advantage of gravity settling and natural flotation. Special sedimentation tanks and clarification tanks are designed to encourage solids to settle so they can be collected as a sludge from the bottom of the tank. Some solids will float naturally to the surface and they can be removed with a skimming device. It is also possible to encourage flotation by introducing finely divided bubbles into the waste stream. The bubbles collect particles as they rise and the combination can be skimmed from the surface. Separated sludges can then be further concentrated by evaporation, filtration, or centrifugation.

Adsorption.   Physical treatment can also be used to remove small concentrations of hazardous substances dissolved in water that would never settle out. One of the most commonly used techniques for removing organics involves the process of *adsorption,* which is the physical adhesion of chemicals onto the surface of a solid. The effectiveness of the adsorbent is directly related to the amount of surface area available to attract the molecules or particles of contaminant. The most commonly used adsorbent is a very porous matrix of *granular activated carbon* (GAC), which has an enormous surface area (on the order of $1000 \ m^2/g$). A single handful of GAC has an internal surface area of about one acre.

Granular activated carbon treatment systems usually consist of a series of large vessels partially filled with adsorbent. Contaminated water enters the top of each vessel, trickles down through the GAC, and is released at the bottom. After a period of time, the carbon filter becomes clogged with adsorbed contaminants

TABLE 6.7　A PARTIAL LIST OF TREATMENT TECHNOLOGIES FOR VARIOUS HAZARDOUS WASTE STREAMS

Treatment process	Hazardous waste streams												Form of waste		
	Corrosives	Cyanides	Halogenated solvents	Nonhalogenated organics	Chlorinated organics	Other organics	Oily wastes	PCBs	Aqueous with metals	Aqueous with organics	Reactives	Contaminated soils	Liquids	Solids/sludges	Gases
Separation/filtration		×	×	×	×	×			×	×			×		
Carbon adsorption									×	×	×		×		×
Air and steam stripping			×	×	×	×				×			×		
Electrolytic recovery									×				×		
Ion exchange	×								×	×			×		
Membranes									×	×			×		
Chemical precipitation	×								×				×		
Chemical oxidation/reduction		×								×			×		
Ozonation		×		×		×					×		×		×
Evaporation			×	×	×	×	×						×	×	
Solidification	×	×										×	×	×	
Liquid injection incineration			×	×	×	×	×						×		×
Rotary kilns			×	×	×	×	×	×				×	×	×	×
Fluidized bed incineration			×	×	×	×	×	×				×	×	×	×
Pyrolysis			×	×	×	×						×	×	×	
Molten glass			×	×	×	×	×			×			×	×	×

Source: Based on Freeman (1989).

and must be either replaced or regenerated. Regeneration can be an expensive, energy-intensive process, usually done off-site. During regeneration, the contaminants are usually burned from the surface of the carbon granules, though in some cases a solvent is used to remove them. Carbon filters that cannot be regenerated due to their contaminant composition must be properly managed for disposal.

Aeration.   For chemicals that are relatively volatile, another physical process, *aeration,* can be used to drive the contaminants out of solution. These stripping systems typically use air, though in some circumstances steam is used. In the most commonly used air stripper, contaminated water is sprayed downward through packing material in a tower, while air is blown upward carrying away the volatiles with it. Such a *packed-tower* can easily remove over 95 percent of the volatile organic compounds (VOCs), including such frequently encountered ones as trichloroethylene, tetrachloroethylene, trichloroethane, benzene, toluene, and other common organics derived from solvents. There is another type of stripper, called an *induced-draft stripper,* which does not use a blower or packing material. In the induced-draft tower, a carefully engineered series of nozzles spray contaminated water horizontally through the sides of a chamber. Air passing through the chamber draws off the volatiles. Induced-draft strippers are cheaper to build and operate, but their performance is much lower than a packed-tower.

By combining air stripping with GAC, many volatile and nonvolatile organic compounds can be removed from water to nondetectable levels. By passing contaminated water first through the air stripper, most of the volatile organics are removed before reaching the GAC system, which extends the life of the carbon before regeneration or replacement is required.

The volatiles removed in an air stripper are, in some circumstances, released directly to the atmosphere. When discharge into the atmosphere is unacceptable, a GAC treatment system can be added to the exhaust air, as shown in Figure 6.11.

Other Physical Processes.   Other physical processes that are sometimes used to treat hazardous wastes include reverse osmosis, ion exchange, and electrodialysis. *Reverse osmosis* devices use pressure to force contaminated water against a semipermeable membrane. The membrane acts as a filter, allowing the water to be pushed through its pores, but restricting the passage of larger molecules that are to be removed. *Ion exchange* is a process wherein ions to be removed from the wastestream are exchanged with ions associated with a special exchange resin. Ion exchange has already been mentioned in the context of water softening where calcium and magnesium ions are replaced with sodium ions from the exchange resin. In the context of hazardous wastes, ion exchange is often used to remove toxic metal ions from solution. *Electrodialysis* uses ion-selective membranes and an electric field to separate anions and cations in solution. In the past, electrodialysis was most often used for purifying brackish water, but it is now finding a role in hazardous waste treatment. Metal salts from plating rinses are sometimes removed in this way, for example.

Chemical Treatment

Chemically treating hazardous waste not only has the potential advantage of converting it to less hazardous forms, but can also produce useful by-products in some circumstances. By encouraging resource recovery, the treatment cost can sometimes be partially offset by the value of the end products produced.

Neutralization. There are many chemical processes that can be used to treat hazardous wastes, and the process decision depends primarily upon the characteristic of the waste. For example, recall that one of RCRA's categories of hazardous waste is anything corrosive, that is, anything having a pH of less than 2 or more than 12.5. Such wastes can be chemically neutralized. Acidic wastewaters are usually neutralized with slaked lime $[Ca(OH)_2]$ in a continuously stirred chemical reactor. The rate of addition of lime is controlled with a feedback control system that monitors pH and adjusts the feed rate accordingly.

Alkaline wastewaters may be neutralized by adding acid directly or by bubbling in gaseous CO_2, forming carbonic acid (H_2CO_3). The advantage of CO_2 is that it is quite often readily available in the exhaust gas from any combustion process at the treatment site. Simultaneous neutralization of acid and caustic waste can be accomplished in the same vessel, as is suggested by Figure 6.12.

Chemical Precipitation. The ability to adjust pH is important not only for waste neutralization, but also because it facilitates other chemical processes that actually remove undesirable substances from the waste stream. For example, a common method for removing heavy metals from a liquid waste is via *chemical precipitation,* which is pH dependent. By properly adjusting pH, the solubility of toxic metals can be decreased, leading to formation of a precipitate that can be removed by settling and filtration.

Frequently, the precipitation involves the use of lime, $Ca(OH)_2$, or caustic (NaOH) to form metal hydroxides. For example, the following reaction suggests the use of lime to form the hydroxide of a divalent metal (M^{2+}):

$$M^{2+} + Ca(OH)_2 \longrightarrow M(OH)_2 + Ca^{2+} \qquad (6.21)$$

Metal hydroxides are relatively insoluble in basic solutions, and, as shown in Figure 6.13, they are *amphoteric;* that is, they have some pH at which their solubility is a minimum. Since each metal has its own optimum pH, it is tricky to control precipitation of a mix of different metals in the same waste. For a waste containing several metals, it may be necessary to use more than one stage of

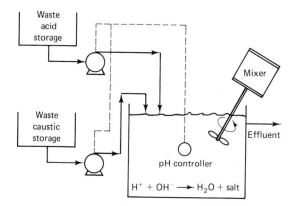

Figure 6.12 Simultaneous neutralization of acid and caustic waste (USEPA, 1987b).

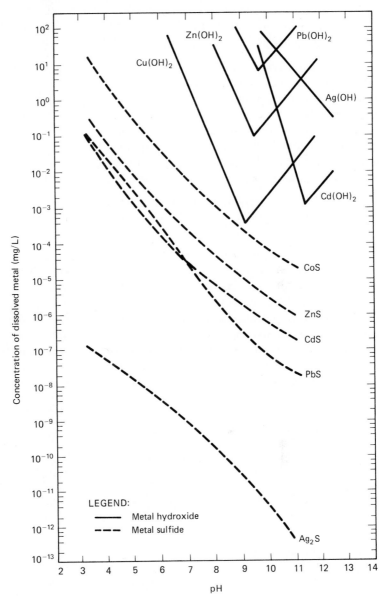

Figure 6.13 Chemical precipitation of metals can be controlled by pH. Metal sulfides are less soluble than metal hydroxides (USEPA, 1980).

precipitation to allow different values of pH to control the removal of different metals.

While hydroxide precipitation using lime is the most common metal removal process, even lower concentrations of metals in the effluent can be obtained by precipitating the metals as sulfides. As can be seen in Figure 6.13, metal sulfides are considerably less soluble than metal hydroxides. A disadvantage of sulfide precipitation is the potential formation of odorous and toxic hydrogen sulfide gas.

Chemical Reduction–Oxidation. *Reduction–oxidation* (redox) reactions provide another important chemical treatment alternative for hazardous wastes. When electrons are removed from an ion, atom, or molecule, the substance is *oxidized;* when electrons are added, it is *reduced.* Both oxidation and reduction occur in the same reaction; hence the abbreviation redox. One of the most important redox treatment processes is the reduction of hexavalent chromium (CrVI) to trivalent chromium (CrIII) in large electroplating operations. Sulfur dioxide is often used as the reducing agent, as shown in the following reactions:

$$3SO_2 + 3H_2O \longrightarrow 3H_2SO_3 \qquad (6.22)$$

$$2CrO_3 + 3H_2SO_3 \longrightarrow Cr_2(SO_4)_3 + 3H_2O \qquad (6.23)$$

The trivalent chromium formed in reaction (6.23) is much less toxic and more easily precipitated than the original hexavalent chromium. Notice that the chromium in reaction (6.23) is reduced from an oxidation state of +6 to +3, while the sulfur is oxidized from +4 to +6.

Another important redox treatment involves the oxidation of cyanide wastes, which are also common in the metal finishing industry. In the following reactions, cyanide is first converted to a less toxic cyanate using alkaline chlorination (pH above 10); further chlorination oxidizes the cyanate to simple carbon dioxide and nitrogen gas. Nearly complete destruction of cyanide results:

$$NaCN + Cl_2 + 2NaOH \longrightarrow NaCNO + 2NaCl + H_2O \qquad (6.24)$$

$$2NaCNO + 3Cl_2 + 4NaOH \longrightarrow 2CO_2 + N_2 + 6NaCl + 2H_2O \qquad (6.25)$$

Wastes that can be treated via redox oxidation include benzene, phenols, most organics, cyanide, arsenic, iron, and manganese; those that can be successfully treated using reduction treatment include chromium (VI), mercury, lead, silver, chlorinated organics like PCBs, and unsaturated hydrocarbons (USEPA, 1988a).

Biological Treatment

Virtually all municipal wastewater treatment plants in the United States and a large number of industrial systems rely on biological treatment processes to decompose organic wastes. Biological treatment systems use microorganisms, mainly bacteria, to metabolize organic material, converting it to carbon dioxide, water, and new bacterial cells. Since biological systems rely on living organisms to transform wastes, considerable care must be exercised to assure conditions are conducive to life. Microbes need a source of carbon and energy, which they can get from the organics that they consume, as well as nutrients such as nitrogen and phosphorus. They are sensitive to pH and temperature and some need oxygen.

As living organisms, they are susceptible to toxic substances, which at first glance makes biological treatment of hazardous wastes seem an unlikely choice.

Surprisingly, though, most hazardous organics are amenable to biological treatment, provided that the proper distribution of organisms can be established and maintained. For any given organic substance, there may be some organisms that will find that substance to be an acceptable food supply, while others may find it toxic. Moreover, organisms that flourish with the substance at one concentration may die when the concentration is increased beyond some critical level. Finally, even though a microbial population may have been established that can handle a particular kind of organic waste, it may be destroyed if the characteristics of the waste are changed too rapidly. If changes are made slowly enough, however, selection pressures may allow the microbial consortium to adjust to the new conditions and thereby remain effective.

Aqueous Waste Treatment. It is convenient to consider biological treatment of various sorts of wastewaters, including leachates from hazardous waste landfills, separately from in situ biological treatment of soils and groundwater. When liquid hazardous wastes can be conveyed to the treatment facility, it is possible to carefully control the characteristics of the waste that reach the biological portion of the facility, increasing the likelihood of a successful degradation process (Figure 6.14).

Biological treatment is just one step in an overall treatment system. An example system would include a chemical treatment stage to oxidize and precipitate some of the toxics, followed by physical treatment to separate the resulting solids from the waste stream. The effluent from the physical treatment step may

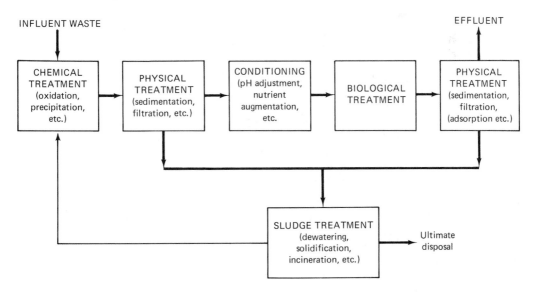

Figure 6.14 General flow diagram for treatment of liquid hazardous waste.

then be conditioned to give it the right pH and nutrient supply needed by the microorganisms in the biological treatment step.

The biological treatment stage itself utilizes processes already described for municipal wastewater treatment plants. After biological treatment, further sedimentation and clarification followed by carbon adsorption can be used to polish the effluent. Inorganic sludges produced during chemical processing and organic sludges from the biotreatment stage are separated from the liquid wastestream and treated. These sludges must be dewatered and disposed of in accordance with RCRA regulations since they are likely to be hazardous themselves.

In Situ Biodegradation.   One way to clean a contaminated aquifer is to pump out the groundwater, treat it, and then either reinject it back into the ground, find some acceptable way to reuse it, or simply "dispose" of the water while gaining minimal environmental or economic benefit from the treated effluent. The latter "pump and dump" schemes have been used extensively in many Superfund sites. Not only does "pump and dump" remediation waste a valuable resource, other problems such as salt water intrusion into the freshwater aquifer (for coastal sites) and land subsidence can result.

Similarly, treatment of contaminated soils above an aquifer has often been accomplished by removing huge quantities of soil from the site, then treating and disposing of them elsewhere.

A promising alternative to moving water or soil to a treatment facility is to, in essence, reverse the process and move the treatment facility to the contaminated water and soil. *In situ biodegradation* is such a process wherein bacteria are used to degrade organic compounds in the soil and groundwater on site. An obvious advantage to in situ biodegradation is that soils and groundwater do not have to be removed, so there is less land disturbance, less wasted water, less risk associated with hazardous waste transportation, and potentially decreased costs. Moreover, there is the potential for aquifer restoration to be more complete since contaminants tend to adsorb onto soil particles, making it difficult to ever completely remove them by pumping.

The technology requires creating conditions underground that will stimulate growth of indigenous, or newly introduced bacteria that have the capability of degrading the organic contaminants of concern. There are basically two approaches to in situ biodegradation. First, the environment of existing microbial populations can be enhanced by adding nutrients and/or oxygen. Nutrients that may be called for include nitrogen, phosphorus, and inorganic salts such as ammonium sulfate, magnesium sulfate, sodium carbonate, and calcium chloride. Significant quantities of these nutrients may have to be added to the affected aquifer before treatment is complete. The supply of oxygen can be enhanced by injecting an oxidant such as hydrogen peroxide or by forcing air through wells with diffusers.

The second approach to in situ biodegradation involves altering the underground microbial population by seeding with new microorganisms that have already been acclimated to the pollutants to be degraded. These new microbes can be picked based on laboratory studies of their effectiveness against the wastes in question. There is also the possibility of genetically altering microorganisms to get strains that will work even better.

In situ biodegradation is a relatively new technique that has been used most frequently to treat soil contaminated with gasoline and diesel. It shows great promise as a way to treat chlorinated solvents such as trichloroethylene (TCE), tetrachloroethylene (PCE), and 1,2-dichloroethylene (DCE), which are among the most commonly found contaminants in underground water supplies. Figure 6.15 shows one version of an in situ bioremediation system.

Waste Incineration

Waste incineration is being strongly advocated by the EPA as its technology of choice for many types of hazardous wastes. Incineration is particularly effective with organic wastes, not only in soils but in other solids, gases, liquids, and slurries (thin mixtures of liquids and solids) and sludges (thick mixtures). Carcinogens, mutagens, teratogens, as well as pathological wastes can all be completely detoxified in a properly operated incinerator. Incinerators are not, however, capable of destroying inorganic compounds, although they can concentrate them in ash, making transportation and disposal more efficient. In addition, metals that volatilize at temperatures below 2000 °F pose a particular problem since, once vaporized, they are difficult to remove using conventional air pollution control equipment.

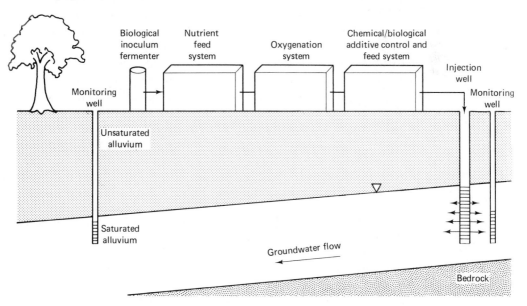

Figure 6.15   In situ bioremediation (USEPA, 1988a).

The principal measure of an incinerator's performance is known as the *destruction and removal efficiency* (DRE). A DRE of 99.99 percent, for example, (commonly called "four nines DRE") means that one molecule of an organic compound is released to the air for every 10 000 molecules entering the incinerator. RCRA requires a minimum DRE of 99.99 percent for most organic compounds, and a DRE of 99.9999 percent (six nines) for dioxins and dibenzofurans. The Toxic Substances Control Act regulations cover thermal destruction of PCBs, and although they are written somewhat differently from RCRA, they in essence require a 99.9999-percent DRE. Incinerator standards also have been written for hydrogen chloride emissions and particulates.

As is the case for all combustion processes, the most critical factors that determine combustion completeness are (1) the temperature in the combustion chamber, (2) the length of time that combustion takes place, (3) the amount of turbulence, or degree of mixing, and (4) the amount of oxygen available for combustion. Controlling these factors, which is crucial to obtaining the high levels of performance required by law, is made especially difficult in hazardous waste incinerators because of the variability of the wastes being burned. In addition to combustion controls, stack gas cleaning systems similar to those that will be described in Chapter 7 are a necessary part of the system. Proper operation and maintenance of these complex incineration systems requires highly trained personnel, diligent and qualified supervisory staff, and an alert governmental agency to assure compliance with all regulations.

While there are a number of types of hazardous waste incinerators, there are only two principal designs that account for most of the existing units in operation: the *liquid injection incinerator* and the *rotary kiln incinerator*. Liquid injection incinerators are the most common, even though they are usable only for gases, liquids, and slurries thin enough to be pumped through an atomizing nozzle. The nozzle emits tiny droplets of waste that are mixed with air and an auxiliary fuel such as natural gas or fuel oil. The resulting gaseous mixture is burned at a very high temperature. The atomizing nozzle used in a liquid injection incinerator must be designed to accommodate the particular characteristics of the expected waste stream, which limits the types of waste that any given incinerator can treat.

The rotary kiln incinerator is more versatile than the liquid injection type, being capable of handling gases, liquids, sludges, and solids of all sorts, including drummed wastes. Figure 6.16 shows a diagram of such an incinerator. The main unit consists of a slightly inclined, rotating cylinder perhaps 2–5 m in diameter and 3–10 m long. Wastes and auxiliary fuel are introduced into the high end of the kiln and combustion takes place while the cylinder slowly rotates. The rotation helps increase turbulence, which improves combustion efficiency. Partially combusted waste gases are passed to a secondary combustion chamber for further oxidation. Rotary kiln incinerators are commercially available as mobile units and fixed installations.

In spite of numerous controls, hazardous waste incinerators have the potential to emit amounts of noxious gases that may be unacceptable to neighbors.

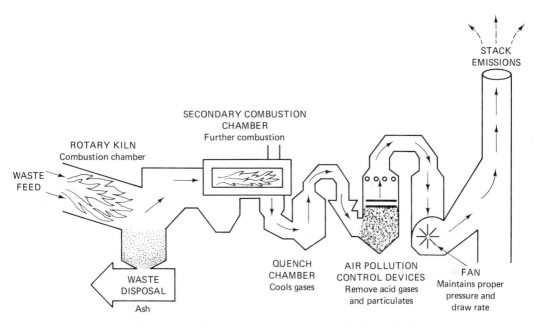

Figure 6.16 A rotary kiln hazardous waste incinerator (USEPA, 1988b).

Emissions may include unburned organic compounds from the original waste, various *products of incomplete combustion* (PICs) formed in the incinerator itself, odors, carbon monoxide, nitrogen and sulfur oxides, hydrogen chloride, and particulates. The unburned ash and sludge from the air pollution control devices are considered hazardous wastes themselves and must be treated as such. If they are transported off-site, then not only are there hazardous materials transported into the facility, but there are some leaving as well. The perception of potentially adverse impacts associated with incineration has made the siting of these facilities an extremely difficult task.

Liquid injection incinerators have been built into two ships, the Vulcanus I and the Vulcanus II, for hazardous waste incineration at sea, but public opposition has prevented their use. Proponents of incineration at sea point out the advantages of incineration, the difficulties associated with land-based siting, and the reductions in human risk that would come with incineration far from any centers of population. Opposition has focused on the chances of hazardous waste spills and the near impossibility of controlling such a spill should it occur. Also, the small amounts of unburned wastes routinely released, as well as those that might be released should the incinerator ever be improperly operated, could have unknown implications for marine life.

A stalemate has developed between the EPA, which is advocating incineration as a way to avoid land disposal (with its almost invariable toxic leakage into the groundwater) and the public, which has, with good reason, learned to be suspicious of any complex and potentially dangerous technology.

Land Disposal

Land disposal techniques include landfills, surface impoundments, underground injection wells, and waste piles. About 5 percent of the hazardous waste that we dispose of on land in the United States is placed in specially designed landfills. About 35 percent is disposed of in diked surface impoundments such as pits, ponds, lagoons, and basins. About 60 percent is disposed of deep underground in underground injection wells. Waste piles, which are noncontainerized accumulations of solid hazardous waste typically used for temporary storage, account for less than 1 percent of our disposal volume (USEPA, 1986b).

Historically, land disposal has been the traditional method of getting rid of hazardous wastes in this country. Unfortunately, many of these disposal sites have been poorly engineered and monitored and the results have sometimes been tragic, as was the case at Love Canal, New York. They have been used extensively in the past because they were the most convenient and inexpensive method of disposal. However, remediation at older sites that have leaked toxics into the soil and groundwater has proven to be tremendously costly and the originally perceived economic advantage of land disposal is now seen to have been short-sighted.

As was mentioned in Chapter 5, the 1984 amendments to RCRA ban unsafe, untreated wastes from land disposal. The amendments require that the EPA assess all hazardous wastes to determine whether they should be banned. If they are banned, the EPA must determine the level of treatment that would be required before land disposal could be allowed. RCRA goes on to provide new restrictions and standards for those land disposal facilities that will be allowed to accept hazardous substances, including (USEPA, 1986b):

- Banning liquids from landfills.
- Banning underground injection of hazardous waste within 1/4-mile of a drinking water well.
- Requiring more stringent structural and design conditions for landfills and surface impoundments, including two or more liners, leachate collection systems above and between the liners, and groundwater monitoring.
- Requiring cleanup or corrective action if hazardous waste leaks from a facility.
- Requiring information from disposal facilities on pathways of potential human exposure to hazardous substances.
- Requiring location standards that are protective of human health and the environment; for example, allowing disposal facilities to be constructed only in suitable hydrogeologic settings.

Landfills. In accordance with these new, more stringent RCRA requirements, the design and operation of hazardous waste landfills has become much more sophisticated. A hazardous waste landfill is now designed as a modular series of three-dimensional control cells. By incorporating separate cells it becomes possible to segregate wastes so that only compatible wastes are disposed of together. Arriving wastes are placed in an appropriate cell and covered at the end of each working day with a layer of cover soil.

Beneath the hazardous wastes there must be a double-liner system to stop the flow of liquids, called *leachate,* from entering the soil and groundwater beneath the site. The upper liner must be a *flexible-membrane lining* (FML), usually made of sheets of plastic or rubber. Commonly used plastics include polyvinyl chloride (PVC), high-density polyethylene (HDPE), and chlorinated polyethylene (CPE). Rubber FMLs include chlorosulfonated polyethylene (CSPE) and ethylene propylene diene monomer (EPDM). Depending on the material chosen for the FML, the thickness is typically anywhere from 0.25 mm (10 mils) to over 2.5 mm (100 mils). The lower liner is usually an FML, but recompacted clay at least 3 ft thick is also considered acceptable.

Leachate that accumulates above each liner is collected in a series of perforated drainage pipes and pumped to the surface for treatment. To help reduce the amount of leachate formed by precipitation seeping into the landfill, a low permeability cap is placed over completed cells. When the landfill is finally closed, a cap that may consist of an FML along with a layer of compacted clay is placed over the entire top, with enough slope to assure drainage away from the wastes.

The landfill must also include monitoring facilities. The groundwater flowing beneath the site should be tested with monitoring wells placed upgradient and downgradient from the site. There may need to be only one upgradient well to test the ''natural'' quality of the groundwater before it flows under the site, but there should be at least three or more monitoring wells placed downgradient to assure detection of any leakage from the site. In addition, the soil under the site, above the water table, should be tested using devices called a suction lysimeters.

A cross section of a completed hazardous waste landfill is shown in Figure 6.17.

Surface Impoundments.   Surface impoundments are excavated or diked areas used to store liquid hazardous wastes. Usually storage is temporary unless the impoundment has been designed to eventually be closed as a landfill. Impoundments have been popular because they have been cheap and because wastes remain accessible, allowing some treatment to take place during storage. Typical treatment technologies used in surface impoundments include neutralization, precipitation, settling, and biodegradation.

Historically, surface impoundments have typically been poorly constructed and monitored. In a survey of 180 000 surface impoundments, the EPA estimated that prior to 1980 only about one-fourth were lined and fewer than 10 percent had monitoring programs (USEPA, 1984). The same survey also found that surface impoundments were usually poorly sited. More than half were located over very thin or very permeable soils that would allow easy transport of leachate to ground-

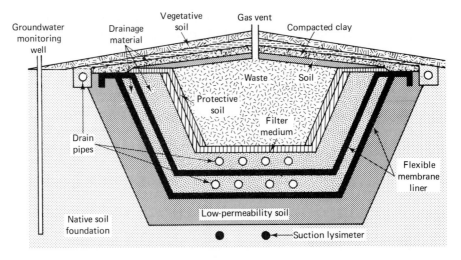

Figure 6.17 Schematic of a hazardous waste landfill.

water. Over three-fourths of the impoundments were located over very thick and permeable aquifers that would allow relatively rapid dispersion of contaminants should they reach the water table. Moreover, about 98 percent of the surface impoundments were located less than one mile from sources of high-quality drinking water.

As a result of these poor siting, construction, and management problems, surface impoundments are the principal source of contamination in a large number of Superfund sites. Recent EPA regulations require new surface impoundments, or expansions to existing impoundments, to have two or more liners, a leachate-collection system, and monitoring programs similar to those required for landfills. However, the legacy of past practices will undoubtedly take billions of dollars and decades of time to remediate.

Underground Injection.   The most popular way to dispose of liquid hazardous wastes has been to force them underground through deep injection wells (Figure 6.18). To help assure that underground drinking water supplies will not become contaminated, injection wells used to dispose of hazardous industrial wastes are required to extend below the lowest formation containing underground sources of drinking water. Typical injection depths are more than 700 m below the surface. Since the main concern with underground injection is the potential for contaminating underground drinking water supplies, the regulation of such systems has come under the Safe Drinking Water Act of 1974.

Unfortunately, a number of hazardous waste injection wells have had leakage problems, so such wells cannot be considered entirely safe. Regulations covering construction, operation, and monitoring of injection wells are becoming more stringent, and, as is the case for all land disposal options, continued reliance on this technology is being discouraged.

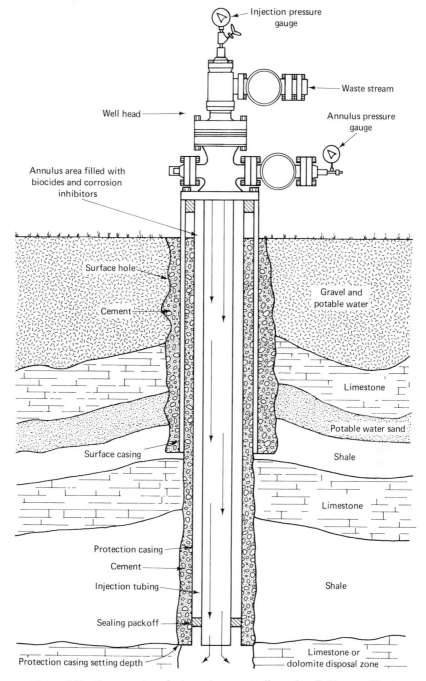

Figure 6.18   Cross section of a hazardous waste disposal well. (*Source:* Wentz, C. A. *Hazardous Waste Management*. Copyright © 1989 by McGraw-Hill, Inc. Used with permission of the publisher.)

PROBLEMS

6.1. A sample of groundwater has 150 mg/L of Ca^{2+} and 60 mg/L of Mg^{2+}. Find the total hardness expressed in milliequivalents per liter (meq/L) and mg/L as $CaCO_3$. Using Table 6.5, how would this water be classified (e.g., soft, hard, . . .)?

6.2. For a solution with pH equal to 9.0, express the concentrations of H^+ and OH^- in meq/L and mg/L as $CaCO_3$.

6.3. A sample of water at pH 10.5 has 39.0 mg/L of CO_3^{2-} and 24.5 mg/L of HCO_3^-.
  a. Ignoring the contribution of $[H^+]$ and $[OH^-]$ to alkalinity, what is the alkalinity as $CaCO_3$?
  b. Including the contribution of $[H^+]$ and $[OH^-]$, find the alkalinity as $CaCO_3$.

6.4. A sample of water has the following concentrations of ions (and the pH is near neutral):

Cations	mg/L	Anions	mg/L
Ca^{2+}	95.0	HCO_3^-	160.0
Mg^{2+}	26.0	SO_4^{2-}	135.0
Na^+	15.0	Cl^-	73.0

  a. What is the total hardness (TH)?
  b. What is the carbonate hardness (CH)?
  c. What is the noncarbonate hardness (NCH)?
  d. What is the alkalinity?
  e. What is the total dissolved solids concentration?
  f. Draw an ion concentration bar graph.

6.5. A sample of water has the following concentrations of ions:

Cations	mg/L	Anions	mg/L
Ca^{2+}	40.0	HCO_3^-	110.0
Mg^{2+}	10.0	SO_4^{2-}	67.2
Na^+	?	Cl^-	11.0
K^+	7.0		

  a. Assuming no other constituents are missing, use an anion–cation balance to estimate the concentration of Na^+.
  b. What is the total hardness (TH)?
  c. Draw an ion concentration bar graph.

REFERENCES

DAVIS, M. L., and D. A. CORNWELL, 1985, *Introduction to Environmental Engineering,* PWS Engineering, Boston, MA.

FREEMAN, H. M. (ed.), 1989, *Standard Handbook of Hazardous Waste Treatment and Disposal,* McGraw Hill, New York.

IRVINE, R. L., and P. A. WILDERER, 1989, Aerobic processes, in *Standard Handbook of Hazardous Waste Treatment and Disposal,* H. M. Freeman (ed.), McGraw Hill, New York.

LINSLEY, R. K., and J. B. FRANZINI, 1979, *Water-Resources Engineering,* 3d ed., McGraw Hill, New York.

MORRISON, A., 1983, In third world villages, a simple handpump saves lives, *Civil Engineering/ASCE* October: 68–72.

ROBECK, G. G., N. A. CLARKE, and K. A. DOSTAL, 1962, Effectiveness of water treatment processes in virus removal, *Journal American Water Works Association,* 54:1275–1290.

TCHOBANOGLOUS, G., and E. D. SCHROEDER, 1985, *Water Quality,* Addison-Wesley, Reading, MA.

USEPA, 1976, *National Interim Primary Drinking Water Regulations,* Office of Water Supply, EPA-570/9-76-003, Washington, DC.

USEPA, 1980, *Summary Report, Control and Treatment Technology for Metal Finishing Industry,* EPA 625/8-80-003, Washington, DC.

USEPA, 1984, *Surface and Impoundment Assessment National Report,* EPA-570/9-84-002, Washington, DC.

USEPA, 1986a, Drinking water in America: An overview, *EPA Journal,* September.

USEPA, 1986b, *Solving the Hazardous Waste Problem: EPA's RCRA Program,* Office of Solid Waste, Washington, DC.

USEPA, 1987a, *The Hazardous Waste System,* Environmental Protection Agency, Office of Solid Waste and Emergency Response, Washington, DC.

USEPA, 1987b, *A Compendium of Technologies Used in the Treatment of Hazardous Wastes,* Center for Environmental Research Information, EPA/625/8-87/014, Washington, DC.

USEPA, 1988a, Technology Screening Guide for Treatment of CERCLA Soils and Sludges, Office of Solid Waste and Emergency Response, EPA/502/2-88/004, Washington, DC.

USEPA, 1988b. *Hazardous Waste Incineration: Questions and Answers,* Office of Solid Waste, EPA/530-SW-88-018, Washington DC.

VIESSMAN, W., Jr., and M. J. HAMMER, 1985, *Water Supply and Pollution Control,* 4th ed., Harper & Row, New York.

WENTZ, C. A., 1989, *Hazardous Waste Management,* McGraw Hill, New York.

CHAPTER 7

Air
Pollution

*The air of Athens is pure, whence the inhabitants have
more piercing apprehension and quicker reflexes than the
rest of Greece. Thebes, on the other hand, is thick and
foggy, its inhabitants dull and slow.*
Cicero

7.1 INTRODUCTION

Air pollution is certainly not a new phenomenon. Indeed, early references to it
date to the Middle Ages, when smoke from burning coal was already considered
such a serious problem that in 1307, King Edward I banned its use in lime kilns in
London. In more recent times, though still decades ago, several serious episodes
focused attention on the need to control the quality of the air we breathe. The
worst of these occurred in London, in 1952. A week of intense fog and smoke
resulted in over 4000 excess deaths that were directly attributed to the pollution.
In the United States the most alarming episode occurred during a 4-day period in
1948 in Donora, Pennsylvania, when 20 deaths and almost 6000 illnesses were
linked to air pollution. At the time, Donora had a population of only 14 000,
making this the highest per capita death rate ever recorded for an air pollution
episode.

Those air pollution episodes were the result of exceptionally high concentrations of sulfur oxides and particulate matter, the primary constituents of *industrial
smog* or *sulfurous smog*. Sulfurous smog is caused almost entirely by combustion

of fossil fuels, especially coal, in stationary sources such power plants and smelters. In contrast, the air pollution problem in many cities is caused by emissions of carbon monoxide, oxides of nitrogen, and various hydrocarbons, that swirl around in the atmosphere reacting with each other and with sunlight to form *photochemical smog*. Although stationary sources also contribute to photochemical smog, the problem is most closely associated with motor vehicles.

Much of the work on air pollution in the last few decades has centered on a small set of compounds, called *criteria pollutants,* that have been identified as contributors to both sulfurous and photochemical smog problems. The sources, transport, effects, and methods of controlling these criteria pollutants will be the focus of this chapter. More recently, the importance of hazardous air pollutants has begun to be appreciated, and one such pollutant, radon gas, is receiving considerable attention. Radon is just one of a number pollutants that we are exposed to inside of buildings. Indoor air pollution is now recognized as one of the most important routes for human exposure to a number of contaminants, including radon, and it will be touched upon in this chapter.

In the next chapter we will discuss the growing problems of global climate change and stratospheric ozone depletion caused by accumulations of carbon dioxide and other trace gases that are not usually thought of as air pollutants since they do not pose direct threats to human health.

7.2 SUMMARY OF THE CLEAN AIR ACT

Initial efforts on the part of the U.S. Congress to address the nation's air pollution problem began with the passage of the Air Pollution Control Act of 1955. Although it provided funding only for research, and not control, it was an important milestone because it opened the door to federal participation in efforts to deal with air pollution. Up until that time, air pollution had been thought to be a state and local problem. This was followed by a series of legislative actions by Congress that included the Clean Air Act of 1963, the Motor Vehicle Air Pollution Control Act of 1965, the Air Quality Act of 1967, the Clean Air Act Amendments of 1970, the Energy Supply and Environmental Coordination Act of 1974, and the Clean Air Act Amendments of 1977, all of which are sometimes lumped together and referred to as simply the *Clean Air Act* (CAA).

Much of the real structure to the Clean Air Act was established in the 1970 Amendments. In those amendments, the Environmental Protection Agency (EPA) was required to establish *National Ambient Air Quality Standards* (NAAQS), and states were required to submit *State Implementation Plans* (SIPs) that would show how they would meet those standards. In addition, the Act required *New Source Performance Standards* (NSPS) to be established that would limit emissions from certain specific types of industrial plants. Emission standards were also written for mobile sources that required manufacturers to reduce emissions from new cars by 90 percent.

There is a difference between these two types of standards that should be noted. Ambient air quality standards are acceptable *concentrations* of pollution in the atmosphere, while emission standards are allowable *rates* at which pollutants can be released from a source.

Air Quality Standards

National Ambient Air Quality Standards have been established by the EPA at two levels: *primary* and *secondary*. Primary standards are required to be set at levels that will protect public health and include an "adequate margin of safety," regardless of whether the standards are economically or technologically achievable. Primary standards must protect even the most sensitive individuals, including the elderly and those with respiratory ailments. NAAQS are, therefore, conceptually different from maximum contaminant levels (MCLs) that have been set for drinking water. Recall that the Safe Drinking Water Act requires EPA to balance public health benefits with technological and economic feasibility in establishing drinking water MCLs.

Secondary air quality standards are meant to be even more stringent than primary standards. Secondary standards are established to protect public welfare (e.g., structures, crops, animals, fabrics). Given the difficulty in achieving primary standards, secondary standards have played almost no role in air pollution control policy and, in fact, they have usually been set at the same levels as primary standards.

National Ambient Air Quality Standards now exist for six *criteria* pollutants: carbon monoxide, lead, nitrogen dioxide, ozone, sulfur dioxide, and particulates with aerodynamic diameter less than or equal to a nominal 10 μm (PM-10). The law requires that the list of criteria pollutants be reviewed periodically and that standards be adjusted according to the latest scientific information. Past reviews have modified both the list of pollutants and their acceptable concentrations. For example, the older standard for a general category of oxidants has been replaced by a more specific ozone standard. Also, the original particulate standard did not refer to the size of particulates, but as of 1987 it does. Lead has been added to the list, and a broad category of hydrocarbons has been dropped. Table 7.1 lists the current (1990) status of the NAAQS.

For a given region of the country to be in compliance with primary ambient air quality standards, the concentrations given in Table 7.1 cannot be exceeded more than once per calendar year. It should be noted that the law allows states to establish standards that are more stringent than the NAAQS, which California has done. California's standards are also shown in Table 7.1.

Table 7.1 gives concentrations in terms of weight per unit volume (μg/m^3 or mg/m^3) at an assumed temperature of 25 °C and 1 atm of pressure. For gases in the list, the concentrations are also expressed in parts per million (ppm), which are volumetric units. The conversion between units was discussed in Section 1.2, and the following example illustrates the procedure.

TABLE 7.1 NATIONAL AMBIENT AIR QUALITY STANDARDS (NAAQS) AND CALIFORNIA STATE STANDARDS

Pollutant	Averaging time	Federal primary	Federal secondary	California[a]	Objective
Carbon monoxide	8 hr	10 mg/m³ (9 ppm)	None	None	Limit carboxyhemoglobin
	1 hr	40 mg/m³ (35 ppm)	None	23 mg/m³	
Nitrogen dioxide	Annual	100 μg/m³ (0.053 ppm)	Same	None	Prevent health risk and improve visibility
	1 hr	None	None	470 μg/m³	
Ozone	1 hr	235 μg/m³ (0.12 ppm)	Same	200 μg/m³	Prevent eye irritation, breathing difficulties
Sulfur dioxide	Annual	80 μg/m³ (0.03 ppm)	None	None	Prevent increase in respiratory disease, plant damage and odor
	24 hr	365 μg/m³ (0.14 ppm)	None	131 μg/m³	
	3 hr	None	1310	None	
	1 hr	None	None	1310 μg/m³	
PM-10	Annual	50 μg/m³	Same	30 μg/m³	Improve visibility and prevent health effects
	24 hr	150 μg/m³	Same	50 μg/m³	
Lead	1 month	None	None	1.5 μg/m³	Prevent health problems
	3 months	1.5 μg/m³	Same	None	

[a] California also has air quality standards for hydrogen sulfide (42 μg/m³, 1 hr), vinyl chloride (26 μg/m³, 24 hr), and ethylene (0.1 ppm, 8 hr; 0.5 ppm, 1 hr).

Example 7.1   Air Quality Standards Expressed in Volumetric Units

California's air quality standard for nitrogen dioxide (NO_2) is 470 μg/m³ (at a temperature of 25 °C and 1 atm of pressure). Express the concentration in ppm.

Solution Recall that one mole of an ideal gas at 1 atm and 25 °C occupies a volume of 24.45 L (24.45×10^{-3} m³).

The molecular weight of NO_2 is

$$\text{mol wt} = 14 + 2 \times 16 = 46$$

so that

$$NO_2 = \frac{24.45 \times 10^{-3} \text{ m}^3/\text{mol} \times 470 \times 10^{-6} \text{ g/m}^3}{46 \text{ g/mol}}$$

$$= 0.25 \times 10^{-6} = 0.25 \text{ ppm}$$

One advantage of volumetric units is that a given concentration expressed in ppm does not change as temperature and pressure vary, whereas with mixed units (μg/m³), it does.

The NAAQS form the basis for an air pollution index, the *Pollutant Standards Index* (PSI), that has been adopted by the EPA and is used by many cities to report to the public an overall assessment of a given day's air quality. The PSI converts air pollution concentrations to a simple number between zero and 500 and assigns a descriptive term such as "good" or "moderate" to that value. A PSI value of 100 indicates that at least one pollutant reached its ambient air quality standard on that day. The PSI also triggers public health warnings on smoggy days. A PSI of 200 corresponds to a First Stage Alert during which time elderly persons with existing heart or lung disease are advised to stay indoors and reduce their physical activity. A Second Stage Alert is called when the PSI reaches 300, at which point the general public is advised to avoid outdoor activity. The descriptive terms corresponding to PSI numbers are given in Table 7.2.

TABLE 7.2 PSI VALUES AND
AIR QUALITY DESCRIPTORS

PSI value	Descriptor
0–50	Good
51–100	Moderate
101–199	Unhealthful
200–299	Very unhealthful
≥300	Hazardous

The actual calculation of a day's PSI value is based on Table 7.3, which shows individual PSI numbers corresponding to various pollutant concentrations. Individual PSI subindexes are computed for each of the pollutants in the table using linear interpolation between the indicated breakpoints. The highest PSI subindex determines the overall PSI. Note the particulate concentration in Table 7.3 is designated as TSP, which stands for *total suspended particulates*. In addition, a column corresponding to the product of TSP and SO_2 is included in the table. These two pollutants act synergistically, that is, the effect of both together is more than the sum of the individual effects. An example PSI calculation is given below.

TABLE 7.3 POLLUTANT STANDARDS INDEX (PSI) BREAKPOINTS

Index	1-hr O_3 $\mu g/m^3$	8-hr CO mg/m^3	24-h TSP $\mu g/m^3$	24-hr SO_2 $\mu g/m^3$	TSP × SO_2 10^3 $(\mu g/m^3)^2$	1-hr NO_2 $\mu g/m^3$
0	0	0	0	0	—	—
50	118	5	75	80	—	—
100	235	10	260	365	—	—
200	400	17	375	800	65	1130
300	800	34	625	1600	261	2260

Source: 40 CFR (Code of Federal Regulations) 58, 1982.

Example 7.2   Determining the PSI

Suppose on a given day the following maximum concentrations are measured:

$$
\begin{array}{ll}
\text{1-hr } O_3 & 250 \ \mu g/m^3 \\
\text{8-hr CO} & 10 \ mg/m^3 \\
\text{24-hr TSP} & 50 \ \mu g/m^3 \\
\text{24-hr } SO_2 & 100 \ \mu g/m^3
\end{array}
$$

Find the PSI and indicate the descriptor that would be used to characterize the day's air quality.

Solution   Using Table 7.3, it can be seen that the ozone level (O_3) yields a subindex over 100; CO yields an index of 100; TSP yields an index of less than 50; and SO_2 yields an index of less than 100. The product of TSP and SO_2 is $50 \times 100 = 5 \times 10^3$ $(\mu g/m^3)^2$, which is below the lowest value in the table, so there is no subindex for that column. The highest subindex therefore corresponds to O_3. To calculate the PSI, we must interpolate. An ozone concentration of 235 $\mu g/m^3$ corresponds to an index of 100, while a value of 400 $\mu g/m^3$ corresponds to 200. By interpolation, the measured ozone concentration of 250 $\mu g/m^3$ yields an overall PSI value of

$$
PSI = 100 + \frac{(250 - 235)}{(400 - 235)} \times (200 - 100) = 109
$$

so the air quality would be described as unhealthful.

Emission Standards

Besides establishing National Ambient Air Quality Standards, the Clean Air Act also requires EPA to establish emission standards for certain industries. *New Source Performance Standards* (NSPS) have been promulgated for a large number of stationary sources such as fossil-fuel-fired power plants, incinerators, Portland cement plants, nitric acid plants, petroleum refineries, sewage treatment plants, and smelters of various sorts. As an example, the NSPS for electric power plants is given in Table 7.4. Notice the allowable emissions are expressed as pounds of pollutant per million BTU of heat input to the power plant. The emission standard for nitrogen is written in terms of total oxides of nitrogen (NO_x), which is composed mostly of nitric oxide (NO) with some nitrogen dioxide (NO_2).

The emission standard for SO_2 is complicated for coal-fired, steam-electric power plants and is best illustrated using Figure 7.1 and an example. In Figure 7.1, a region of allowable SO_2 emissions is identified. The sulfur content (percent sulfur) of the fuel and the fuel's energy content (Btu/lb) are parameters used to determine a point on one of the indicated arcs. A line drawn from the origin to the sulfur/heat point crosses the "admissible region" boundary at one point, corresponding to the allowable SO_2 emission rate.

TABLE 7.4 NEW SOURCE PERFORMANCE STANDARDS (NSPS) FOR STEAM ELECTRIC
POWER PLANTS (lb/10^6 Btu HEAT INPUT)

Particulates	0.03 lb/10^6 Btu of heat input (13 g/10^6 kJ)
Nitrogen oxides (NO$_x$)	
Gas-fired	0.20 lb/10^6 Btu (86 g/10^6 kJ)
Oil-fired	0.30 lb/10^6 Btu (130 g/10^6 kJ)
Coal-fired[a]	0.60 lb/10^6 Btu (260 g/10^6 kJ)
Sulfur dioxide (SO$_2$)	
Gas or Oil	0.20 lb/10^6 Btu (86 g/10^6 kJ) for gas or oil.
Coal-fired	Allowable emission rates are based on the sulfur content and heating value of the fuel. Controls must reduce emissions by at least 70%. If emission reduction is less than 90%, emissions cannot exceed 0.6 lb/10^6 Btu; above 90%, emissions cannot exceed 1.2 lb/10^6 Btu. See Figure 7.1.

[a] Anthracite or bituminous coal.
Source: 40 CFR 60, 1982.

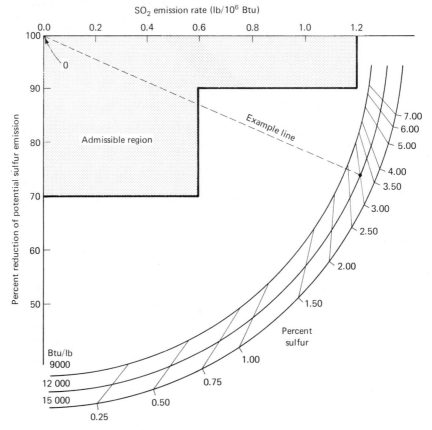

Figure 7.1 Graphical representation of new source performance standard for SO$_2$ emissions from coal-fired power plants. (*Source:* Molburg, 1980.)

Example 7.3   Sulfur Emissions from a New Coal-Fired Power Plant

A new 1000 megawatt (MW) coal-fired power plant is 40 percent efficient. The coal to be burned contains 3 percent sulfur and has a heat content of 12 000 Btu/lb. Find the allowable emission rate of SO_2 and the percent reduction in SO_2 that an emission control system will have to achieve.

Solution   First we must determine the rate at which heat enters the power plant. Since it is 40 percent efficient, and it delivers 1000 MW of electricity (1.0×10^6 kW), the input heat rate is

$$Q_{in} = \frac{Q_{out}}{\eta} = \frac{1.0 \times 10^6 \text{ kW}}{0.40} = 2.5 \times 10^6 \text{ kW}$$

or, in terms of Btu/day

$$Q_{in} = 2.5 \times 10^6 \text{ kW} \times \frac{3412 \text{ Btu}}{\text{kWhr}} \times \frac{24 \text{ hr}}{\text{day}} = 2.05 \times 10^{11} \text{ Btu/day}$$

From Figure 7.1, at 3 percent sulfur and 12 000 Btu/lb of coal, the maximum allowable emission rate is 0.6 lb $SO_2/10^6$ Btu and controls must achieve an SO_2 reduction of about 87 percent. Total daily emissions would be

$$SO_2 \text{ emission rate} = \frac{0.6 \text{ lb}}{10^6 \text{ Btu}} \times \frac{2.05 \times 10^{11} \text{ Btu}}{\text{day}} = 123,000 \text{ lb/day}$$

which is still a sizeable amount.

The Clean Air Act Amendments of 1977

The goal of the 1970 Amendments was to attain clean air by 1975, as defined by the NAAQS, with allowable extensions in certain circumstances until 1977. For a number of reasons, only about one-third of the air quality control regions in the nation were meeting the standards by 1977. This forced Congress to readdress the problem, establishing the Clean Air Act Amendments of 1977. Besides extending the deadlines, the 1977 Amendments had to deal with two important questions. First, what measures should be taken in *nonattainment areas* that were not meeting the standards? Second, should air quality in regions where the air is cleaner than the standards be allowed to degrade toward the standards, and if so, by how much?

For nonattainment areas, the 1970 Act appeared to prohibit any increase in emissions whatsoever, which would have eliminated industrial expansion and severely curtailed local economic growth. To counter this, the EPA adopted a policy of *emission offsets*. To receive a construction permit, a major new source of pollution in a nonattainment area must first find ways to reduce emissions from existing sources. The reductions, or offsets, must exceed the anticipated emissions from the new source. The net effect of this offset policy is that progress is made toward meeting air quality standards in spite of new emission sources being added to the airshed.

Offsets can be obtained in a number of ways. For example, emissions from existing sources in the area might be reduced by installing better emission controls on equipment that may, or may not, be owned by the permit seeker. In some cases, a permit seeker may simply buy out existing emission sources and shut them down. Emission offsets can be "banked" for future use or they can be sold or traded to other companies for whatever the market will bear. In addition to offsets, new sources in nonattainment areas must use emission controls that yield the *Lowest Achievable Emission Rate* (LAER) for the particular process. LAER technology is based on the most stringent emission rate achieved in practice by similar sources, regardless of the economic cost or energy impacts.

The 1970 Amendments were not specific about regions that were cleaner than ambient standards required, and in fact appeared to allow air quality to deteriorate to those standards. The 1977 Amendments settled the issue of whether or not this would be allowed by establishing the concept of *prevention of significant deterioration* (PSD) in attainment areas. Attainment areas are put into one of three classes and the amount of deterioration allowed is determined by the class. Class I areas include National Parks and Wilderness Areas, and almost no increase in pollution is allowed. Moderate deterioration is allowed in class II areas, and even greater amounts are allowed in class III areas. In PSD areas, *best available control technology* (BACT) is required on major new sources. BACT is less stringent than LAER, as it does allow consideration of economic, energy, and environmental impacts of the technology.

Table 7.5 shows the incremental increase in pollutant concentration (for various averaging times) allowed for each PSD class in an attainment area. As of 1988, PSD increments had been established only for sulfur dioxide and total suspended particulates (TSP). The nitrogen dioxide increments shown in the table is a proposal. These increments are actually the maximum amounts allowed. If adding the increment to an area's established baseline pollutant concentration results in a value that exceeds the ambient air quality standard, then the increment

TABLE 7.5 MAXIMUM ALLOWABLE PSD INCREMENTS NOT TO BE EXCEEDED MORE THAN ONCE PER YEAR (μg/m^3)

Pollutant	Class I	Class II	Class III
Sulfur dioxide			
Annual	2	20	40
24-hr	5	91	182
3-hr	25	512	700
Total suspended particulates			
Annual	5	19	37
24-hr	10	37	75
Nitrogen dioxide[a]			
Annual	2.5	25	50

[a] Proposed increment.

is reduced accordingly. To demonstrate compliance with these PSD increments and with air quality standards in general, mathematical models predicting ambient pollutant concentrations must be used for any proposed new source. Such models, which use meteorological and stack emission data to predict air quality impacts, will be described in Section 7.5.

In addition to the six *criteria* pollutants for which ambient air quality standards have been written, the Clean Air Act also regulates emissions of hazardous air pollutants. These pollutants include carcinogens, mutagens, and toxic chemicals. As of 1988, the *National Emission Standards for Hazardous Air Pollutants* (NESHAP) program had designated eight pollutants (asbestos, benzene, beryllium, coke oven emissions, inorganic arsenic, mercury, radionuclides, and vinyl chloride) as being hazardous, and another 21 chemicals were under consideration.

Hazardous substances in air are regulated not only by the Clean Air Act but also by the Resource Conservation and Recovery Act (RCRA) and the Comprehensive Environmental Response, Compensation, and Liability Act (CERCLA or Superfund), as was described in Chapter 5.

7.3 CRITERIA POLLUTANTS

As a result of the Clean Air Act, most of the monitoring of emissions, concentrations, and effects of air pollution has been directed toward the criteria pollutants. The EPA has published many volumes of *Air Quality Criteria* documents, which summarize all of the pertinent literature on each of the criteria pollutants. It is from this information that standards have been written. The documents are an invaluable source of detailed information.

Table 7.6 summarizes 1986 U.S. pollutant emissions by source category and compares the total to that released in 1970. The emissions correspond to the criteria pollutants, except *particulates* refers to all particulates regardless of size. Also, a class of pollutants known as volatile organic compounds (VOCs) substitutes for ozone. Ozone is a *secondary* pollutant; that is, it is not actually emitted

TABLE 7.6 U.S. EMISSION ESTIMATES, 1986 (10^{12} g/yr)

Source	Particulates	SO_x	NO_x	VOC	CO	Lead
Transportation	1.4	0.9	8.5	6.5	42.6	0.0035
Stationary source fuel combustion	1.8	17.2	10.0	2.3	7.2	0.0005
Industrial processes	2.5	3.1	0.6	7.9	4.5	0.0019
Solid waste disposal	0.3	0.0	0.1	0.6	1.7	0.0027
Miscellaneous	0.8	0.0	0.1	2.2	5.0	0.0000
Total (1986)	6.8	21.2	19.3	19.5	60.9	0.0086
Total (1970)	18.5	28.4	18.1	27.5	98.7	0.2038

Source: USEPA (1988a).

but rather is formed by reactions that take place in the atmosphere. Volatile organic compounds are inputs to the photochemical reactions that produce ozone, so they are used as indicators of the potential for ozone formation.

Perhaps the most important feature in Table 7.6 is the significant decrease in emissions in all categories (except for NO_x) between the year in which the 1970 Amendments to the Clean Air Act were passed and 1986. As will be seen later, NO_x has been a particularly hard pollutant to control, especially in motor vehicles. Early controls that reduced hydrocarbon and carbon monoxide emissions in cars actually increased NO_x emissions.

The transportation sector, which includes highway vehicles, aircraft, railroads, and ships, contributes by far the most carbon monoxide, and is a major source of VOCs and NO_x as well. Stationary source fuel combustion emissions come from electric utilities, industrial boilers, and space conditioning equipment in buildings. This category is the dominant source of SO_x and a major source of NO_x. The industrial processes category includes all noncombustion operations that occur in such fields as steel, plastics, and cement production, as well as petroleum refining, transport, and storage. As can be seen, industrial processes are significant sources of particulates and VOCs (much of the VOC emissions are caused by evaporation).

In spite of significant progress that has been made as a result of the Clean Air Act, EPA estimated that as of 1986, 75 million people still were living in counties with measured air quality levels that violated the NAAQS for ozone, 41.7 million were living in air that violated NAAQS for particulates (based on the old measurement technique, which did not account for particle size), 41.4 million for carbon monoxide, 7.5 million for nitrogen dioxide, 4.5 million for lead, and 0.9 million for sulfur dioxide (USEPA, 1988b) (see Figure 7.2).

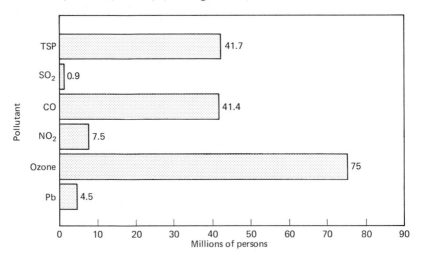

Figure 7.2   Number of persons living in counties with air quality levels above the primary National Ambient Air Quality Standards in 1986 (USEPA 1988b).

Carbon Monoxide

Carbon monoxide is a colorless, odorless, tasteless gas that is by far the most abundant of the criteria pollutants, as Table 7.6 indicates. It is produced when carbonaceous fuels are burned under less than ideal conditions. Incomplete combustion, yielding CO instead of CO_2, results when any of the following four variables are not kept sufficiently high: (1) oxygen supply, (2) flame temperature, (3) gas residence time at high temperature, and (4) combustion chamber turbulence. These parameters are generally under much tighter control in stationary source combustion facilities than in motor vehicles, and CO emissions are correspondingly less. For example, power plants that are designed and managed for maximum combustion efficiency, produce less than 1/2 percent of all CO emissions in spite of the fact that they consume about 30 percent of our fossil fuel.

About 70 percent of CO emissions are from the transportation sector, with almost all of that coming from vehicles. Hourly atmospheric concentrations of CO often reflect city driving patterns. Peaks occur on weekdays during the morning and late afternoon rush hours, while on weekends there is typically but one lower peak in the late afternoon. Personal exposure to CO is very much determined by the proximity of heavy motor vehicle traffic, with some occupational groups such as cab drivers, police, and parking lot attendants receiving far higher than average doses. Figure 7.3, which shows trends in CO emissions from 1940 to 1986, clearly shows the impact of the Clean Air Act, especially on the transportation sector.

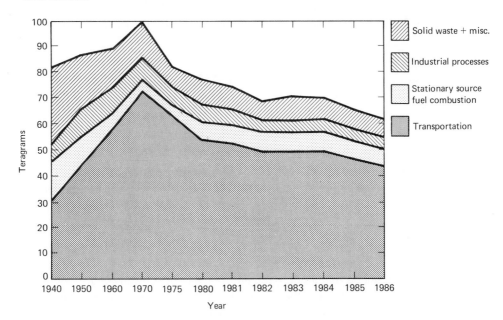

Figure 7.3   Trends in carbon monoxide emissions, 1940–1986 (USEPA, 1988a).

At levels of CO that occur in urban air, there are apparently no detrimental effects on materials or plants; those levels can, however, adversely affect human health. Carbon monoxide interferes with the blood's ability to carry oxygen to the cells of the body. When inhaled, it readily binds to hemoglobin in the bloodstream to form carboxyhemoglobin (COHb). Carbon monoxide, in fact, has a much greater affinity for hemoglobin than does oxygen, so that even small amounts of CO can seriously reduce the amount of oxygen conveyed throughout the body. With the bloodstream carrying less oxygen, brain function is affected and heart rate increases in an attempt to offset the oxygen deficit.

The usual way to express the amount of carboxyhemoglobin in the blood is as a percentage of the saturation level, %COHb. The amount of COHb formed in the blood is related to three factors: the CO concentration, the length of time exposed, and the breathing rate. The following expression has been used to relate %COHb to these three factors:

$$\%\text{COHb} = 0.005[\text{CO}]^{0.85} (\alpha t)^{0.63} \qquad (7.1)$$

where %COHb = carboxyhemoglobin as a percentage of saturation
  [CO] = carbon monoxide concentration in ppm
  α = a physical activity level coefficient
  t = exposure time in minutes

For sedentary activity, $\alpha = 1$, while for heavy work, a factor of 3 is sometimes used (USHEW, 1970; Dimitriades and Whisman, 1971). Equation 7.1 breaks down for large values of time, corresponding to the bloodstream reaching saturation.

Example 7.4   Federal Standard for CO

Estimate the %COHb expected for a 1-hr exposure to 35 ppm (the federal standard) for sedentary activities, and for heavy work.

Solution   Using Eq. 7.1, we find for sedentary activities with $\alpha = 1$:

$$\%\text{COHb} = 0.005 \times (35)^{0.85} \times (60)^{0.63} = 1.35 \text{ percent}$$

assuming $\alpha = 3$ for heavy activity,

$$\%\text{COHb} = 0.005 \times (35)^{0.85} \times (3 \times 60)^{0.63} = 2.7 \text{ percent}$$

Physiological effects can be noted at small percentages of COHb, increasing in severity as the concentration increases. Individuals with heart conditions are the most sensitive to COHb since the heart must work harder in an attempt to offset the reduction in oxygen. Studies of patients with angina pectoris (a heart condition characterized by chest pain) have shown an earlier than usual onset of pain during exercise when levels are as low as 2 percent COHb. In Example 7.4, it was calculated that an individual engaged in heavy activity for one hour, breathing CO at the federal ambient air quality standard of 35 ppm, would be likely to reach 2.7 percent COHb. That is above the level at which health effects have been noted, and the federal standard has been criticized as a result. California's CO standard has been set lower (20 ppm) in an attempt to assure less than 2 percent COHb.

The reduction of oxygen in the bloodstream also affects the brain's ability to perceive and react. At 2.5 percent COHb, studies have shown an impairment in time-interval discrimination. (Subjects were less able to distinguish the duration of a tone signal). Studies have also shown that at 5 percent, psychomotor response times may be affected and patients with heart disease experience increased physiological stress. At 10 percent COHb, most people will experience dizziness and headache. Concentrations above 50 percent can be lethal.

Carbon monoxide concentrations in urban areas frequently range from 5 to 50 ppm, and measurements made on congested highways indicate that drivers can be exposed to CO concentrations of 100 ppm. Of course, smokers create their own CO problem. Cigarette smoke contains more than 400 ppm CO and smokers frequently have COHb levels between 5 and 10 percent. Tobacco smoke in bars and restaurants often raises indoor CO levels to 20–30 ppm, which is close to the 1-hr ambient standard (Wadden and Scheff, 1983).

Figure 7.4 illustrates the accumulation of COHb with time. It takes approximately 3 or 4 hr for COHb to reach 50 percent of its saturation value. Fortunately, COHb is removed from the bloodstream when clean air is breathed. Healthy subjects clear about half of the CO from their blood in 3–4 hours, so adverse effects are usually temporary.

Oxides of Nitrogen

Although 7 oxides of nitrogen are known to occur, NO, NO_2, NO_3, N_2O, N_2O_3, N_2O_4, and N_2O_5, the only two that are important in the study of air pollution are nitric oxide (NO) and nitrogen dioxide (NO_2). There are two sources of nitrogen

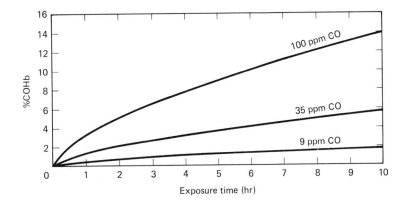

Figure 7.4   Percent COHb plotted from (7.1) for sedentary activity. 35 ppm CO is the federal 1-hr standard; 9 ppm is the 8-hr standard. 100 ppm is not uncommon in heavy traffic.

oxides (or NO$_x$) during the combustion of fossil fuels. *Thermal NO$_x$* are created when nitrogen and oxygen in the combustion air are heated to a high enough temperature (above about 1000 K) to oxidize the nitrogen. *Fuel NO$_x$* result from the oxidation of nitrogen compounds that are chemically bound in the fuel molecules themselves. Different fuels have different amounts of nitrogen in them, with natural gas having almost none and some coal having as much 3 percent by weight. Both thermal NO$_x$ and fuel NO$_x$ can be significant contributors to the total NO$_x$ emissions, but fuel NO$_x$ often is the dominant source.

Almost all NO$_x$ emissions are in the form of NO, which has no known adverse health effects at concentrations found in the atmosphere. However, NO can oxidize to NO$_2$, which, in turn, may react with hydrocarbons in the presence of sunlight to form photochemical smog. That is injurious. Nitrogen dioxide also reacts with the hydroxyl radical (HO) in the atmosphere to form nitric acid (HNO$_3$), which is washed out of the atmosphere as acid rain. The direct health effects of NO$_2$ itself are still uncertain. At higher concentrations than are normally found in the atmosphere, it is an acute irritant. Prolonged exposure to relatively low concentrations, typical of some polluted environments, has been linked to increased bronchitis in children. It also can cause damage to plants, and when converted to nitric acid it leads to corrosion of metal surfaces. NO is a colorless gas, but NO$_2$ gives smog its reddish-brown color.

Reductions in NO$_x$ emissions have been harder to come by than reductions in other criteria pollutants. In fact, emissions were relatively flat through the 1980s, as Figure 7.5 indicates. This is an accomplishment, considering the rapid

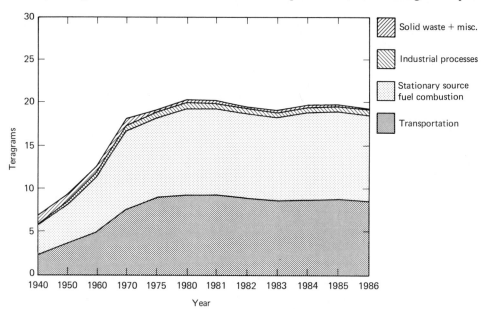

Figure 7.5   Trends in nitrogen oxide emissions, 1940–1986 (USEPA, 1988a).

growth rates that preceded the 1970 Clean Air Act Amendments. As will be discussed further when mobile source controls are introduced, modifications to the combustion process that improve emissions of carbon monoxide tend to make the NO_x problem worse, and vice versa. To control CO, it helps to increase the combustion air supply and to raise the temperature. To control NO_x, the opposite is true.

Photochemical Smog and Ozone

When oxides of nitrogen, various hydrocarbons, and sunlight come together, they can initiate a complex set of reactions that produce a number of secondary pollutants known as photochemical oxidants. Ozone (O_3) is the most abundant of the photochemical oxidants, and it is the one for which an ambient air quality standard has been written. Although it is responsible for many of the undesirable properties of photochemical smog, from chest constriction and irritation of the mucous membrane in people, to the cracking of rubber products and damage to vegetation, it is not itself a cause of the eye irritation that is our most common complaint about smog. Eye irritation is caused by other components of photochemical smog, principally formaldehyde, peroxybenzoyl nitrate (PBzN), peroxyacetyl nitrate (PAN), and acrolein.

In the very simplest of terms, we can express the formation of photochemical smog as

$$\text{Hydrocarbons} + NO_x + \text{Sunlight} \longrightarrow \text{Photochemical smog} \qquad (7.2)$$

The reaction in (7.2) gives us only the simplest overview. We can add a few details to give a sense of the key reactions involved, but a complete analysis is far beyond the scope of this book.

The NO–NO_2–O_3 Photochemical Reaction Sequence

Consider some of the important reactions involving NO_x without the complications associated with the added hydrocarbons. We can begin with the formation of NO during combustion (for simplicity, we shall just show the thermal NO_x reaction):

$$N_2 + O_2 \longrightarrow 2NO \qquad (7.3)$$

The nitric oxide thus emitted can oxidize to NO_2:

$$2NO + O_2 \longrightarrow 2NO_2 \qquad (7.4)$$

If sunlight is available, NO_2 can photolyze, and the freed atomic oxygen can then help to form ozone, as the following pair of reactions suggests:

$$NO_2 + h\nu \longrightarrow NO + O \qquad (7.5)$$

$$O + O_2 + M \longrightarrow O_3 + M \qquad (7.6)$$

where $h\nu$ represents a photon ($\lambda < 0.38\ \mu$m), and M represents a molecule (usually O_2 or N_2 since they are most abundant in air) whose presence is necessary to absorb excess energy from the reaction. Without M, the ozone would have too much energy to be stable, and it would dissociate back to O and O_2.

Ozone can then convert NO back to NO_2:

$$O_3 + NO \longrightarrow NO_2 + O_2 \qquad (7.7)$$

This set of reactions creates a cycle that is represented in Figure 7.6. If these were the only reactions involved, and if they were to reach steady state, the rates of production and destruction of each of the key players, NO, NO_2, and O_3, would be balanced, and there would be no net change in their concentrations over time.

Figure 7.6, even without consideration of the effect of hydrocarbons, helps explain the sequence of stages through which atmospheric NO, NO_2, and O_3 progress on a typical smoggy day. The diagram suggests that we might expect NO concentrations to rise as early morning traffic emits its load of NO. Then, as the morning progresses, we would expect to see a drop in NO and a rise in NO_2 as NO gets converted to NO_2. As the sun's intensity increases toward noon, the rate of photolysis of NO_2 increases; thus NO_2 begins to drop while O_3 rises. Ozone is so effective in its reaction with NO (7.7) that as long as there is O_3 present, NO concentrations do not rise through the rest of the afternoon even though there may be new emissions.

The nitrogen dioxide photolytic cycle helps provide an explanation for the sequence: NO to NO_2 to O_3, observable in both laboratory smog chambers and in cities such as Los Angeles (see Figure 7.7). A careful analysis of the reactions, however, predicts O_3 concentrations that are much lower than those frequently found on smoggy days. If only the NO_2 photolytic cycle is involved, O_3 cannot accumulate in sufficient quantity to explain actual measured data. By introducing certain types of hydrocarbons into the cycle, however, the balance of production and destruction of O_3 can be upset, allowing more O_3 to accumulate and thus explaining the discrepancy.

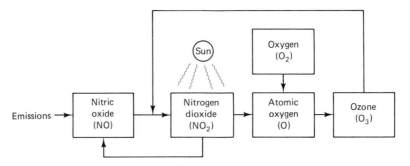

Figure 7.6   Simplified atmospheric nitrogen photolytic cycle.

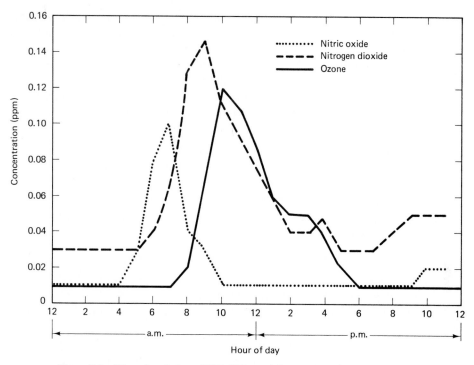

Figure 7.7   Diurnal variation of NO, NO_2, and O_3 concentrations in Los Angeles on July 19, 1965 (USHEW, 1970).

Hydrocarbons and NO_x

Expanding the nitrogen dioxide photolytic cycle to include hydrocarbons and other organics helps explain the increase in ozone above what would be predicted by the NO_x cycle alone. It also enables us to account for some of the other objectionable organic compounds found in photochemical smog.

The chemistry of photochemical smog is exceedingly complex. A multitude of organic chemicals are introduced to the atmosphere when fuels burn or volatiles evaporate, and many more are produced in the atmosphere as a result of chemical reactions. To help describe some of these reactions, let us begin with a very brief explanation of some of the nomenclature and notation used in organic chemistry, building upon what was introduced in Chapter 2.

As described in Section 2.4, the *alkanes* are hydrocarbons in which each carbon forms single bonds with other atoms. The alkane series is the familiar methane (CH_4), ethane (C_2H_6), propane (C_3H_8), and so on, and is represented in general by ($C_n H_{2n+2}$). If one of the hydrogens is removed from an alkane, the resulting *free radical* is called an *alkyl*. The alkyls then form a series beginning with methyl ($CH_3\cdot$), ethyl ($C_2H_5\cdot$), and so on, where the dot indicates an unpaired electron. We could represent an alkyl with the general expression $C_nH_{2n+1}\cdot$, but it is more convenient to call it simply $R\cdot$.

A basic chemical unit that comes up over and over again in the study of photochemical smog is a *carbonyl,* a carbon atom with a double bond joining it to an oxygen as shown below. A carbonyl with one bond shared with an alkyl, R·, and the other with a hydrogen atom, forms an *aldehyde.* Aldehydes can thus be written as RCHO. The simplest aldehyde is *formaldehyde* (HCHO), which corresponds to R· being just a single hydrogen atom. A more complex aldehyde is acrolein. Both formaldehyde and acrolein are eye-irritating components of photochemical smog.

$$
\begin{array}{cccc}
\quad\;\; \text{O} & \quad\;\; \text{O} & \quad\;\; \text{O} & \qquad\;\; \text{O} \\
\quad\;\; \| & \quad\;\; \| & \quad\;\; \| & \qquad\;\; \| \\
\text{—C—} & \text{R—C—H} & \text{H—C—H} & \text{CH}_2\!\!=\!\!\text{CH—C—H} \\
\text{carbonyl} & \text{aldehyde} & \text{formaldehyde} & \text{acrolein}
\end{array}
$$

An important key to understanding atmospheric organic chemistry is the OH radical, which is formed when atomic oxygen reacts with water:

$$O + H_2O \longrightarrow 2OH\cdot \tag{7.8}$$

The OH radical is extremely reactive, and its atmospheric concentration is so low that it has been difficult to detect. Nevertheless, it plays a key role in many reactions, including the oxidations of NO_2 to nitric acid and CO to CO_2.

$$OH\cdot + NO_2 \longrightarrow HNO_3 \tag{7.9}$$

$$OH\cdot + CO \longrightarrow CO_2 + H\cdot \tag{7.10}$$

It was mentioned above that the NO_2 photolytic cycle, by itself, underpredicts the observed concentrations of O_3. As that cycle is described in (7.5) to (7.7), the availability of NO_2 affects the rate of production of O_3, while the availability of NO affects the rate of destruction of O_3. The balance of O_3 production and destruction can be upset if there are other reactions that will enhance the rate of conversion of NO to NO_2. Any reactions that will help convert NO to NO_2 will increase O_3 concentrations both by reducing the amount of NO available to destroy O_3, and increasing the amount of NO_2 available to create O_3.

The following three reactions provide one explanation for the way that hydrocarbons can enhance the rate of conversion of NO to NO_2 and hence increase O_3 concentrations (see Seinfeld (1986), for example, for a much more complete description). Letting RH be a hydrocarbon, we have

$$RH + OH\cdot \longrightarrow R\cdot + H_2O \tag{7.11}$$

$$R\cdot + O_2 \longrightarrow RO_2\cdot \tag{7.12}$$

$$RO_2\cdot + NO \longrightarrow RO\cdot + NO_2 \tag{7.13}$$

These reactions are not complete, though they do show how hydrocarbons can help convert NO to NO_2, and thus increase O_3. As written, an already scarce OH· is required to start the chain, and it appears to be destroyed in the process. Unless there is some way to rejuvenate that OH·, these reactions could not continue for long. The following pair of reactions shows not only one way that OH· is regenerated, but also, in the process, how another NO is converted to NO_2. As well, an aldehyde can be formed from this process:

$$RO· + O_2 \longrightarrow HO_2· + R'CHO \qquad (7.14)$$

$$HO_2· + NO \longrightarrow NO_2 + OH· \qquad (7.15)$$

The net effect of reactions 7.11–7.15 is that one hydrocarbon molecule converts two molecules of NO to NO_2 and produces an aldehyde. The removal of NO slows the rate at which O_3 is removed, while the addition of NO_2 increases the rate at which it is produced, thus leading to higher levels of O_3 in the air. These are summarized in Figure 7.8.

Example 7.5   Ethane to Acetaldehyde

Suppose the hydrocarbon that begins reactions 7.11–7.15 is ethane (C_2H_6). Write the sequence of reactions.

Solution   The hydrocarbon RH that appears in (7.11) is ethane (C_2H_6), so the free radical R· is simply $C_2H_5·$. Reaction 7.11 thus becomes

$$C_2H_6 + OH· \longrightarrow C_2H_5· + H_2O$$

Reactions 7.12–7.14 are

$$C_2H_5· + O_2 \longrightarrow C_2H_5O_2·$$

$$C_2H_5O_2· + NO \longrightarrow C_2H_5O· + NO_2$$

$$C_2H_5O· + O_2 \longrightarrow HO_2· + CH_3CHO$$

Notice that R′ in Equation 7.14 is thus $CH_3·$. Finally, (7.15) is

$$HO_2· + NO \longrightarrow NO_2 + OH·$$

The CH_3CHO produced is an aldehyde called *acetaldehyde*. As we shall see, it plays a role in the formation of the eye irritant peroxyacetyl nitrate (PAN).

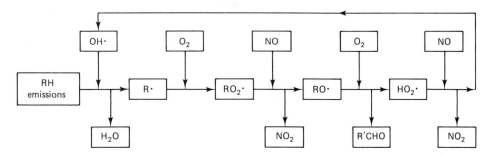

Figure 7.8   One way that hydrocarbons can cause NO to convert to NO_2. Reducing NO slows the removal of O_3, while increasing NO_2 increases the production of O_3, so this cycle, when combined with Figure 7.6, helps account for elevated atmospheric O_3 levels.

There are other ways that NO can be converted to NO_2. Carbon monoxide, for example, can do it too, in reactions that are similar to those given above. Start with CO and OH·, as in (7.10):

$$OH· + CO \longrightarrow CO_2 + H· \tag{7.10}$$

The hydrogen atom then quickly combines with O_2 to form the hydroperoxyl radical, $HO_2·$:

$$H· + O_2 \longrightarrow HO_2· \tag{7.16}$$

Now, going back to (7.15), NO is converted NO_2:

$$HO_2· + NO \longrightarrow NO_2 + OH· \tag{7.15}$$

So we see one way that another of our criteria pollutants, CO, can contribute to the photochemical smog problem. By increasing the rate at which NO is converted to NO_2, CO aids in the accumulation of O_3.

We can also extend these relationships to show the formation of another of the eye irritants, peroxyacetyl nitrate (PAN):

$$CH_3C(O)O_2NO_2$$

peroxyacetyl nitrate (PAN)

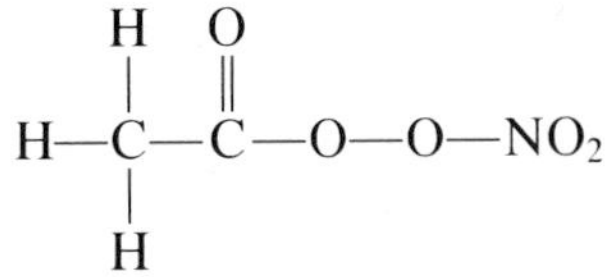

Acetaldehyde (CH_3CHO), which Example 7.5 indicated can be formed from ethane emissions, can react with OH· with much the same relationship as given in (7.11):

$$CH_3CHO + O_2 + OH· \longrightarrow CH_3C(O)O_2· + H_2O \tag{7.17}$$

The resulting acetylperoxy radical can react with NO_2 to create PAN:

$$CH_3C(O)O_2· + NO_2 \longrightarrow CH_3C(O)O_2NO_2 \tag{7.18}$$

It must be pointed out once again, that the above reactions are only a very limited description of the complex chemistry going on each day in the atmosphere over our cities.

Sources and Effects of Photochemical Oxidants

Since ozone and other photochemical oxidants are not primary pollutants, there is no direct way to specify emissions. An annual emissions inventory is kept by the EPA, however, for the precursors to photochemical oxidants, namely nitrogen oxides and volatile organic compounds (VOCs). We have already considered recent trends in NO_x, and Figure 7.9 shows a similar graph for VOC emissions. A significant drop came during the first few years after the 1970 Clean Air Act Amendments were passed, but reductions since then have been modest. As can be seen, the two major source categories for VOCs are transportation and industrial processes. When all precursors are considered, especially if an adjust-

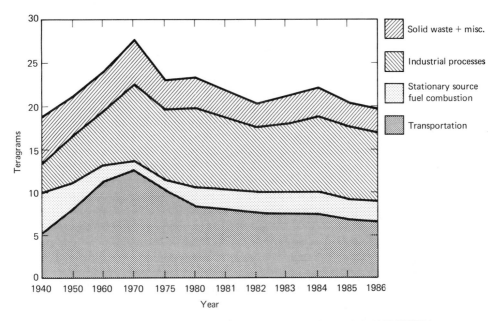

Figure 7.9   Trends in volatile organic compound emissions, 1940–1986 (USEPA, 1988a).

ment is made for the sources' proximity to cities, the automobile is the dominant source of photochemical smog. Conditions in Los Angeles, for example, are ideal for smog formation: the near total dependence on the automobile for transportation results in high hydrocarbon and NO_x emissions; there are long lasting atmospheric inversions that restrict the vertical dispersion of pollutants; a ring of mountains nearly surrounds the city on three sides, reducing the horizontal dispersion; and there is an abundance of sunshine to power the photochemical reactions.

Photochemical smog is known to cause many annoying respiratory effects such as coughing, shortness of breath, airway constriction, headache, chest tightness, and eye, nose, and throat irritation. It also has been correlated with decreased athletic performance. No convincing relationship has been observed, however, between short-term variations in photochemical oxidants and daily mortality or hospital admissions. Although it aggregates symptoms among patients with existing respiratory diseases, photochemical smog has not been shown to be a cause of such diseases.

Ozone has been shown to cause damage to tree foliage and to reduce growth rates of certain sensitive tree species. It also reduces yields of major agricultural crops, such as corn, wheat, soybeans, and peanuts. Ozone alone is thought to be responsible for about 90 percent of all of the damage that air pollutants cause to agriculture, with a total economic cost that has been estimated at 6–7 percent of U.S. agricultural productivity (OTA, 1984).

Particulate Matter

Atmospheric *particulate matter* is defined to be any dispersed matter, solid or liquid, in which the individual aggregates are larger than single small molecules (about 0.0002 μm in diameter), but smaller than about 500 μm. As a category of criteria pollutant, particulate matter is extremely diverse and complex, since size and chemical composition, as well as atmospheric concentration, are important characteristics.

A number of terms are used to categorize particulates, depending on their size and phase (liquid or solid). The most general term is *aerosol,* which applies to any tiny particles, liquid or solid, dispersed in the atmosphere. Solid particles are called *dusts* if they are caused by grinding or crushing operations. Solid particles are called *fumes* if they are formed when vapors condense. Liquid particles may be called *mist* or, more loosely, *fog. Smoke* and *soot* are terms used to describe particles composed primarily of carbon that result from incomplete combustion. *Smog* is a term that was derived from smoke and fog, originally referring to particulate matter, but now describing air pollution in general.

Although particles may have very irregular shapes, their size can be described by an equivalent *aerodynamic diameter* determined by comparing them with perfect spheres having the same settling velocity. The particles of most interest have aerodynamic diameters in the range of 0.1 to 10 μm (roughly the size of bacteria). Particles smaller than these undergo random (Brownian) motion and, through coagulation, generally grow to sizes larger than 0.1 μm. Particles larger than 10 μm settle quite quickly. A 10-μm particle, for example, has a settling velocity of approximately 20 cm/min.

The ability of the human respiratory system to defend itself against particulate matter is, to a large extent, determined by the size of the particles. To help understand these defense mechanisms, consider the illustration in Figure 7.10. The upper respiratory system consists of the nasal cavity and the trachea, while the lower respiratory system consists of the bronchial tubes and the lungs themselves. Each bronchus divides over and over again into smaller and smaller branches, terminating with a large number of tiny air sacs called alveoli.

Large particles that enter the respiratory system can be trapped by the hairs and lining of the nose. Once captured, they can be driven out by a cough or sneeze. Smaller particles that make it into the tracheobronchial system can be captured by mucus, worked back to the throat by tiny hairlike *cilia,* and removed by swallowing or spitting. Particles larger than about 10 μm are quite effectively removed in the upper respiratory system by these defense mechanisms. Smaller particles, however, are often able to traverse the many turns and bends in the upper respiratory system without being captured on the mucous lining. These particles may make it into the lungs, but depending on their size, they may or may not be deposited there. Some particles are so small that they tend to follow the airstream into the lungs and then right back out again. Particles roughly between 0.5 and 10 μm may be small enough to reach the lung, and large enough to be

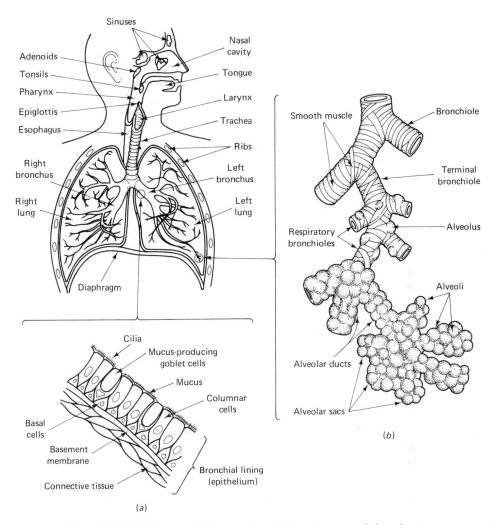

Figure 7.10   The human respiratory system: (*a*) the system as a whole and a cross section of the bronchial lining showing the cilia; (*b*) details of the lower respiratory system terminating in the alveoli. (*Source:* Williamson, S. J. *Fundamentals of Air Pollution,* © 1973 by Addison-Wesley Publishing Co. Reprinted by permission of Addison-Wesley Publishing Co., Inc., Reading, MA.)

deposited there by sedimentation. Sedimentation is most effective for particles between 2 and 4 μm.

The original NAAQS for particulates did not take size into account. Larger particles could dominate the weight per unit volume measure, but be unimportant in terms of human health risk. With the PM-10 standard promulgated in 1987, however, only particles smaller than 10 μm, which are capable of reaching the lungs, are measured.

Elevated particulate concentrations in the atmosphere, especially in conjunction with oxides of sulfur, have been linked to rises in the number of hospital visits for upper respiratory infections, cardiac disorders, bronchitis, asthma, pneumonia, emphysema, and the like. In addition, some particulates are especially dangerous because of their toxicity. Many carbonaceous particles, especially those containing polycyclic aromatic hydrocarbons (PAHs), are suspected carcinogens.

Particulate emissions have decreased substantially in the past few decades. As Figure 7.11 shows, particulate emissions in 1950 were estimated to be about 25 million metric tons, while in 1986 they were about 7 million tons. Reductions between 1940 and 1970 were due more to the replacement of coal-burning locomotives and fewer forest wildfires than to any particular change in emissions from fuel combustion. However, since the Clean Air Act of 1970, tremendous reductions in combustion emissions have been achieved, especially by the electric utilities. In 1970 utilities burned 320 million tons of coal, releasing 2.3 million tons of particulates. In 1986, 685 million tons of coal were burned and only 0.4 million tons of particulates were emitted (USEPA, 1988a).

Oxides of Sulfur

Over 80 percent of anthropogenic sulfur oxide emissions are the result of fossil fuel combustion in stationary sources. Of that, almost 85 percent is released from electric utility power plants. Only about 2 percent comes from highway vehicles.

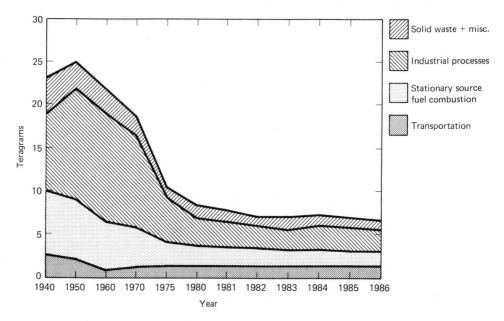

Figure 7.11   Trends in particulate emissions, 1940–1986 (USEPA, 1988a).

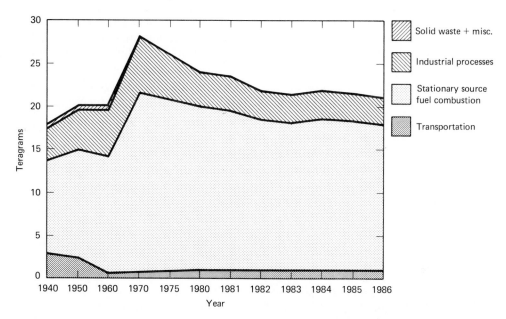

Figure 7.12   Trends in sulfur oxide emissions, 1940–1986 (USEPA, 1988a).

The only significant noncombustion sources of sulfur emissions are associated with petroleum refining, copper smelting, and cement manufacture. As Figure 7.12 indicates, total sulfur oxide emissions have been reduced by about 25 percent since 1970 as a result of fuels with lower sulfur content, flue gas scrubbing, and more extensive controls on sulfuric acid plants and smelters.

Oil and coal generally contain appreciable quantities of sulfur (roughly 0.5–6 percent), either in the form of inorganic sulfides or as organic sulfur. When these fuels are burned, the sulfur is released mostly as sulfur dioxide (SO_2), but also with small amounts of sulfur trioxide (SO_3). Sulfur dioxide, once released, can convert to SO_3 in a series of reactions which, once again, involve a free radical such as OH·

$$SO_2 + OH \cdot \longrightarrow HOSO_2 \cdot \qquad (7.19)$$

$$HOSO_2 \cdot + O_2 \longrightarrow SO_3 + HO_2 \cdot \qquad (7.20)$$

The $HO_2 \cdot$ radical can then react with NO to return the initial OH·, as in (7.15). Sulfur trioxide reacts very quickly with H_2O to form sulfuric acid, which is the principal cause of acid rain.

$$SO_3 + H_2O \longrightarrow H_2SO_4 \qquad (7.21)$$

Sulfuric acid molecules rapidly become particles by either condensing on existing particles in the air or by merging with water vapor to form $H_2O-H_2SO_4$ droplets. Often a significant fraction of particulate matter in the atmosphere consists of such sulfate (SO_4^{2-}) aerosols.

The transformation from SO_2 gas to sulfate particles (SO_4) is gradual, taking a matter of days. During that time, sulfur pollution may be deposited back onto the land or into water, either in the form of SO_2 or sulfate. In either form, sulfur pollution can be deposited by removal during precipitation (wet deposition), or by slow, continuous removal processes that occur without precipitation (dry deposition). Figure 7.13 suggests the effects of time and distance on the conversion and deposition of sulfur.

Figure 7.14 shows the contours of pH for wet deposition in the United States. Recall from Chapter 2 that natural rainfall would have a pH value between 5 and 5.6, and anything less is loosely called "acid rain." A comparison between Figure 7.14 and Figure 4.15 shows large areas of the eastern United States and Canada that unfortunately have the worst acid rain falling on lakes that are inherently the most sensitive to acidification. About two-thirds of U.S. coal consumption, the source of most sulfur emissions, is east of the Mississippi, a fact that correlates well with the acidity of rainfall in the eastern half of the United States, as shown in Figure 7.14.

Most sulfate particles in urban air have an effective size of less than 2 μm, with most of them being in the range of 0.2 to 0.9 μm. Their size is comparable to the wavelengths of visible light, and their presence greatly affects visibility. Their size also allows deep penetration into the respiratory system.

Sulfur dioxide is highly water soluble, much more so than any of the other criteria pollutants. As a result, when it is inhaled it is most likely to be absorbed in the moist passages of the upper respiratory tract, the nose, and upper airways. Other gases, being less soluble, are more likely to reach the terminal air sacs of the

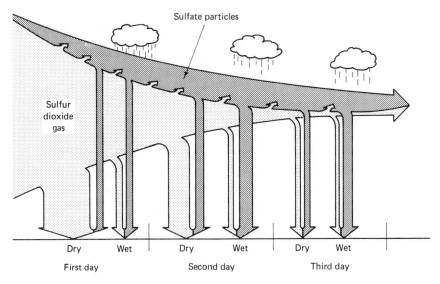

Figure 7.13 The effects of time and distance on conversion and deposition of sulfur pollution (OTA, 1984).

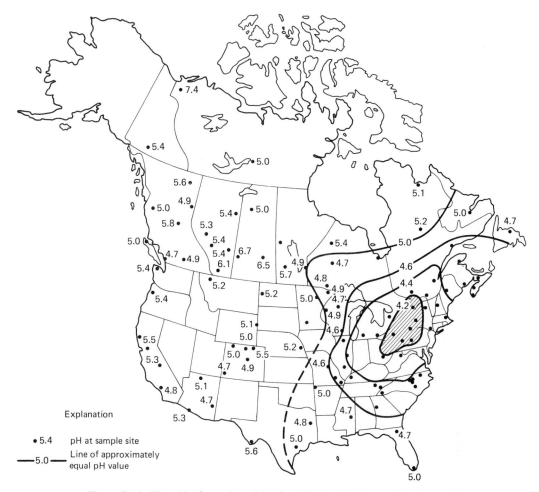

Figure 7.14 The pH of wet deposition in 1982 (precipitation-weighted annual average). (*Source:* Interagency Task Force on Acid Precipitation, 1983.)

lungs. When sulfur is entrained in an aerosol, however, the aerodynamic properties of the particles themselves affect the area of deposition, and it is possible for sulfur oxides to reach far deeper into the lungs. The combination of particulate matter and sulfur oxides can then act synergistically, with the effects of both together being much more detrimental than either of them separately. In fact, in every major air pollution episode, the combination of sulfur oxides and particulates has been implicated as a cause of the excess mortality observed. The magnitude of the health risk posed by current levels of sulfates and other particulates has been estimated at 50 000 premature deaths (2 percent of total deaths) per year in the United States and Canada (OTA, 1984). A summary of adverse health effects of SO_2 is shown in Figure 7.15.

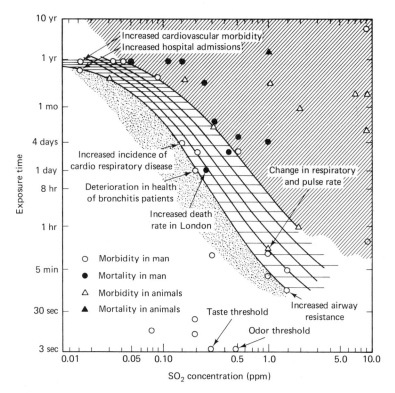

Figure 7.15 Effects of sulfur oxides on human health. Shaded area represents range of exposures in which excess deaths have been reported. The grid area represents the exposure range in which significant health effects have been reported. The speckled area represents the range of exposures where health effects are suspected. (USHEW, 1967.)

Sulfur dioxide can damage vegetation, although aerosols of sulfuric acid appear to be less toxic to plants. Sulfurous pollutants can discolor paint, corrode metals, and cause organic fibers to weaken. Airborne sulfates significantly reduce visibility and discolor the atmosphere. Most of the visibility impairment in the eastern United States is caused by sulfates, while in the west it is more often nitrogen oxides and dust.

Prolonged exposure to sulfates causes serious damage to building marble, limestone, and mortar, as the carbonates in these materials are replaced by sulfates. The reaction between limestone ($CaCO_3$) and sulfuric acid shows such a replacement:

$$CaCO_3 + H_2SO_4 \longrightarrow CaSO_4 + CO_2 + H_2O \qquad (7.22)$$

The calcium sulfate (gypsum) produced by this reaction is water soluble and easily washes away, leaving a pitted, eroded surface. Many of the world's historic buildings and statues are rapidly being degraded due to this exposure. It is common now in such monuments as the Acropolis in Greece for the original outdoor statuary to be moved into air conditioned museums, leaving plaster replicas in their place.

Lead

Most lead emissions in the past have been from motor vehicles burning gasoline containing the antiknock additive, tetraethyl lead $(C_2H_5)_4Pb$. Major reductions in transportation sector emissions had already occurred before the NAAQS standard for lead was promulgated in 1978. The automobile industry had chosen catalytic converters to meet the motor vehicle emission restrictions mandated by the Clean Air Act of 1970. Catalytic converters are quickly rendered inoperable with leaded fuels, so essentially all new cars must use unleaded gasoline. In addition, since many motor vehicles without converters continue to burn leaded fuels, 1984 EPA rules required refiners to lower the lead content of leaded gasoline from 1.1 to 0.1 g/gal. Emissions from stationary sources have also dropped due to control programs oriented toward attaining both the particulate and lead ambient air quality standards. Between 1970, when data on lead first began to be carefully assembled, and 1986, lead emissions have dropped over 95 percent, as shown in Figure 7.16.

Lead is emitted to the atmosphere primarily in the form of inorganic particulates. Much of this is removed from the atmosphere by settling in the immediate vicinity of the source. Airborne lead may affect human populations by direct

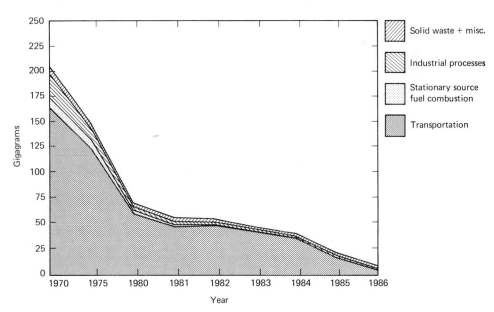

Figure 7.16 Trends in lead emissions, 1970–1986 (USEPA, 1988a).

inhalation, in which case people living nearest to highways are at greatest risk, or it can be ingested after the lead is deposited onto foodstuffs. Most of human exposure to airborne lead is the result of inhalation. It has been estimated that about one-third of the lead particles inhaled are deposited in the respiratory system, and that about half of those are absorbed by the bloodstream.

Measurements made in actual communities suggest that an increase in airborne lead concentration of 1 μg/m^3 results in an increase of about 1–2 μg per deciliter (μg/dL) in blood lead levels (the NAAQS standard has been set at 1.5 μg/m^3). Lead poisoning can cause aggressive, hostile, and destructive behavioral changes, as well as learning disabilities, seizures, severe and permanent brain damage, and even death. Children and pregnant women are at greatest risk. Blood lead levels associated with neurobehavioral changes in children appear to begin at 50–60 μg/dL. Encephalopathy, with possible severe brain damage or death, occurs at levels somewhat above 80 μg/dL (USEPA, 1977). In 1977, the EPA estimated that 600 000 children in the United States had blood levels above 40 μg/dL due to exposure to lead from all environmental sources. These sources include air emissions, drinking water (lead can be leached out of lead solder used in copper piping systems), and ingestion of lead in food and leaded paint.

7.4 AIR POLLUTION AND METEOROLOGY

Obviously, air quality at a given site varies tremendously from day to day, even though emissions may remain relatively constant. The determining factors have to do with the weather: how strong the winds are, what direction they are blowing, the temperature profile, how much sunlight is available to power photochemical reactions, and how long it has been since the last strong winds or precipitation were able to clear the air. Air quality is dependent on the dynamics of the atmosphere, the study of which is called *meteorology*.

Adiabatic Lapse Rate

The ease with which pollutants can disperse vertically into the atmosphere is largely determined by the the rate of change of air temperature with altitude. For some temperature profiles the air is *stable,* that is, air at a given altitude has physical forces acting on it that make it want to remain at that elevation. Stable air discourages the dispersion and dilution of pollutants. For other temperature profiles, the air is unstable. In this case rapid vertical mixing takes place that encourages pollutant dispersal and increases air quality. Obviously, vertical stability of the atmosphere is an important factor that helps determine the ability of the atmosphere to dilute emissions; hence, it is crucial to air quality.

Let us investigate the relationship between atmospheric stability and temperature. It is useful to imagine a "parcel" of air being made up of a number of air molecules with an imaginary boundary around them. If this parcel of air moves upward in the atmosphere, it will experience less pressure, causing it to expand and cool. On the other hand, if it moves downward, more pressure will compress the air and its temperature will increase. This heating or cooling of a gas as it is compressed or expanded should be a familiar concept. Pumping up a bicycle tire, for example, warms the valve on the tire, while letting the air out cools it off.

As a starting point, we need a relationship that expresses an air parcel's change of temperature as it moves up or down in the atmosphere. As it moves, we can imagine its temperature, pressure, and volume changing, and we might imagine its surroundings adding or subtracting energy from the parcel. If we make small changes in these quantities, and apply both the ideal gas law and the first law of thermodynamics, it is relatively straightforward to derive the following expression, (see Problem 7.10):

$$dQ = C_p \, dT - V \, dP \tag{7.23}$$

where dQ = heat added to the parcel per unit mass (J/kg)

C_p = specific heat at constant pressure, that is, the amount of heat required to raise the temperature of 1 kg of air by 1 °C while holding its pressure constant (=1005 J/kg-°C)

dT = incremental temperature change (K)

V = volume per unit mass (m³/kg)

dP = incremental pressure change in the parcel (Pa)

Let us make the quite accurate assumption that as the parcel moves, there is no heat transferred across its boundary, that is, that this process is *adiabatic*. This means that $dQ = 0$; so we can rearrange (7.23) as

$$\frac{dT}{dP} = \frac{V}{C_p} \tag{7.24}$$

Equation 7.24 gives us an indication of how atmospheric temperature would change with air pressure, but what we are really interested in is how it changes with altitude. To do that we need to know how pressure and altitude are related. Consider a static column of air with cross section A, as shown in Figure 7.17. A horizontal slice of air in that column of thickness dz and density ρ, will

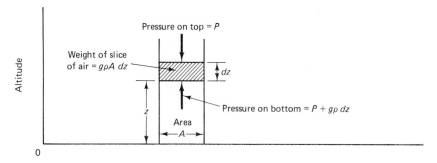

Figure 7.17 A column of air in static equilibrium used to determine the relationship between air pressure and altitude.

have mass $\rho A\ dz$. If the pressure at the top of the slice due to the weight of air above it is $P(z + dz)$, then the pressure at the bottom of the slice, $P(z)$, will be $P(z + dz)$ plus the added weight per unit area of the slice itself:

$$P(z) = P(z + dz) + \frac{g\rho A\ dz}{A}$$

where g is the gravitational constant. We can write the incremental pressure dP for an incremental change in elevation, dz as

$$dP = P(z + dz) - P(z) = -g\rho\ dz \qquad (7.25)$$

Expressing the rate of change in temperature with altitude as a product, and substituting in (7.24) and (7.25), gives

$$\frac{dT}{dz} = \frac{dT}{dP}\frac{dP}{dz} = \frac{V}{C_p}(-g\rho) \qquad (7.26)$$

However, since V is volume per unit mass and ρ is mass per unit volume, the product $V\rho = 1$, and the expression simplifies to

$$\frac{dT}{dz} = \frac{-g}{C_p} \qquad (7.27)$$

The negative sign indicates that temperature decreases with increasing altitude. Substituting the constant $g = 9.806$ m/s^2, and the constant-volume specific heat of dry air at room temperature, $C_p = 1005$ J/kg-°C into (7.27) yields

$$\frac{dT}{dz} = \frac{-9.806\ \text{m/s}^2}{1005\ \text{J/kg-°C}} \times \frac{1\ \text{J}}{\text{kg-m}^2/\text{s}^2} = -0.00976\ °\text{C/m} \qquad (7.28)$$

This is a very important result. When its sign is changed to keep things simple, $-dT/dz$ is given a special name: the *dry adiabatic lapse rate*, Γ.

$$\Gamma = -\frac{dT}{dz} = 9.76\ °\text{C/km} = 5.4\ °\text{F}/1000\ \text{ft} \qquad (7.29)$$

Equation 7.29 tells us that moving a parcel of dry air up or down will cause its temperature to change by 9.76 °C/km, or essentially 1 °C/100 m. This temperature profile will be used as a reference against which actual ambient air temperature gradients will be compared. As we shall see, if the actual air temperature decreases faster with increasing elevation than the adiabatic lapse rate, the air will be unstable and rapid mixing and dilution of pollutants will occur. Conversely, if the actual air temperature drops more slowly than the adiabatic lapse rate, the air will be stable and pollutants will concentrate.

Equation 7.29 was derived by assuming that our parcel of air could be treated as an ideal gas that could be moved without heat transfer taking place between it and its surroundings. Both assumptions are very good. It was also assumed that the air was dry, but this may or may not be a good assumption. If air has some water vapor in it, C_p changes slightly from the value assumed, but not enough to warrant a correction. On the other hand, if enough water vapor is present that condensation occurs when the parcel is raised and cooled, latent heat will be released. The added heat means a saturated air parcel will not cool as rapidly as a dry one. Unlike the dry adiabatic lapse rate, the wet adiabatic lapse rate is not a constant. This is because the amount of moisture that air can hold before reaching saturation and condensation begins is a function of temperature. A reasonable average value of wet adiabatic lapse rate is about 6 °C/km, although near the cold poles it will be higher, and in wet, tropical areas it will be lower.

Atmospheric Stability

Our interest in lapse rates is based on the need to understand atmospheric stability, which is, in turn, a crucial factor in the atmosphere's ability to dilute pollution. For a number of reasons such as wind, sunlight, and geographical features, the actual rate of change of air temperature with altitude (*ambient* lapse rate) is almost always different from the 1 °C/100 m *adiabatic* lapse rate just calculated.

Consider a parcel of air at some given altitude. It has the same temperature and pressure as the air surrounding it. Our test for atmospheric stability will be based on the following thought experiment. If we imagine raising the parcel of air slightly, it will experience less atmospheric pressure, so it will expand. Because it will have done work on its environment (by expanding), the internal energy in the parcel will be reduced so its temperature will drop. Assuming the parcel is raised fast enough to be able to ignore any heat transfer between the surrounding air and the parcel, the cooling will follow the adiabatic lapse rate. After raising the parcel, note its temperature, and compare it with the temperature of the surrounding air. If the parcel, at this higher elevation, is now colder than its surroundings, it will be denser than the surrounding air and will want to sink back down again. That is, whatever it was that caused the parcel to start to move upward will immediately be opposed by conditions that make the parcel want to go back down again. The atmosphere is said to be stable. If, however, raising the parcel causes its temperature to be greater than the surrounding air, it will be less dense than the surrounding air and it will experience buoyancy forces that will encourage it to keep moving upward. The original motion upward will be reinforced and the parcel will continue to climb. This is an unstable atmosphere.

Consider Figure 7.18*a*, which shows an ambient temperature profile for air that cools less rapidly with altitude than the adiabatic lapse rate. Imagine a 20 °C parcel of air at 1000 m to be just like the air surrounding it. If that parcel is raised by 100 m, it will cool adiabatically by about 1 °C, so it will now reside at an elevation of 1100 m with temperature 19 °C. The surrounding air at 1100 m is shown to be 19.5 °C. The parcel of air is cooler and denser than its surroundings, so it sinks. In other words, nudging a parcel of air up creates forces that want to return it to the original elevation. Suppose it drops 100 m below its initial elevation, warming adiabatically to 21 °C. At 900 m, the parcel is 21 °C and the surrounding air is 20.5 °C. The parcel is warmer than the surrounding air so it is

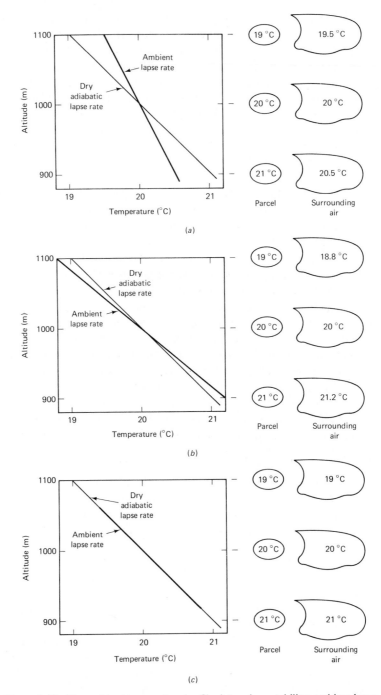

Figure 7.18 The ambient temperature profile determines stability: ambient lapse rate (*a*) less than Γ (stable); (*b*) greater than Γ (unstable); (*c*) equal to Γ (neutral).

buoyant and wants to move upward (warm air rises). In other words, nudging the parcel down creates forces that want to push it back up to the original elevation. This temperature profile therefore corresponds to a *stable* atmosphere. The ambient lapse rate is said to be *subadiabatic*.

In Figure 7.18*b*, the ambient temperature cools more rapidly than the adiabatic lapse rate. If we once again imagine a parcel of air at 1000 m and 20 °C that is for one reason or another nudged upward, it will find itself warmer than the surrounding air. At its new elevation, it experiences the same pressure as the air around it, but it is warmer, so it will be buoyant and continue to climb. Conversely, a parcel starting at 1000 m and 20 °C that starts moving downward will get cooler and denser than its surroundings. It will continue to sink. The ambient air is *unstable,* and the ambient lapse rate is said to be *superadiabatic*.

If the ambient lapse rate is equal to the adiabatic lapse rate, as shown in Figure 7.18*c*, moving the parcel upward or downward results in its temperature changing by the same amount as its surroundings. In any new position, it experiences no forces that make it either continue its motion or return to its original elevation. The parcel likes where it was, and it likes its new position too. Such an atmospheric is said to be *neutrally stable*.

Figure 7.19 summarizes this conclusion. For ambient lapse rates greater than the adiabatic lapse rate (ambient cooling faster with increasing altitude than adiabatic), the atmosphere is unstable and vertical motions are enhanced. For ambient lapse rates less than the adiabatic lapse rate, the atmosphere is stable and vertical motions are suppressed. An extreme case is one in which temperatures increase with altitude. Such *temperature inversions* yield a very stable air mass. In Chapter 8, it will be noted that the absorption of incoming solar energy by oxygen and ozone in the stratosphere creates a temperature inversion there. That temperature inversion causes the stratosphere to be extremely stable, oftentimes trapping pollutants for many years.

Temperature Inversions

Temperature inversions represent the extreme case of atmospheric stability, creating a virtual lid on the upward movement of pollution. There are several causes

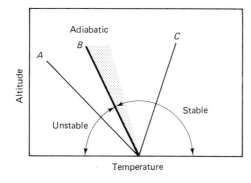

Figure 7.19 Temperature profiles to the left of the adiabatic lapse rate correspond to an unstable atmosphere (line *A*); profiles to the right are stable (line *C*). The dry adiabatic lapse rate is line *B*. The speckled area is meant to suggest slopes that correspond to the wet adiabatic lapse rate.

of inversions, but only two are of major importance from an air quality stand-point. The first, *radiation inversions,* are caused by nocturnal cooling of the earth's surface, especially on clear winter nights. The second, *subsidence inversions,* are the result of the compressive heating of descending air masses in high pressure zones. There are other causes of inversions, but they are less important. There are *frontal* inversions when a cold air mass passes under a warm air mass. These are short-lived, however, and tend to be accompanied by precipitation that cleanses the air. There are also inversions associated with geographical features of the landscape. Warm air passing over a cold body of water, for example, creates an inversion. There are also inversions in valleys when cold air rolls down the canyons at night under warmer air that might exist aloft.

Radiation Inversions

The surface of the earth cools down at night by radiating energy toward space. On a cloudy night, the earth's radiation tends to be absorbed by water vapor in the atmosphere, which in turn reradiates some of that energy back to the ground. On a clear night, however, the surface more readily radiates energy to space; thus, the ground can cool much more rapidly. As the ground cools, the temperature of the air in contact with the surface also drops. As is often the case on clear winter nights, the temperature of the air just above the ground becomes colder than the air above it, creating an inversion. Radiation inversions begin to form at about dusk. As the evening progresses, the inversion extends to a higher and higher elevation, reaching perhaps a few hundred meters before morning comes and the sun warms the ground again. Figure 7.20 shows the development of a radiation inversion through the night, followed by the erosion of the inversion that takes place the next day.

Radiation inversions occur close to the ground, mostly during the winter, and last for only a matter of hours. They often begin at about the time traffic builds up in the early evening, which traps auto exhaust at ground level and causes

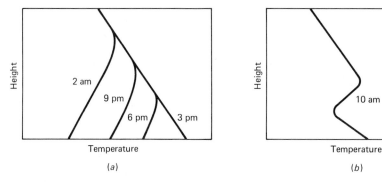

Figure 7.20 Development of a radiation inversion (*a*), and the subsequent erosion of the inversion (*b*). The times are representative only. The breakup of the inversion in the morning leads to a process called fumigation.

elevated concentrations of pollution for commuters. Without sunlight, photo-chemical reactions cannot take place, so the biggest problem can be the accumulation of carbon monoxide. In the morning, as the sun warms the ground and the inversion begins to break up, pollutants that have been trapped in the stable air mass are suddenly brought back to earth in a process known as *fumigation*. Fumigation can cause a short-lived, high concentration of pollution at ground level.

Radiation inversions are important in another context besides air pollution. Fruit growers in places like California have long known that their crops are in greatest danger of frost damage on winter nights when the skies are clear and a radiation inversion sets in. Since the air even a few meters up is warmer than the air at crop level, one way to help protect sensitive crops on such nights is to simply mix the air with large, motor-driven fans.

Subsidence Inversions

While radiation inversions are mostly a short-lived, ground-level, wintertime phenomenon, the other important cause of inversions, subsidence, creates quite the opposite characteristics. Subsidence inversions may last for months on end, occur at higher elevations, and are more common in summer than winter.

Subsidence inversions are associated with high-pressure weather systems, known as *anticyclones*. Air in the middle of a high pressure zone is descending, while on the edges, it is rising. Air near the ground moves outward from the center, while air aloft moves toward the center from the edges. The result is a massive vertical circulation system. As air in the center of the system falls, it experiences greater pressure and is compressed and heated. If its temperature at elevation z_1 is T_1, then as it falls to elevation z_2 it will be heated adiabatically to $T_2 = T_1 + \Gamma(z_1 - z_2)$. As is often the case, this compressive heating warms the descending air to a higher temperature than the air below, whose temperature is dictated primarily by conditions on the ground. The result is an inversion, located anywhere from several hundred meters above the surface to several thousand meters, that lasts as long as the high-pressure weather system persists.

Since subsiding air is getting warmer, it is more and more able to hold water vapor as it descends. Without sources of new moisture, its relative humidity drops; thus, there is little chance for clouds to form. The result is that high-pressure zones create clear, dry weather with lots of sunshine during the day and clear skies at night. Clear skies allow solar warming of the earth's surface. This helps create superadiabatic conditions under the inversion during the daytime and, hence, good mixing. At night, the surface can cool quickly by radiation, which may result in a radiation inversion located under the subsidence inversion as shown in Figure 7.21.

Some anticyclones and their accompanying subsidence inversions drift across the continents (west to east in midlatitudes) so that at any given spot they may appear to come and go with some frequency. On the other hand, other

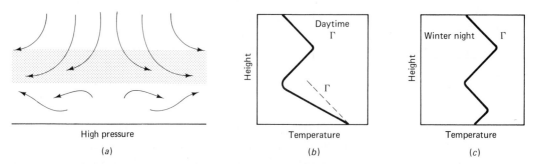

Figure 7.21 Subsidence inversions. (*a*) As air descends it is warmed adiabatically creating clear skies and an upper level inversion (dots); (*b*) during the day, air under the inversion may be unstable due to solar warming of the surface; (*c*) radiation inversions may form at night under the subsidence inversion, especially in winter.

anticyclones are semipermanent in nature and can cause subsidence inversions to last for months at a time. These semipermanent highs are the result of the general atmospheric circulation patterns shown in Figure 7.22.

At the equinox, the equator is directly under the sun, and the air there will be heated, become buoyant, and rise. As that air approaches the tropopause, it begins to turn, some heading north and some south. An eighteenth century meteorologist, George Hadley, postulated that the air would continue to the poles before descending. In actuality, it descends at a latitude of about 30° and then returns to the equator, forming what are known as a *Hadley cell*. In similar fashion, though not as distinct, there are Hadley cells, linked as in a chain, between 30° and 60° latitude, and between 60° and the poles. The descending air at 30° creates a persistent high-pressure zone with corresponding lack of clouds or rainfall that is a contributing factor to the creation of the world's great deserts.

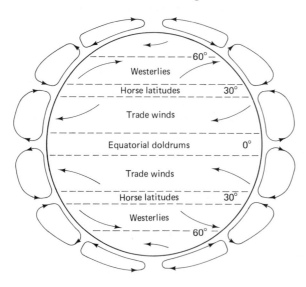

Figure 7.22 Idealized general air circulation patterns, drawn at the equinox. High-pressure zones are common around 30° latitude, producing clear skies and subsidence inversions.

The deserts of southern California, the southwestern United States, the Sahara, the Chilean desert, the Kalahari in South Africa, and the great deserts in Australia, are all located at roughly 30° latitude for this reason. Conversely, rising air near the equator and near 60° latitude tends to be moist after passing over oceans. As it rises and cools, the moisture condenses, causing clouds and rainfall. For this reason, some of the wettest regions of the world are in those bands of latitude.

The air that moves along the surface of the earth to close a Hadley cell is affected by Coriolis forces, causing it to curve toward the west if it is moving toward the equator, or toward the east if it is moving toward the poles. The resulting winds between roughly 30° and 60° are known as the *westerlies,* and between 30° and the equator they are the *trade winds.* Near the equator, there is little wind since the air is mostly rising. That band is called the *equatorial doldrums.* Similarly, the surface air is relatively calm around 30° latitude, forming a band called the *horse latitudes.* (Apparently, early explorers were sometimes forced to lighten the load by throwing horses overboard to avoid being becalmed as they sailed to the New World.)

These important bands of latitudes, the doldrums and horse latitudes, move up and down the globe as the seasons change. Figure 7.22 shows global circulation patterns at an equinox, when the sun is directly over the equator. In summer, the sun moves northward, as do the persistent high- and low-pressure zones associated with these bands. In the winter, the sun is directly overhead somewhere in the Southern Hemisphere and the bands move southward. There is, for example, a massive high pressure zone off the coast of California that moves into place over Los Angeles (latitude 34°) and San Francisco (latitude 38°) in the spring and remains there until late fall. That is the principal reason for California's sunny climate, as well as its smog. Clear skies assure plenty of sunlight to power photochemical reactions, the lack of rainfall eliminates that atmospheric cleansing mechanism, and prolonged subsidence inversions concentrate the pollutants. In the case of Los Angeles, there is also a ring of mountains around the city that tends to keep winds from blowing the smog away.

The global circulation patterns of Figure 7.22 are, of course, idealized. The interactions between sea and land, the effects of storms and other periodic disturbances, and geographical features such as high mountain ranges all make this model useful only on a macroscale. Even such a simplified global model, however, does help to explain a number of significant features of the world's climate and some aspects of regional air pollution problems.

7.5. ATMOSPHERIC DISPERSION

The Clean Air Act specifies new source emission standards, and it specifies ambient air quality standards. The connecting link between the two is the atmosphere. How do pollutants behave once they have been emitted, and how may we predict their concentrations in the atmosphere? How can the limitations imposed by a Prevention of Significant Deterioration policy be shown to be satisfied for new sources in an attainment area? How can we predict the improvement in air quality that must be achieved when new sources are proposed for a nonattainment area? To help answer questions like these, computer models that use such information as predicted emissions, smokestack heights, wind data, atmospheric temperature profiles, ambient temperatures, solar insolation, and local terrain features have been developed.

Maximum Mixing Depth

The amount of air available to dilute pollutants is related to the wind speed and to the extent to which emissions can rise into the atmosphere. We now know the importance of the atmospheric temperature profile in determining its stability. This, in turn, affects the vertical dispersion of pollutants. It would seem useful, then, to have some measure of the thickness of the air layer near the surface of the earth in which we could expect to have turbulence and effective dilution taking place.

Imagine a parcel of air at ground level being warmed by convection on a sunny day. Its buoyancy will make it rise, and we want to know how high it will tend to go. As the parcel moves upward it cools adiabatically at about 1 °C/ 100 m. If its temperature is warmer than that of the surrounding air, it will continue to rise; if it is less, it will fall. At some point, if its temperature becomes equal to that of the surrounding air, it will stop moving. If we have a temperature profile of the atmosphere, then we can determine the elevation to which our parcel will ascend by simply projecting the ground temperature upward, at the adiabatic lapse rate, until it crosses the ambient temperature profile. The crossing point determines what is called the *mixing depth* (sometimes called the *mixing height*) of the atmosphere.

While we could imagine making this projection on a moment-to-moment basis, the results would not be worth that much effort. In practice, what is usually done is to obtain a single, daily temperature profile, usually at night or in the early morning, using a balloon equipped to transmit the necessary data. The dry adiabatic temperature line is then drawn from the surface, usually with a starting point equal to the average maximum surface temperature for the month. The elevation where these lines cross is known as the mean *maximum mixing depth., as shown in Figure 7.23.

The product of the maximum mixing depth and the average wind speed within the mixing depth is sometimes used as an indicator of the atmosphere's dispersive capability. This product is known as the *ventilation coefficient* (m²/s). Values of ventilation coefficient less than about 6000 m²/s are considered indicative of high air pollution potential (Portelli and Lewis, 1987).

Wind speeds generally increase with height, and it is sometimes helpful to be able to estimate wind at an elevation higher than the standard 10-m weather

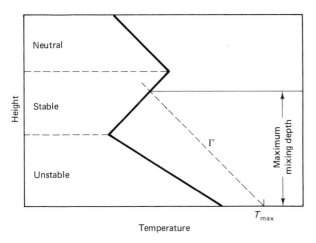

Figure 7.23 The maximum mixing depth is obtained by projecting the dry adiabatic lapse rate line to the point of intersection with the atmospheric temperature profile. Notice that the mixing depth does not necessarily correspond to the layer of unstable air.

station anemometer. The following power law expression is frequently used for elevations less than a few hundred meters above the ground:

$$\frac{u_1}{u_2} = \left(\frac{z_1}{z_2}\right)^p \tag{7.30}$$

where u_1, u_2 = wind speed at the higher and lower elevation, respectively
$\quad\quad z_1$, z_2 = higher and lower elevation, respectively
$\quad\quad\quad p$ = a dimensionless parameter that varies with atmospheric stability

There is considerable uncertainty involved in estimating an appropriate value for the parameter p. The more stable the atmosphere, the higher the windspeed becomes as elevation increases, and the higher p is. The local terrain also affects p. The rougher the terrain, the more the windspeed near the surface will differ from the windspeed higher up. So, increasing surface roughness requires higher values for p.

Table 7.7 gives values for p that are recommended by the EPA for rough surfaces in the vicinity of the measurement anemometer (Peterson, 1978). For smooth terrain, flat fields, or over bodies of water, the values of p given in Table 7.7 should be multiplied by a factor of 0.6. The *stability class* indicators in the table will be further clarified in Table 7.8, when we compute the characteristics of smokestack plumes.

Example 7.6   Ventilation Coefficient

Suppose the atmospheric temperature profile is isothermal (constant temperature) at 20 °C and the estimated maximum daily surface temperature is 25 °C. The weather station anemometer is at a height of 10.0 m in the city (rough terrain). It indicates an average windspeed of 3.0 m/s. Estimate the mixing depth and the ventilation coefficient.

TABLE 7.7 WIND PROFILE EXPONENT p FOR ROUGH
TERRAIN[a]

Stability class	Description	Exponent, p
A	Very unstable	0.15
B	Moderately unstable	0.15
C	Slightly unstable	0.20
D	Neutral	0.25
E	Slightly stable	0.40
F	Stable	0.60

[a] For smooth terrain, multiply p by 0.6; see Table 7.8 for
further descriptions of the stability classifications used
here.
Source: Peterson (1978).

Solution The dry adiabatic lapse rate is 1 °C/100 m, so projecting that lapse rate
from 25 °C at the surface until it reaches the 20 °C isothermal means the mixing depth
will be 500 m as shown in Figure 7.24. Since the temperature profile is isothermal,
let's choose the ''slightly stable'' stability class, with $p = 0.4$, to use in (7.30). We
need to estimate the average windspeed in the 500 m mixing depth. For a quick
estimate, we might use the windspeed at the half-way point, 250 m:

$$\frac{u_1}{3.0} = \left(\frac{250}{10}\right)^{0.4} = 3.6$$

$$u_1 = 3.0 \times 3.6 = 10.9 \text{ m/s} \qquad \text{at 250 m}$$

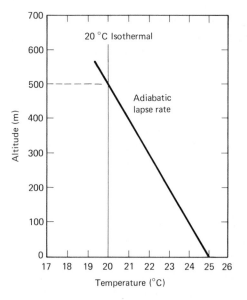

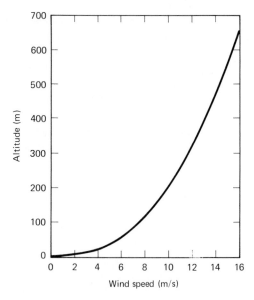

Figure 7.24 Temperature and windspeed for Example 7.6.

which would give a ventilation coefficient of 10.9 m/s $\times$ 500 m = 5450 m^2/s. To be more careful, we could find the average over the 500 m with an integral

$$\bar{u}_1 = \frac{1}{500} \int_0^{500} u_1 \, dz = \frac{1}{500} \int_0^{500} 3.0 \left(\frac{z}{10}\right)^{0.4} dz = 10.2 \text{ m/s}$$

The ventilation coefficient is then 10.2 m/s $\times$ 500 m = 5100 m^2/s. Given the uncertainties in (7.30), the extra work required to do the integration may not be justified.

Smokestack Plumes and Atmospheric Lapse Rates

The atmospheric temperature profile affects the dispersion of pollutants from a smokestack, as shown in Figure 7.25. If a smokestack were to emit pollutants into a neutrally stable atmosphere, we might expect the plume to be relatively symmetrical, as shown in (*a*). The term used to describe this plume is *coning*. In (*b*), the atmosphere is very unstable, and there is rapid vertical air movement, both up and down, producing a *looping* plume. In (*c*), the stable atmosphere greatly restricts the dispersion of the plume in the vertical direction, although it still spreads horizontally. The result is called a *fanning* plume. In (*d*), emissions from a smokestack that is under an inversion layer head downward much more easily than upward. The resulting *fumigation* can lead to greatly elevated downwind, ground-level concentrations.

When the stack is above an inversion layer, as in Figure 7.25*e*, mixing in the upward direction is uninhibited, but downward motion is greatly restricted by the inversion's stable air. Such *lofting* plumes are helpful in terms of exposure to people at ground level. Thus, a common approach to air pollution control has been to build taller and taller stacks to emit pollutants above inversions. An unfortunate consequence of this approach has been that pollutants released from tall stacks are then able to travel great distances, so that effects such as acid deposition can be felt hundreds of miles from the source.

The Point-Source Gaussian Plume Model

At the heart of most every computer program that attempts to relate emissions to air quality is the assumption that the time-averaged pollutant concentration downwind from a source can be modeled using a normal, or Gaussian, distribution curve (for a brief discussion of the Gaussian curve, see Section 3.3). The basic Gaussian dispersion model applies to a single *point source,* such as a smokestack, but it can be modified to account for *line sources* (such as emissions from motor vehicles along a highway), or *area sources* (one can model these as a large number of point sources). To begin, consider just a single point source such as is shown in Figure 7.26. The coordinate system has been set up to show a cross section of the plume, with z representing the vertical direction and x being the distance directly downwind from the source. If we were to observe the plume at any particular instant, it might have some irregular shape, such as the outline of the looping

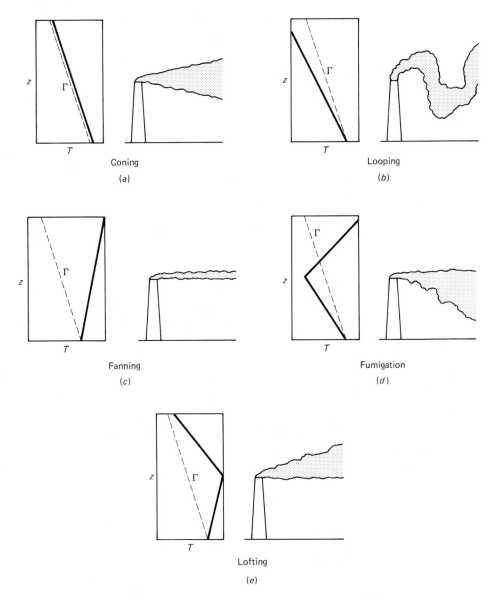

Figure 7.25 Effect of atmospheric lapse rates and stack heights on plume behavior. The dashed line is the dry adiabatic lapse rate for reference. (*Source:* Williamson, S. J. *Fundamentals of Air Pollution,* © 1973 by Addison-Wesley Publishing Co. Reprinted by permission of Addison-Wesley Publishing Co., Inc., Reading, MA.)

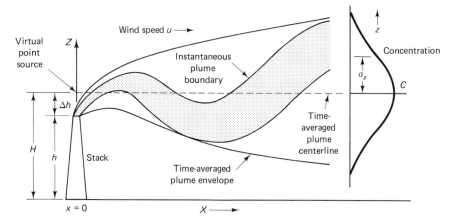

Figure 7.26   The instantaneous plume boundary and a time-averaged plume envelope.

plume shown. A few minutes later, however, the plume might have an entirely different boundary. If we were to set up a camera and leave the shutter open for a while, we could imagine getting a photograph of a time-averaged plume envelope such as that shown in the figure.

Since stack emissions have some initial upward velocity and buoyancy, it might be some distance downwind, before the plume envelope might begin to look symmetrical about a centerline. The centerline would be somewhat above the actual stack height. The highest concentration of pollutants would be along this centerline, with lower and lower values as we get further and further away. The Gaussian plume model assumes that the pollutant concentration follows a normal distribution in both the vertical plane, as is shown in the figure, and in the horizontal direction, not shown. It also treats emissions as if they came from a virtual point source along the plume centerline, at an effective stack height H.

The Gaussian-point source dispersion equation relates average steady-state pollutant concentrations to the source strength, wind speed, effective stack height, and atmospheric conditions. Its form can be derived from basic considerations involving gaseous diffusion in three-dimensional space. The derivation, however, is beyond the scope of this book (see, for example, Wark and Warner, 1981). It is important to note the following assumptions that are incorporated into the analysis:

- The rate of emissions from the source is constant.
- The wind speed is constant both in time and with elevation.
- The pollutant is conservative; that is, it is not lost by decay, chemical reaction, or deposition. When it hits the ground, none is absorbed and all is reflected.
- The terrain is relatively flat, open country.

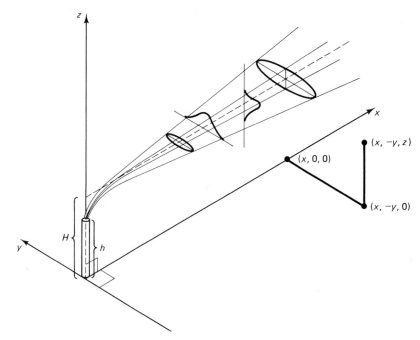

Figure 7.27   Plume dispersion coordinate system, showing Gaussian distributions in the horizontal and vertical directions (Turner, 1970).

The three-dimensional coordinate system established in Figure 7.27 has the stack at the origin, with distance directly downwind given by x, distance off the downwind axis specified by y, and elevation given by z. Since our concern is going to be only with receptors (people and ecosystems) at ground level, the form of the Gaussian plume equation given here is less general than it can be, and applies only for $z = 0$:

$$C(x, y) = \frac{Q}{\pi u \sigma_y \sigma_z} \left(\exp \frac{-H^2}{2\sigma_z^2} \right) \left(\exp \frac{-y^2}{2\sigma_y^2} \right) \qquad (7.31)$$

where $C(x, y)$ = concentration at ground-level at the point (x, y), $\mu g/m^3$

       x = distance directly downwind, m

       y = horizontal distance from the plume centerline, m

       Q = emission rate of pollutants, $\mu g/s$

       H = effective stack height, m, ($H = h + \Delta h$, where h = actual stack height, and Δh = plume rise)

       u = average wind speed at the effective height of the stack, m/s

       σ_y = horizontal dispersion coefficient (standard deviation), m

       σ_z = vertical dispersion coefficient (standard deviation), m

Before we get into the details of (7.31), several features are worth noting. Ground-level pollution concentration is directly proportional to the source strength Q, so it is easy to determine how much source reduction is necessary to achieve a desired decrease in downwind concentration. The units of Q, by the way, have been given in μg so concentrations will be in the usual μg/m^3. Also note that ground-level concentrations decrease for higher stacks, although the relationship is not linear. Finally, downwind concentration appears to be inversely proportional to the wind speed, which is what might be expected intuitively. In actuality, the inverse relationship is slightly modified by the dependence of the plume rise Δh on wind speed. Higher wind speeds reduce the effective height of the stack H, which keeps ground-level pollution from dropping quite as much as a simple inverse relationship would imply.

Finally, and most importantly, although the Gaussian plume equation is based on both theory and actual measured data, it is still, at best, a crude model. Predictions based on the model should be assumed to be accurate to within perhaps ± 50 percent. In spite of that uncertainty, however, it is still very useful since it has almost universal acceptance, it is easy to use, and it allows comparison between the estimates by different modelers in varying situations.

The Gaussian Dispersion Coefficients

The two dispersion coefficients, σ_y and σ_z, need explanation. These are really just the standard deviations of the horizontal and vertical Gaussian distributions, respectively (about 68 percent of the area under a Gaussian curve is within $\pm 1\sigma$ of the mean value). Smaller values for a dispersion coefficient mean the Gaussian curve is narrower, with a higher peak, while larger values mean the opposite. The further downwind we go from the source, the larger these coefficients become. This causes the Gaussian curves to spread further and further. Not only are these coefficients a function of downwind distance, they also depend, in a complex way, on atmospheric stability.

The most common procedure for estimating the dispersion coefficients was introduced by Pasquill (1961), modified by Gifford (1961), and adopted by the U.S. Public Health Service (Turner, 1970); it is presented here as Figure 7.28. The parameters A–F in Figure 7.28 represent stability classifications based on qualitative descriptions of prevailing environmental conditions. Table 7.8 describes these parameters. For example, a clear summer day, with the sun higher than 60° above the horizon and wind speeds less than 2 m/s (at an elevation of 10 m), creates a *very unstable* atmosphere with stability classification A. The opposite extreme is classification F, which is labeled *stable,* and corresponds to a clear night (less than 3/8 of the sky covered by clouds), with winds less than 3 m/s. It should be noted that the stability classifications used in Table 7.7 to estimate wind speeds at elevations other than the standard 10-m anemometer height, are the same as the A–F classifications used here.

It is often easier to use a computer to work with (7.31), in which case the graphical presentation of dispersion coefficients in Figure 7.28 is inconvenient. A reasonable fit to those graphs can be obtained using the following equations

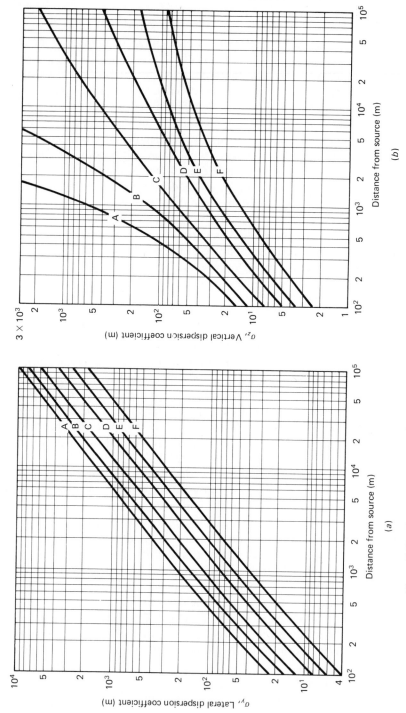

Figure 7.28 Gaussian dispersion coefficients as a function of distance downwind: (a) horizontal coefficient σ_y; (b) vertical coefficient σ_z. (Turner, 1970).

TABLE 7.8 ATMOSPHERIC STABILITY CLASSIFICATIONS

Surface wind speed[a] (m/s)	Day solar insolation			Night cloudiness[e]	
	Strong[b]	Moderate[c]	Slight[d]	Cloudy (≥4/8)	Clear (≤3/8)
<2	A	A–B[f]	B	E	F
2–3	A–B	B	C	E	F
3–5	B	B–C	C	D	E
5–6	C	C–D	D	D	D
>6	C	D	D	D	D

[a] Surface wind speed is measured at 10 m above the ground.

[b] Corresponds to clear summer day with sun higher than 60° above the horizon.

[c] Corresponds to a summer day with a few broken clouds, or a clear day with sun 35–60° above the horizon.

[d] Corresponds to a fall afternoon, or a cloudy summer day, or clear summer day with the sun 15–35° above the horizon.

[e] Cloudiness is defined as the fraction of sky covered by clouds.

[f] For A–B, B–C, or C–D conditions, average the values obtained for each.

Note: A, Very unstable; B, moderately unstable; C, slightly unstable; D, neutral; E, slightly stable; F, stable. Regardless of wind speed, class D should be assumed for overcast conditions, day or night.

Source: Turner (1970).

(Martin, 1976):

$$\sigma_y = ax^{0.894} \tag{7.32}$$

and

$$\sigma_z = cx^d + f \tag{7.33}$$

where the constants a, c, d, and f are given in Table 7.9 for each stability classification. The downwind distance x must be expressed in kilometers to yield σ_y and σ_z in meters. For convenience, a few values for the dispersion coefficients calculated using (7.32) and (7.33) are listed in Table 7.10.

Example 7.7   A Power Plant Plume

A new 40-percent efficient, 1000-MW power plant burns coal with 3 percent sulfur and a heat content of 12 000 Btu/lb. It emits SO_2 at the legally allowable rate from a stack with an effective height equal to 300 m. Predict the ground-level concentration of SO_2 4 km directly downwind. Winds at 10 m height are 2.5 m/s and it is a cloudy summer day.

Solution   To calculate the allowable emission rate, we could use Figure 7.1, which summarizes the new source performance standards for SO_2. This power plant, however, is the same one that was analyzed in Example 7.3 where SO_2 emissions were

TABLE 7.9 VALUES OF THE CONSTANTS, *a, c, d,* AND *f* FOR USE IN (7.32) AND (7.33)[a]

Stability	a	$x \leq 1$ km				$x \geq 1$ km		
		c	d	f		c	d	f
A	213	440.8	1.941	9.27		459.7	2.094	−9.6
B	156	106.6	1.149	3.3		108.2	1.098	2.0
C	104	61.0	0.911	0		61.0	0.911	0
D	68	33.2	0.725	−1.7		44.5	0.516	−13.0
E	50.5	22.8	0.678	−1.3		55.4	0.305	−34.0
F	34	14.35	0.740	−0.35		62.6	0.180	−48.6

[a] The computed values of σ will be in meters when x is given in kilometers.
Source: Martin (1976).

found to be 123 000 lb/day. Converting that to μg/s gives

$$Q = 123\ 000 \text{ lb/day} \times \frac{1000 \text{ g}}{2.2 \text{ lb}} \times \frac{10^6 \ \mu\text{g}}{\text{g}} \times \frac{1 \text{ day}}{86\ 400 \text{ s}} = 6.47 \times 10^8 \ \mu\text{g/s}$$

From Table 7.8 and the given wind speed and solar conditions, we find that the stability classification is C. Note that Table 7.8 uses the wind speed as measured at the standard 10-m anemometer height. The wind speed in the Gaussian plume equation requires that we estimate the wind at the effective stack height. To estimate winds at 300 m we can use (7.30), which requires an estimate of the wind parameter p. From Table 7.7 for stability class C, we determine an appropriate p to be 0.20. Winds at 300 m then can be estimated to be

$$u_1 = u_2 \left(\frac{z_1}{z_2}\right)^p = 2.5 \left(\frac{300}{10}\right)^{0.20} = 4.9 \text{ m/s}$$

At 4 km downwind, Table 7.10 indicates the dispersion coefficients are $\sigma_y = 359$ m and $\sigma_z = 216$ m. Plugging these into (7.31), with $y = 0$ (since the concern is directly downwind) gives

$$C(x, y) = \frac{Q}{\pi u \sigma_y \sigma_z} \left(\exp \frac{-H^2}{2\sigma_z^2}\right)\left(\exp \frac{-y^2}{2\sigma_y^2}\right)$$

$$C(4, 0) = \frac{6.47 \times 10^8 \ \mu\text{g/s}}{\pi \times 4.9 \text{ m/s} \times 359 \text{ m} \times 216 \text{ m}} \left[\exp \frac{-(300)^2}{2 \times (216)^2}\right]$$

$$= 206 \ \mu\text{g/m}^3$$

So, the power plant in Example 7.7 would add 206 μg/m^3 to whatever SO_2 level is already there from other sources. For perspective, let us compare these emissions to ambient air quality standards. Table 7.1 indicates that the annual average SO_2 concentration must be less than 80 μg/m^3. This power plant by itself would cause pollution to greatly exceed that standard if these atmospheric conditions prevailed over the full year. The 24-hr SO_2 standard is 365 μg/m^3, so that under the conditions stated, this plant would not violate that standard.

TABLE 7.10 DISPERSION COEFFICIENTS (m) FOR SELECTED DISTANCES DOWNWIND (km), COMPUTED WITH (7.32) AND (7.33)

Distance x (km)	Stability class and σ_y						Stability class and σ_z					
	A	B	C	D	E	F	A	B	C	D	E	F
0.2	51	37	25	16	12	8	29	20	14	9	6	4
0.4	94	69	46	30	22	15	84	40	26	15	11	7
0.6	135	99	66	43	32	22	173	63	38	21	15	9
0.8	174	128	85	56	41	28	295	86	50	27	18	12
1	213	156	104	68	50	34	450	110	61	31	22	14
2	396	290	193	126	94	63	1953	234	115	51	34	22
4	736	539	359	235	174	117		498	216	78	51	32
8	1367	1001	667	436	324	218		1063	406	117	70	42
16	2540	1860	1240	811	602	405		2274	763	173	95	55
20	3101	2271	1514	990	735	495		2904	934	196	104	59

Example 7.7 opens the door to many interesting questions. How does the concentration vary as distance downwind changes? What would be the effect of changes in wind speed or stability classification? How would we utilize statistical data on wind speed, wind direction, and atmospheric conditions to be sure that all air quality standards will be met? If they will not be met, what are the alternatives? Some examples come to mind. We might raise the stack height (adding to the acid deposition problem); we might increase the efficiency of the scrubber to clean flue gases; or we might use coal with a lower sulfur content. Perhaps energy conservation efforts might reduce the size of the power plant required to meet projected needs. Clearly, to do a proper siting analysis for a new source, such as the power plant in the problem, would require quite a complex study. Although we will not carry out such calculations here, we do have the crucial starting point for such a study, namely the Gaussian plume model.

Downwind Ground-Level Concentration

The ground-level concentrations directly downwind are of interest, since pollution will be highest along that axis. With $y = 0$, (7.31) simplifies to

$$C(x, 0) = \frac{Q}{\pi u \sigma_y \sigma_z} \left(\exp \frac{-H^2}{2\sigma_z^2} \right) \tag{7.34}$$

Plotting (7.34) by hand is somewhat tedious, especially if we want to do a sensitivity analysis to see how the results change with changing stack heights and atmospheric conditions. It is much easier to work this out with a computer, and it is especially simple if a spreadsheet program with graphics capability is used. Using Example 7.7 as a base case, the effect on downwind concentration of changes in stability classification with fixed effective stack height, has been plotted in Figure 7.29(b). The effect of changing effective stack height with fixed stability classification is shown in Figure 7.29a.

The impact of changing stability classification is perhaps unexpected. It is not the stable atmosphere that produces the highest peak downwind concentration. Rather, it is the least stable atmosphere that does. The turbulence in an unstable atmosphere brings the plume to earth very quickly, resulting in high peak values near the stack. Downwind, however, concentrations drop off very quickly. This may be a satisfactory situation as long as any populations or ecosystems that might be damaged by the pollution are more than a few kilometers away. The stable atmosphere, on the other hand, has a much lower peak. However, beyond a few kilometers, the concentration is higher than for an unstable atmosphere and continues to be appreciable for a considerable distance downwind. Although the conclusions are correct, the figure overstates the case somewhat because it assumes fixed effective stack height. In reality, the plume rise is itself a function of stability class; thus, less stable atmospheres have higher effective stack height, producing somewhat lower ground-level concentrations than shown.

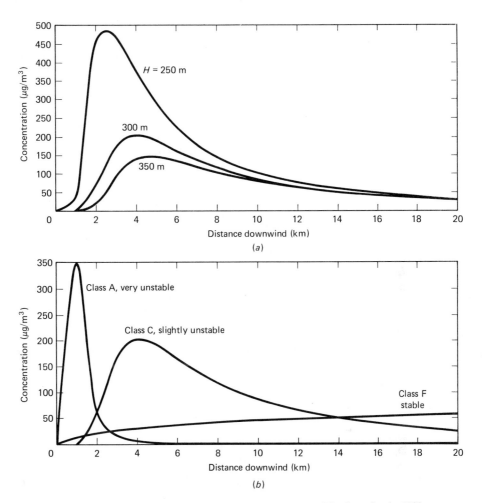

Figure 7.29   The effect of variations in key parameters on SO_2 plume for the 1000-MW coal plant of Example 7.7. (*a*) Effect of effective stack height for constant stability classification, and (*b*) effect of stability classification given a constant effective stack height.

The plume is also quite sensitive to changes in effective stack height, as can be seen in Figure 7.29*a*. Lowering the effective height from 300 to 250 m causes the peak to more than double in concentration. While these heights are enormous, remember that they are not the actual height of the stack itself, which is usually considerably lower. It is not uncommon, for example, for the actual stack height to be less than half of the effective height. There are some tremendously tall stacks, however. The highest one in the world, a smelter in Subdury, Ontario, is as tall as the Empire State Building (380 m).

An obvious question is, How can the peak downwind concentration be predicted from (7.34)? Unfortunately, it is not possible to derive a mathematical solution. One way to predict the peak is to simply plot curves of the sort shown in Figure 7.29 using a computer. With the ready availability of personal computers and spreadsheet programs, that has become an easy enough way to deal with the problem. For hand calculations, however, this approach would be far too tedious. Turner (1970) has derived curves that use the stability classification and effective stack height as parameters to find the distance downwind to the maximum concentration (x_{max}). A normalized concentration $(Cu/Q)_{max}$ is also obtained, from which the maximum concentration can be found using the following:

$$C_{max} = \frac{Q}{u} \left(\frac{Cu}{Q} \right)_{max} \tag{7.35}$$

The curves are presented in Figure 7.30, and the following example illustrates their use.

Example 7.8   Peak Downwind Concentration

For the 1000-MW, coal-fired power plant of Example 7.7, use Figure 7.30 to determine the distance downwind to reach maximum SO_2 concentration. Then find that concentration.

Solution   The stability classification is C and the effective stack height is 300 m. From Figure 7.30, the distance downwind, x_{max}, is just under 4 km (which agrees with Figure 7.29). Reading down from the point on the figure corresponding to stability class C and effective stack height $H = 300$ m, $(Cu/Q)_{max}$ looks to be about 1.5×10^{-6} m^{-2}. Wind speed at the effective height of the stack was found in Example 7.7 to be $u = 4.9$ m/s, and Q was 6.47×10^8 μg/s, so that

$$C_{max} = \frac{Q}{u} \left(\frac{Cu}{Q} \right)_{max}$$

$$= \frac{6.47 \times 10^8 \ \mu g/s}{4.9 \ m/s} \times 1.5 \times 10^{-6} \ m^{-2} = 198 \ \mu g/m^3$$

which is fairly close to the 206 μg/m^3 calculated at 4 km in Example 7.7. It is quite difficult to read Figure 7.30, so high accuracy should not be expected.

Plume Rise

So far, we have dealt only with effective stack height H in the calculations. The difference between the actual stack height h and the effective height H is called the plume rise Δh. Plume rise is caused by a combination of factors, the most important ones being the buoyancy and momentum of the exhaust gases, and the stability of the atmosphere itself. Buoyancy results when exhaust gases are hotter than the ambient air, or when the molecular weight of the exhaust is lower than that of air (or a combination of both factors). Momentum is caused by the mass and velocity of the gases as they leave the stack.

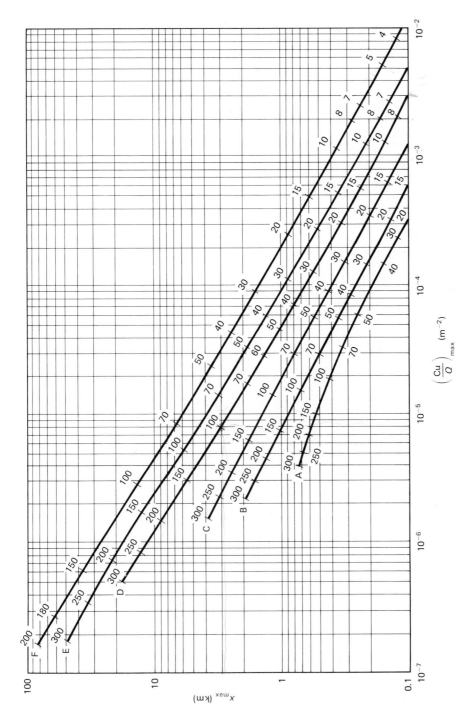

Figure 7.30 To determine the peak downwind plume concentration, enter the graph at the appropriate stability classification and effective stack height (numbers above the lines, in meters) and then move across to find the distance to the peak, and down, to find a parameter from which the peak concentration can be found (Turner, 1970).

Unfortunately, a number of techniques have been proposed in the literature for dealing with plume rise, and they tend to yield very different results. The EPA recommends a model based on work by Briggs (1972), and so we will use it, although with certain restrictions to keep the presentation relatively simple. Although plume rise depends on both momentum and buoyancy, we will assume that the exhaust gases have molecular weight close enough to that of air (taken to be 28.9 g/mol) that we can neglect the buoyancy due to density differences.

It is relatively straightforward to develop a relationship between the buoyancy force on a parcel of exhaust gas and its temperature, exit velocity, and stack diameter (for example, see Perkins, 1974). From such an analysis, the following *buoyancy flux parameter* emerges:

$$F = gr^2 v_s \left(1 - \frac{T_a}{T_s}\right) \tag{7.36}$$

where F = buoyancy flux parameter, m^4/s^3
    g = gravitational acceleration, 9.8 m/s^2
    r = inside radius of the stack, m
    v_s = stack gas exit velocity, m/s
    T_s = stack gas temperature, K
    T_a = ambient temperature, K

For *neutral* or *unstable* conditions in the atmosphere (stability categories A–D), the following equation can be used to estimate plume rise:

$$\Delta h = \frac{1.6 F^{1/3} x_f^{2/3}}{u} \tag{7.37}$$

where Δh = plume rise, m
    u = wind speed at stack height, m/s
    x_f = distance downwind to point of final plume rise, m

Use $x_f = 120 F^{0.4}$ if $F \geq 55$ m^4/s^3

    $x_f = 50 F^{5/8}$ if $F < 55$ m^4/s^3

For *stable,* windy conditions (stability categories E and F), use the following

$$\Delta h = 2.4 \left(\frac{F}{uS}\right)^{1/3} \tag{7.38}$$

where S is a stability parameter with units of s^{-2} given by

$$S = \frac{g}{T_a} \left(\frac{dT_a}{dz} + \Gamma\right) \tag{7.39}$$

Recall that Γ is the adiabatic lapse rate, +0.01 °C/m. The derivative dT_a/dz represents the actual rate of change of ambient temperature with altitude (note that a positive value means temperature is increasing with altitude).

Example 7.9   Plume Rise

A 750-MW coal-fired power plant has a 250-m stack with inside radius 4 m. The exit velocity of the stack gases is estimated at 15 m/s, at a temperature of 140 °C (413 K). Ambient temperature is 25 °C (298 K) and winds at stack height are estimated to be 5 m/s. Estimate the effective height of the stack if (a) the atmosphere is stable with temperature increasing at the rate of 2 °C/km, (b) the atmosphere is slightly unstable, class C.

Solution   First, find the buoyancy parameter F from (7.36):

$$F = gr^2 v_s \left(1 - \frac{T_a}{T_s}\right)$$

$$F = 9.8 \text{ m/s}^2 \times (4 \text{ m})^2 \times 15 \text{ m/s} \times \left(1 - \frac{298}{413}\right) = 655 \text{ m}^4/\text{s}^3$$

a. With the atmosphere stable, we need to use (7.38) and (7.39):

$$S = \frac{g}{T_a}\left(\frac{dT_a}{dz} + \Gamma\right)$$

$$= \frac{9.8 \text{ m/s}^2}{298 \text{ K}}(0.002 + 0.01)\text{K/m} = 0.0004/\text{s}^2$$

$$\Delta h = 2.4 \left(\frac{F}{uS}\right)^{1/3}$$

$$= 2.4 \left(\frac{655 \text{ m}^4/\text{s}^3}{5 \text{ m/s} \times 0.0004/\text{s}^2}\right)^{1/3} = 165 \text{ m}$$

So, the effective stack height is $H = h + \Delta h = 250 + 165 = 415$ m

b. With an unstable atmosphere, class C, we need to use (7.37). Since $F > 55$ m^4/s^3, the distance downwind to the point of final plume rise that should be used is

$$x_f = 120F^{0.4} = 120 \times (655)^{0.4} = 1600 \text{ m}$$

$$\Delta h = 1.6F^{1/3}x_f^{2/3}/u = 1.6(655)^{1/3}(1600)^{2/3}/5 = 380 \text{ m}$$

and the effective stack height is $H = 250 + 380 = 630$ m.

The Briggs model, then, predicts a very substantial plume rise relative to the actual stack height that is quite sensitive to atmospheric conditions.

Downwind Concentration under a Temperature Inversion

The Gaussian plume equation, as presented thus far, applies to an atmosphere where the temperature profile is a simple straight line. If, as is often the case, there is an inversion above the effective stack height, then the basic Gaussian equation must be modified to account for the fact that the vertical dispersion of pollutants is limited by the inversion.

If the pollutants are assumed to reflect off the inversion layer, just as they were assumed to reflect off the ground in the basic Gaussian equation, then an estimate of the concentration at any point downwind would require an analysis of these multiple reflections. That complexity can be avoided if we are willing to restrict our predictions of plume concentration to distances far enough downwind that the summation of these multiple reflections converges into a closed form solution. Beyond that distance, the air is considered to be completely mixed under the inversion, with uniform concentrations from ground level to the bottom of the inversion layer.

Turner (1970) suggests the following modified Gaussian equation to estimate concentrations downwind under an inversion. It is derived based on the assumption that the downwind distance from the source is at least twice the distance to where the plume first interacts with the inversion layer.

$$C(x, y) = \frac{Q}{(2\pi)^{1/2}u\sigma_y L}\left(\exp\frac{-y^2}{2\sigma_y^2}\right) \qquad \text{for } x \geq 2X_L \qquad (7.40)$$

where L = elevation of the bottom of the inversion layer (m)

X_L = the distance downwind where the plume first encounters the inversion layer

Notice that (7.40) is applicable only for distances $x \geq 2X_L$ (see Figure 7.31). That distance, X_L, occurs at the point where the vertical dispersion coefficient, σ_z, is equal to

$$\sigma_z = 0.47(L - H) \qquad (\text{at } x = X_L) \qquad (7.41)$$

Once σ_z is found from (7.41), the distance X_L can be estimated using Figure 7.28*b*, or using Eq. 7.33 and Table 7.9. For distances $x \leq X_L$, the standard Gaussian plume equation (7.31) can be used to estimate downwind concentrations. For $X_L \geq 2X_L$, (7.40) applies. For distances between X_L and $2X_L$, concentrations can be estimated by interpolating between the values computed for $x = X_L$ and $x = 2X_L$.

Example 7.10 Concentration under an Inversion Aloft

Consider a stack with effective height 100 m, emitting SO$_2$ at the rate of 200 g/s. Winds at 10-m elevation are blowing at 5.5 m/s and at the stack height they are 10

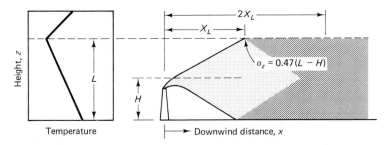

Figure 7.31 Plume dispersion under an elevated inversion. Equation 7.40 applies for distances greater than $2X_L$ downwind, where X_L occurs at the point where $\sigma_z = 0.47(L - H)$.

m/s. It is a clear summer day with the sun nearly overhead and an inversion layer starting at 300 m. Estimate the ground-level SO_2 concentration at a distance downwind twice that where reflection begins to occur from the inversion.

Solution   Equation 7.41 gives σ_z at the distance X_L,

$$\sigma_z = 0.47\,(L - H) = 0.47 \times (300 - 100) = 94 \text{ m}$$

To find X_L we can use Figure 7.28b, but first we need the stability classification below the inversion. From Table 7.8, a clear summer day with 5.5 m/s wind speed corresponds to stability class C. Entering Figure 7.28b on the vertical axis at $\sigma_z = 94$ m, then going across to the class C line and dropping down to the horizontal axis leads to an estimate for X_L of about 1600 m. (Note the difficulty in reading this value from the figure; we could also use (7.33) with Table 7.9 to obtain an estimate analytically.)

To find the concentration at a distance $x = 2X_L = 2 \times 1600$ m $= 3200$ m, we can use (7.40), but first we need to estimate σ_y at that point. Using Figure 7.28a at $x = 3200$ m, and class C, we can estimate σ_y to be about 300 m (again, we could have used Eq. 7.32 with Table 7.9). The concentration directly downwind corresponds to $y = 0$, so

$$C(2X_L, 0) = \frac{Q}{(2\pi)^{1/2}u\sigma_y L} = \frac{200 \text{ g/s}}{(2\pi)^{1/2}\,10 \text{ m/s} \times 300 \text{ m} \times 300 \text{ m}}$$

$$= 89 \times 10^{-6} \text{ g/m}^3 = 89 \text{ } \mu\text{g/m}^3$$

A Line-Source Dispersion Model

In some circumstances it is appropriate to model sources distributed along a line as if they formed a continuously emitting, infinite line source. Examples of line sources that might be modeled this way include motor vehicles traveling along a straight section of highway, agricultural burning along the edge of a field, or a line of industrial sources on the banks of a river. For simplicity, we will consider only the case of an infinite-length source at ground level, with winds blowing perpendicular to the line. Under these specialized circumstances, the ground-level concentration of pollution at distance x from the line source can be described by the following:

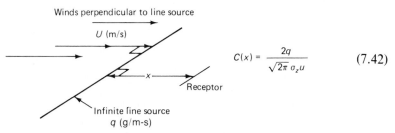

$$C(x) = \frac{2q}{\sqrt{2\pi}\,\sigma_z u} \tag{7.42}$$

where q is an emission rate per unit of distance along the line (g/m-s).

Example 7.11   CO Near a Freeway

The federal emission standards for CO from new vehicles is 3.4 g/mile. Suppose a highway has 10 vehicles per second passing a given spot, each emitting 3.4 g/mile of

CO. If the wind is perpendicular to the highway and blowing at 5 mph (2.2 m/s) on an overcast day, estimate the ground-level CO concentration 200 m from the freeway.

Solution   We need to estimate the CO emission rate per meter of freeway,

$$q = 10 \text{ vehicles/s} \times 3.4 \text{ g/veh-mile} \times 1 \text{ mile}/1609 \text{ m} = 0.021 \text{ g/m-s}$$

and we need the vertical dispersion coefficient σ_z. Since this is an overcast day, the footnotes of Table 7.8 indicate we should use stability classification D. Table 7.10 indicates that at a distance of 200 m (0.2 km) for class D, σ_z is 9 m. Substituting these values into (7.42) gives

$$C(0.2 \text{ km}) = \frac{2 \times 0.021 \text{ g/m-s} \times 10^3 \text{ mg/g}}{(2\pi)^{1/2} \times 2.2 \text{ m/s} \times 9 \text{ m}} = 0.845 \text{ mg/m}^3$$

The 1-hr CO air quality standard is 40 mg/m³, so this estimate is well within the air quality standard.

Area-Source Models

For distributed sources, there are a number of approaches that can be taken to estimate pollutant concentrations. If there are a modest number of point sources, it is reasonable to use the point-source Gaussian plume equation for each source to predict its individual contribution. Then, by superposition, find the total concentration at a given location by summing the individual contributions. Multiple use of the Gaussian line-source equation is another approach. By dividing an area into a series of parallel strips, then treating each strip as a line source, the total concentration on any strip can be estimated.

A much simpler, more intuitive approach can be taken to estimate pollutant concentrations over an area (such as a city) by using the box model concepts introduced in Section 1.3. Consider the airshed over an urban area to be represented by a rectangular box, such as is shown in Figure 7.32, with base dimensions L and W and height H. The box is oriented so that wind, with speed u, is normal to one side of the box. The height of the box is determined by atmospheric

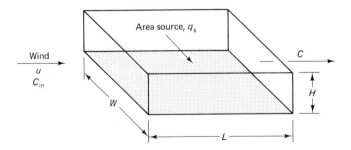

Figure 7.32   Box model for an airshed over a city. Emissions per unit area are given by q_s, pollutants are assumed to be uniformly mixed in the box with concentration C, while upwind of the box the concentration is C_{in}.

conditions, and we could consider it to be just the mixing depth. Emissions per unit area will be represented by q_s (g/m²-s).

Consider the air blowing into the box on the upwind side to have pollutant concentration C_{in} and, for simplicity, assume that no pollution is lost from the box along the sides parallel to the wind or from the top. We will also assume that the pollutants are rapidly and completely mixed in the box, creating a uniform average concentration C. Finally, we will treat the pollutants as if they are conservative; that is, they do not react, decay, or fall out of the airstream. All of these restrictions can be modified in more sophisticated versions of a box model.

Working with pollutant mass, the amount of pollution in the box is the volume of the box times the concentration, $LWHC$. The rate at which air is entering and leaving the box is the area of either end times the wind speed, WHu, so the rate at which pollution is entering the box is $WHuC_{in}$. The rate that it leaves the box is $WHuC$. If we assume the pollutant is conservative, then we can write the following mass balance for the box:

$$\begin{pmatrix} \text{Rate of change of pollution} \\ \text{in the box} \end{pmatrix} = \begin{pmatrix} \text{Rate of pollution} \\ \text{entering the box} \end{pmatrix} - \begin{pmatrix} \text{Rate of pollution} \\ \text{leaving the box} \end{pmatrix}$$

or

$$LWH \frac{dC}{dt} = q_s LW + WHuC_{in} - WHuC \qquad (7.43)$$

where C = pollutant concentration in the airshed
$\quad C_{in}$ = concentration in the incoming air
$\quad q_s$ = emission rate per unit of area
$\quad H$ = mixing height
$\quad L$ = length of airshed
$\quad W$ = width of airshed
$\quad u$ = average wind speed against one edge of the box

The steady-state solution to (7.43) can be obtained by simply setting $dC/dt = 0$, so that

$$C(\infty) = \frac{q_s L}{uH} + C_{in} \qquad (7.44)$$

which looks reasonable. If the air entering the box is clean, the steady-state concentration is proportional to the emission rate and inversely proportional to the ventilation coefficient (the product of mixing depth and wind speed). If it is not clean, we just add the effect of the incoming concentration. We can also solve Eq. 7.43 to obtain the time-dependent increase in pollution above the city. Letting $C(0)$ be the concentration in the airshed above the city (the box) at time $t = 0$, the solution becomes

$$C(t) = \left(\frac{q_s L}{uH} + C_{in} \right)(1 - e^{-ut/L}) + C(0)e^{-ut/L} \qquad (7.45)$$

If we assume the incoming wind blows no pollution into the box and if the initial concentration in the box is zero, then Eq. 7.45 simplifies to

$$C(t) = \frac{q_s L}{uH} (1 - e^{-ut/L}) \tag{7.46}$$

When $t = L/u$, the exponential function becomes e^{-1} and the concentration reaches about 63 percent of its final value. That value of time has various names. It is called the *time constant*, the *ventilation time*, or the *residence time*.

Example 7.12 Evening Rush Hour Traffic

Suppose that within a square city, 15 km on a side, there are 200 000 cars on the road, each being driven 30 km between 4 pm and 6 pm, and each emitting 3 g/km of CO. It is a clear winter evening with a radiation inversion that restricts mixing to estimated 20.0 m, and the wind is bringing clean air in at a steady rate of 1.0 m/s along an edge of the city. Use a box model to estimate the CO concentration at 6 pm if there was no CO in the air at 4 pm, and the only source of CO is cars. Assume that CO is conservative and that there is complete and instantaneous mixing in the box.

Solution The emissions per m^2, q_s, would be

$$q_s = \frac{200\ 000 \text{ cars} \times 30 \text{ km/car} \times 3 \text{ g/km}}{(15 \times 10^3 \text{ m})^2 \times 3600 \text{ s/hr} \times 2 \text{ hr}} = 1.1 \times 10^{-5} \text{ g/m}^2\text{-s}$$

The concentration after 2 hr (7200 s) would be

$$C(t) = \frac{q_s L}{uH} (1 - e^{-ut/L})$$

$$C(2 \text{ hr}) = \frac{1.1 \times 10^{-5} \text{ g/m}^2\text{-s} \times 15 \times 10^3 \text{ m}}{1.0 \text{ m/s} \times 20.0 \text{ m}} \left[1 - \exp\left(\frac{1.0 \text{ m/s} \times 7200 \text{ s}}{15\ 000 \text{ m}}\right)\right]$$

$$= 3.2 \times 10^{-3} \text{ g/m}^3 = 3.2 \text{ mg/m}^3$$

which is considerably below both the 1-hr NAAQS for CO of 40 mg/m^3, and the 8-hr standard of 10 mg/m^3. Any CO that was already in the air at 4 pm would, of course, increase this estimate. The time constant, $L/u = 15\ 000 \text{ m}/(1 \text{ m/s}) = 15\ 000 \text{ s} = 4.2$ hr, suggests that in these 2 hr, the concentration is well below what it would become if these conditions were to continue.

7.6 INDOOR AIR QUALITY

So far, our attention has been directed toward ambient air quality. However, people tend to spend more time indoors than out and, in many circumstances, the air we breathe indoors is even more polluted than outdoor air. Some of the pollutants are the same ones that we are now familiar with from our study of ambient air quality. Combustion that takes place inside of homes and buildings to heat water, cook, and provide space heating can produce elevated levels of carbon monoxide and nitrogen oxides, for example. Another example would be certain photocopying machines that emit ozone. Other pollutants are somewhat unique to the indoor air quality problem, such as formaldehyde emissions from particleboard, plywood, urea-formaldehyde foam insulation, various adhesives, and other building materials; asbestos used for fireproofing and insulation; and various volatile organics emitted from household cleaning products. Many pollutants, such as cigarette smoke or radon gas, if they are emitted outdoors have plenty of dilution air so that people tend not to be exposed to hazardous levels of contamination. Indoors, however, these pollutants can be concentrated, leading all too often to harmful exposure levels.

Table 7.11 summarizes some of these pollutants, their sources, and exposure guidelines if any are available. Although the list is extensive, and each pollutant is important, tobacco smoke and radon gas deserve special attention. Smoking, for example, is blamed for over half a million deaths per year in the United States, while radon may be causing as many as 20 000 deaths per year. Radon was introduced in Chapter 1, and more will be said about it in this section.

TABLE 7.11 SOURCES AND EXPOSURE GUIDELINES OF INDOOR AIR CONTAMINANTS

Pollutant and indoor sources	Guidelines, average concentrations
Asbestos and other fibrous aerosols Friable asbestos; fireproofing, thermal and acoustic insulation, decoration. Hard asbestos: vinyl floor and cement products.	0.2 fibers/mL for fibers longer than 5 μm
Carbon monoxide Kerosene and gas space heaters, gas stoves, wood stoves, fireplaces, smoking.	10 mg/m^3 for 8 hr, 40 mg/m^3 for 1 hr
Formaldehyde Particleboard, paneling, plywood, carpets, ceiling tile, urea-formaldehyde foam insulation, other construction materials.	120 μg/m^3
Inhalable particulate matter Smoking, vacuuming, wood stoves, fireplaces.	55–110 μg/m^3 annual, 150–350 μg/m^3 for 24 hr
Nitrogen dioxide Kerosene and gas space heaters, gas stoves.	100 μg/m^3 annual
Ozone Photocopying machines, electrostatic air cleaners.	235 μg/m^3/hr once a year
Radon and radon progeny Diffusion from soil, groundwater, building materials	0.01 working levels annual
Sulfur dioxide Kerosene space heaters.	80 μg/m^3 annual, 365 μg/m^3 24-hr
Volatile organics Cooking, smoking, room deodorizers, cleaning sprays, paints, varnishes, solvents, carpets, furniture, draperies.	None available

Source: Nagda et al. (1987).

Tobacco smoke is especially important since it contains numerous known or suspected carcinogens, including benzene, hydrazine, benzo-α-anthracene, benzo-α-pyrene (BaP), and nickel. Smoke particles are small, averaging about 0.2 μm, so they are easily carried into the deepest regions of the lungs. A single cigarette gives off on the order of 10^{12} particles, most of which are released while the cigarette is simply smoldering in the air (*sidestream* smoke) rather than when a smoker takes a puff (*mainstream* smoke). Smokers, of course, voluntarily expose themselves to the greatest concentrations, although nonsmokers can be exposed to significant amounts as well. Other indoor air pollutants arising from tobacco smoke include carbon monoxide, nicotine, nitrosamines, acrolein and other aldehydes.

Another potentially important source of indoor air pollution is caused by wood-burning stoves and fireplaces. Wood combustion produces CO, NO_x, hydrocarbons, and respirable particles, and some emissions that are suspected carcinogens, such as benzo-α-pyrene. While most wood smoke is released to the outdoors, where it has become a particularly significant cause of ambient air pollution in some areas, some fraction remains inside.

Infiltration and Ventilation

Just as with outdoor air, the amount of air available to dilute pollutants is an important indicator of likely contaminant concentrations. Indoor air can be exchanged with outdoor air by any combination of three mechanisms: infiltration, natural ventilation, and forced ventilation. *Infiltration* is the term used to describe the natural air exchange that occurs between a building and its environment when doors and windows are closed. That is, it is leakage that occurs through various cracks and holes that exist in the building envelope. *Natural ventilation* is the air exchange that occurs when windows or doors are purposely opened to increase air circulation, while *forced ventilation* occurs when mechanical air handling systems induce air exchange using fans or blowers.

Large amounts of energy are lost when conditioned air (heated or cooled) that leaks out of buildings is replaced by outside air that must be mechanically heated or cooled to maintain desired interior temperatures. It is not uncommon, for example, for one-third of a home's space heating and cooling energy requirements to be caused by unwanted leakage of air, that is, infiltration. Nationwide, about 10 percent of total U.S. energy consumption is accounted for by infiltration, which translates into tens of billions of dollars worth of wasted energy each year. Since infiltration is quite easily, and cheaply, controlled, it is not surprising that tightening buildings has become a popular way to help save energy. Unfortunately, in the process of saving energy, we may be exacerbating indoor air quality problems.

Air leaks in and out of buildings through numerous cracks and openings in the building envelope. The obvious cracks around windows and doors are the

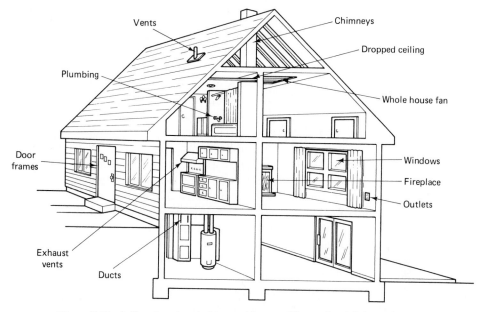

Figure 7.33   Infiltration sites in homes. (*Source:* Diamond and Grimsrud, 1984.)

usual ones we try to plug with caulk and weatherstripping, but there are many less obvious, but potentially more important, leakage areas, such as those created when plumbing, ducts, and electrical wiring penetrate walls, ceilings, and floors; fireplaces without dampers; ceiling holes created around recessed light fixtures, attic access hatches, and any other bypasses created between heated spaces and the attic; gaps where foundations are connected to walls; exhaust vents in bathrooms and kitchens; and leaky ductwork in homes with forced-air heating systems. Figure 7.33 shows some of these infiltration sites, while Table 7.12 indicates the relative importance of typical leakage rates in typical American homes.

TABLE 7.12   DISTRIBUTION OF AIR
LEAKAGE FOR A TYPICAL U.S. RESIDENCE

Component	Range (%)	Average (%)
Walls/floor	18–50	35
Ceiling	3–30	18
Heating system	3–28	15
Windows/doors	6–22	15
Fireplace	0–30	12
Vents	2–12	5

Source: Diamond and Grimsrud (1984).

Infiltration is driven by pressure differences between the inside of the building and the outdoor air. These pressure differences can be caused by wind, or by inside-to-outside temperature differences. Wind blowing against a building creates higher pressure on one side of the building than the other, inducing infiltration through cracks and other openings in the walls. Temperature-induced infiltration (usually referred to as the *stack effect*) is influenced less by holes in the walls than by various openings in the floors and ceilings. In the winter, warm air in a building wants to rise, exiting through breaks in the ceiling and drawing in colder air through floor openings. Thus, infiltration rates are influenced not only by how fast the wind is blowing and how great the temperature difference is between inside and out, but also by the locations of the leaks in the building envelope. Greater leakage areas in the floor and ceiling encourage stack-driven infiltration, while leakage areas in vertical surfaces encourage wind-driven infiltration.

Moreover, while it is usually assumed that increasing the infiltration rate will enhance indoor air quality, that may not be the case in one important circumstance, namely for radon that is emitted from the soil under a building. For radon, wind-driven infiltration helps reduce indoor concentrations by allowing relatively radon-free fresh air to blow into the building. Stack-driven infiltration, which draws air through the floor, may actually encourage new radon to enter the building, negating the cleaning that infiltration usually causes. Figure 7.34 illustrates these important differences.

Infiltration rates may be expressed in units such as m³/hr or cubic feet per minute (cfm), but more often the units are given in air changes per hour (ach). The air exchange rate in air changes per hour is simply the number of times per hour that a volume of air equal to the volume of space in the house is exchanged with outside air. Typical average infiltration rates in American homes range from about 0.5 to 1 ach, with newer houses being more likely to have rates at the lower end of the scale, while older homes are closer to the top end. Some very poorly built houses have rates as high as 3–4 ach. Carefully constructed new homes today can quite easily be built to achieve infiltration rates that are as low as 0.1 ach by using continuous plastic sheet "vapor barriers" in the walls, along with careful application of foam sealants and caulks to seal cracks and holes. At such low infiltration rates, moisture and pollutant buildup can be serious enough to require extra ventilation, and the trick is to get that ventilation without throwing away the heat that the outgoing stale air contains.

One way to get extra ventilation with minimal heat loss is with a mechanical *heat-recovery ventilator* (HRV), in which the warm, outgoing stale air transfers much of its heat to the cold, fresh air being drawn into the house. Another simpler and cheaper approach is to provide mechanical ventilation systems that can be used intermittently in the immediate vicinity of concentrated sources of pollutants. Exhaust fans in bathrooms and range hoods over gas stoves, for example, can greatly reduce indoor pollution, and by using them only as necessary, heat losses can be modest.

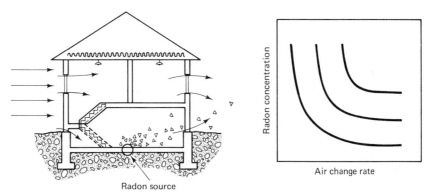

Radon source

Note: Air is driving radon gas out of the house to the right.

(*a*) Infiltration dominated by wind effects.

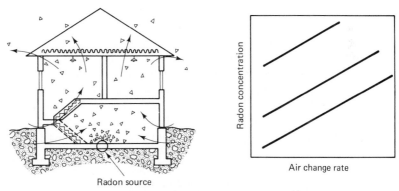

Radon source

Note: Radon gas is being drawn in from the bottom of the house.

(*b*) Infiltration dominated by stack effects.

Figure 7.34 Wind-driven (*a*) and stack-driven (*b*) infiltration may cause opposite effects on indoor radon concentrations. (*Source:* Reece, 1988.)

An Indoor Air Quality Model

It is quite straightforward to apply the box model concepts developed earlier to the problem of indoor air quality. The simple model we will use treats the building as a single, well-mixed box, with sources and sinks for the pollutants in question. If necessary, the simple model can be expanded to include several boxes, each characterized by uniform pollutant concentrations. A two-box model, for example, is sometimes used for radon estimates, where one box is used to model radon concentrations within the living space of a dwelling and the other models the air space beneath the house.

Consider the simple, one-box model of a building shown in Figure 7.35. There are sources of pollution within the building that can be characterized by

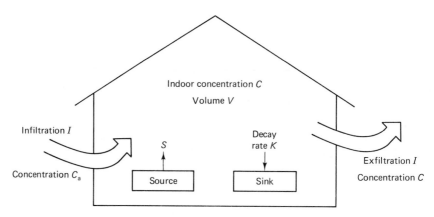

Figure 7.35   Box model for indoor air pollution.

various emission rates. In addition, ambient air entering the building may bring new sources of pollution, which adds to whatever may be generated inside. Those pollutants may be removed from the building by infiltration or ventilation, or they may be nonconservative and decay with time. In addition, if there is a mechanical air cleaning system, some pollutants may be removed as indoor air is passed through the cleaning system and returned. To help keep the model simple, we will ignore such mechanical filtration.

A basic mass balance for pollution in the building, assuming well-mixed conditions, is

$$\begin{pmatrix} \text{Rate of increase} \\ \text{in the box} \end{pmatrix} = \begin{pmatrix} \text{Rate of pollution} \\ \text{entering the box} \end{pmatrix}$$

$$- \begin{pmatrix} \text{Rate of pollution} \\ \text{leaving the box} \end{pmatrix} - \begin{pmatrix} \text{Rate of decay} \\ \text{in the box} \end{pmatrix}$$

$$V \frac{dC}{dt} = S + C_a I V - C I V - K C V \qquad (7.47)$$

where V = volume of conditioned space in building (m³/air change)
 I = air exchange rate (ach)
 S = pollutant source strength (mg/hr)
 C = indoor concentration (mg/m³)
 C_a = ambient concentration (mg/m³)
 K = pollutant decay rate or reactivity (1/hr)

The steady-state solution can be found by setting $dC/dt = 0$, yielding

$$C(\infty) = \frac{S/V + C_a I}{I + K} \qquad (7.48)$$

A general solution is

$$C(t) = \frac{S/V + C_a I}{I + K} (1 - e^{-(I+K)t}) + C(0)e^{-(I+K)t} \tag{7.49}$$

where $C(0)$ = the initial concentration in the building.

Some of the pollutants that we might want to model, such as CO and NO, can be treated as if they were conservative; that is, they do not decay with time or have significant reactivities; thus, $K = 0$. A special case for conservative pollutants, corresponding to negligible ambient concentration ($C_a = 0$), and zero initial indoor concentration, is

$$C(t) = \frac{S}{IV} (1 - e^{-It}) \tag{7.50}$$

Example 7.13   A Portable Kerosene-Fired Heater

An unvented, portable, radiant heater, fueled with kerosene, is tested under controlled laboratory conditions. After running the heater for 2 hr in a test chamber with a 46.0-m³ volume and an infiltration rate of 0.25 ach, the concentration of CO reaches 20 mg/m³. Initial CO in the lab is zero, and the ambient CO level is negligible throughout the run. Treating CO as a conservative pollutant, find the rate at which the heater emits CO. If the heater were to be used in a small home to heat 120 m³ of space having 0.4 ach, predict the steady-state concentration.

Solution   Rearranging (7.50) to solve for the source rate S gives

$$S = \frac{IVC(t)}{1 - e^{-It}} = \frac{0.25 \text{ ach} \times 46.0 \text{ m}^3/\text{ac} \times 20 \text{ mg/m}^3}{1 - e^{-0.25/\text{hr} \times 2 \text{ hr}}} = 585 \text{ mg/hr}$$

In the small home, the steady-state concentration would be

$$C(\infty) = \frac{S}{IV} = \frac{585 \text{ mg/hr}}{0.4 \text{ ach} \times 120 \text{ m}^3/\text{ac}} = 12.2 \text{ mg/m}^3$$

which exceeds the 8-hr ambient standard of 10 mg/m³ for CO.

Portable kerosene-fired heaters, such as the one introduced in the example, became quite popular soon after the energy crises of the 1970s. They are designated as either being of the radiant type or convective type. Radiant heaters heat objects in a direct line of sight by radiation, while convective heaters heat the air, which, in turn, heats objects in the room by convection. Radiant heaters operate at a lower temperature, so their NO$_x$ emissions are reduced, but their CO emissions are higher than the hotter running convective types. Both types of kerosene heaters have high enough emission rates to cause concern if used in enclosed spaces such as a home or trailer. Some average emission rates for these heaters, as well as for unvented gas ranges, are given in Table 7.13, and estimates of appropriate decay constants (or reactivity constants) are given in Table 7.14.

TABLE 7.13 SOME MEASURED POLLUTANT EMISSION RATES
FOR VARIOUS SOURCES

Source	Pollutant emission rate, (mg/hr)			
	CO	NO_x^a	SO_2	HCHO
Gas range				
Oven	1900	52	0.9	23
One top burner	1840	83	1.5	16
Kerosene heater[b]				
Convective	71	122	—	1.1
Radiant	590	15	—	4.0
One cigarette				
(sidestream smoke)[c] (mg)	86	0.05	—	1.44

[a] NO_x reported as N.

[b] New portable heaters, warm-up emissions not included, SO_2 not measured.

[c] *Source:* National Research Council (1981).

Sources: Traynor et al. (1981, 1982).

Radon

By far the most serious indoor air pollutant is radon gas. Radon gas and its radioactive daughters are believed to be causing between 5000 and 20 000 lung cancer deaths per year in the United States alone. Recall from Section 2.5, that radon-222 is a radioactive gas, with a half-life of 3.8 days, that is part of a natural decay chain beginning with uranium and ending with lead. Radon is chemically inert, but its short-lived decay products—polonium, lead, and bismuth—are chemically active and easily become attached to inhaled particles that can lodge in the lungs. In fact, it is the alpha-emitting polonium, formed as radon-222 decays, that causes the greatest lung damage.

TABLE 7.14 DECAY CONSTANTS
OR REACTIVITIES, K

Pollutant	K (1/hr)
CO	0.0
NO	0.0
NO_x (as N)	0.15
HCHO	0.4
SO_2	0.23
Particles (<0.5 μm)	0.48
Radon	7.6×10^{-3}

Source: Traynor et al. (1981).

Radon can be emitted from some earth-derived building materials such as brick, concrete, and tile. It can also exist in groundwater, to be released when that water is aerated, for example, during a shower, (the radon risk in water is from inhalation of the released gas, not from drinking the water itself). Radon has also been detected in some natural gas supplies, so modest amounts may be released during cooking. By far the most important source of radon, however, is soils and rocks that contain radium. The radon gas formed as radium decays can work its way to the surface. From there it can enter a house through the floor, especially if there is stack driven infiltration. Figure 7.36 shows major radon entry routes into a home from soil, building materials, and groundwater.

Radon-rich, high-permeability soils are considered to be causing the highest levels of indoor radon concentrations. The emanation rate from soil seems to range from about 0.1 pCi/m²-s to over 100 pCi/m²-s, with a value of 1 pCi/m²-s being fairly typical. This large variation in emission rates is the principal cause of variations in indoor radon concentrations measured across the country. The infiltration rate is also a key variable, but a much less important one. When emission rates are low, as is the case for most of the United States, homes can be tightly constructed for energy efficiency without concern for elevated levels of radon. On the other hand, in areas where soil radon emissions are high, tight building construction techniques can exacerbate the radon problem.

The following example shows how we can estimate the radon concentration in a home, using the indoor air pollution model of the previous section.

Example 7.14   Indoor Radon Concentration

Suppose the soil under a single-story house emits 1.0 pCi/m²-s of radon gas. As a worst case, assume that all of this gas finds its way through the floor and into the house. The house has 250 m² of floor space, an average ceiling height of 2.6 m , and an air change rate of 0.9 ach. Estimate the steady-state concentration of radon in the house, assuming the ambient concentration is negligible.

Solution   We can find the steady-state concentration using (7.48). The decay rate of radon is given in Table 7.14 as $K = 7.6 \times 10^{-3}$/hr, or we could have calculated that value from (3.7), which shows the relationship between K and half-life.

$$C(\infty) = \frac{S/V + C_a I}{I + K}$$

$$= \frac{(1 \text{ pCi/m}^2\text{s} \times 3600 \text{ s/hr} \times 250 \text{ m}^2)/(250 \text{ m}^2 \times 2.6 \text{ m})}{0.9/\text{hr} + 7.6 \times 10^{-3}/\text{hr}}$$

$$= 1.5 \times 10^3 \text{ pCi/m}^3 = 1.5 \text{ pCi/L}$$

This example house has a radon concentration of 1.5 pCi/L, which is apparently fairly typical in the United States. The EPA is suggesting a radon guideline of 4 pCi/L as a level at which residents might consider taking some remedial action, and 8 pCi/L as a level at which action is recommended.

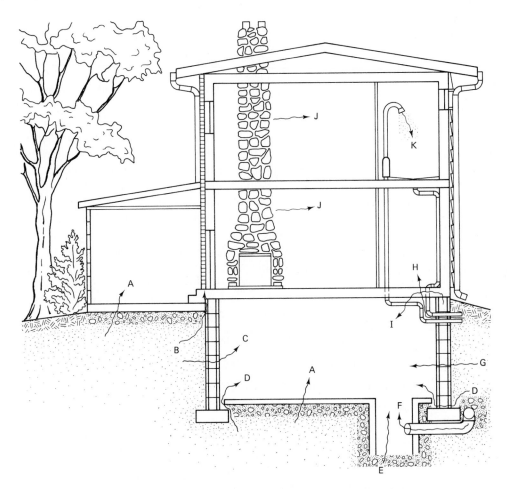

Figure 7.36 Major radon entry points into a home: A, cracks in concrete slabs; B, spaces behind brick veneer walls that rest on hollow-block foundation; C, pores and cracks in concrete blocks; D, floor–wall joints; E, exposed soil, as in a sump; F, weeping (drain) tile, if drained to open sump; G, mortar joints; H, loose-fitting pipe penetrations; I, open tops of block walls; J, building materials, such as some rock; K, water (from some wells). (USEPA, 1987.)

A key question, of course, is how these levels of radon translate into risks of lung cancer. The answer is based in large part on historical data for underground uranium miners who have been exposed to high concentrations of radon and for whom the incidence rate of lung cancer is much higher than that of the general population. Translating data from miners, however, to risks in residences introduces large uncertainties. Working miners, for example, would be expected to have higher breathing rates than people resting at home. They also would be expected to be exposed to much greater concentrations of particulate matter, along with the radon. Much of the danger associated with radon is caused by radon daughters that attach themselves to small particulate matter that is inhaled deep into the lungs. The combination of cigarette smoke and radon is, as a result, an especially deadly combination.

While we are describing radon concentrations in terms of picocuries per liter, there is another unit, the *working level,* that is sometimes used. One working level (WL) is defined as 100 pCi/L of radon-222 in equilibrium with its progeny. This strange sounding unit has its roots in the historical lung cancer data of miners. One *working level month* (WLM) is the exposure that miners would experience if exposed to one working level for 173 hr (the hours in one work month). Miners exposed to 1 WLM have an increased risk of lung cancer of about 300 in one million.

Nero (1986) estimates that people living in an average 1.5-pCi/L house all their lives will have an exposure of about 0.3 WLM per year, or a total of about 20 WLM. Their lung cancer risk is estimated to be about 0.3 percent, give or take a factor of two or three. While that risk is small relative to voluntary actions such as driving or smoking, it is extremely large compared to allowable risks for other carcinogens that are regulated by the EPA. As was mentioned in Chapter 5, standards for most carcinogens are established at levels that control lifetime risk to 0.01–0.00001 percent. At 4 pCi/L, the risk is comparable to 200 chest x rays per year. Between 10 and 20 pCi/L, the lung cancer risk is equivalent to that faced by a one-pack-a-day smoker (USEPA, 1986).

Available techniques to help reduce indoor radon concentrations depend somewhat on what type of floor construction has been employed. Many houses in the United States are built over basements that may or may not be heated; other homes, especially in the west, are built over crawl spaces or on concrete slabs. With every type of construction, mitigation begins with efforts to seal any and all cracks and openings between the floor and the soil beneath. For houses with basements or crawl spaces, increased ventilation of those areas, either by natural or mechanical means can be very effective. If the basement is heated, a heat recovery ventilator may be called for to avoid wasting excessive amounts of energy. One of the most widely used radon reduction techniques uses pipes that penetrate a slab floor and fans that suck radon gas out of the aggregate beneath the floor.

Since stack-driven infiltration can draw radon gas out of the soil, efforts should be taken to seal spaces in ceilings as well as floors. Avoiding recessed light fixtures that create openings directly into the attic space, and sealing attic access stairs can help. Providing combustion air directly from the outside to fireplaces or woodstoves, rather than letting it come from interior spaces, can also reduce the vacuum effect created when hot flue gases are released. Several of these mitigation measures are illustrated in Figure 7.37.

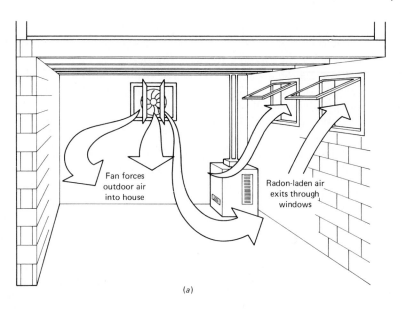

Fan forces outdoor air into house

Radon-laden air exits through windows

(a)

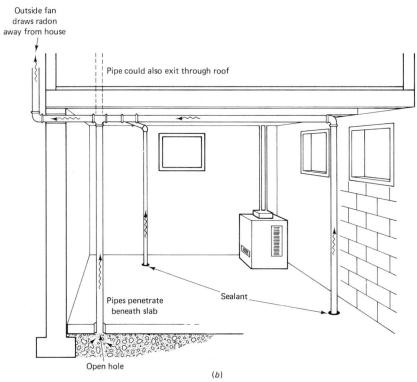

Outside fan draws radon away from house

Pipe could also exit through roof

Pipes penetrate beneath slab

Sealant

Open hole

(b)

Figure 7.37 Radon reduction methods: (*a*) forced ventilation in a basement; (*b*) subslab suction; (*c*) reducing the vacuum effect (USEPA, 1987).

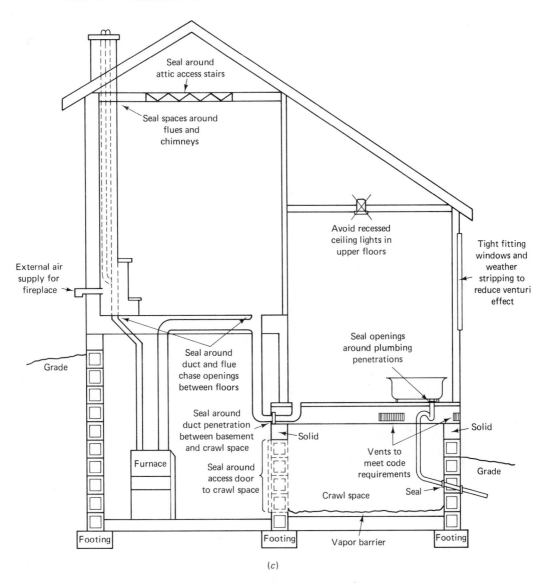

Figure 7.37   (*Continued*)

7.7 EMISSION CONTROLS

Since most air pollutants are produced during combustion, one of the most important, but most overlooked, approaches to reducing emissions is simply to reduce the consumption of fossil fuels. There are two broad approaches that can be taken to reduce fossil fuel consumption: (1) increasing energy efficiency, and, (2) substituting other, less polluting, energy sources for fossil fuels. In the United States, energy efficiency improvements, such as better insulated homes and offices, more efficient appliances and automobiles, improved lighting systems in buildings, and more efficient manufacturing processes, can reduce the energy required to run our society by at least half. The second approach is to decrease our reliance on fossil fuels by increasing power production from solar, wind, hydroelectric, geothermal, and nuclear energy sources.

To the extent that fossil fuels continue to be used, there are three general approaches that can be used to reduce emissions:

1. *Precombustion controls* reduce the emission potential of the fuel itself. Examples include switching to fuels with less sulfur or nitrogen content in power plants, or burning ethanol or methanol in automobiles instead of gasoline. Also, especially in coal-fired power plants, fuel can be physically or chemically treated to remove some of the sulfur or nitrogen before combustion.

2. *Combustion controls* reduce emissions by improving the combustion process itself. Examples include new burners in power plants that reduce NO_x emissions, and new fluidized bed boilers that reduce both NO_x and SO_x. In automobiles, modifications can be made to conventional internal combustion engines, or completely different engine designs might be developed, such as gas turbines or Stirling engines.

3. *Postcombustion controls* capture emissions after they have been formed but before they are released to the air. On power plants, these may be combinations of particulate collection devices and flue-gas desulfurization techniques, used after combustion but before the exhaust stack. On automobiles, most commonly, these are catalytic converters put on the exhaust pipe itself.

Although most air pollution is caused by fossil-fuel combustion, other processes, such as evaporation of volatile organic substances, grinding, and forest fires, can be important as well. Our focus, however, will be on just the two most critical sources: coal combustion in power plants, and automobile emissions.

Emission Controls for Coal-Fired Power Plants

Coal-fired power plants emit great quantities of sulfur oxides, nitrogen oxides, and particulates. Most early emission controls were designed to reduce particulates using postcombustion equipment such as baghouses and electrostatic precipitators. More recently, especially as a result of increased awareness of the seriousness of global acid deposition, attention has shifted somewhat toward control of nitrogen and sulfur oxides using redesigned combustors for NO_x reductions and scrubbers for SO_x control.

Before going into any of the details on these, and other emission control techniques, it is useful to introduce a typical conventional coal-fired power plant, as shown in Figure 7.38. In this plant, coal that has been crushed in a pulverizer is burned to make steam in a boiler for the turbine–generator system. The steam is then condensed, in this case using a cooling tower to dissipate waste heat to the atmosphere, and the condensate is then pumped back into the boiler. The flue gas from the boiler is sent to an electrostatic precipitator, which adds a charge to the particulates in the gas stream so that they can be attracted to electrodes that collect them. Next, a wet scrubber sprays a limestone slurry over the flue gas, precipitating the sulfur and removing it in a sludge of calcium sulfite or calcium sulfate, which then must be treated and disposed of.

Precombustion Controls

One of the two precombustion control techniques used on coal-fired power plants to control sulfur emissions is *fuel switching*. As the name suggests, it involves either substituting low sulfur coal for fuel with higher sulfur content, or blending the two. Fuel switching can reduce emissions by anywhere from 30 to 90 percent, depending on the sulfur content of the fuel currently being burned.

The sulfur content of coal is sometimes expressed as a percentage, or, quite often, in terms of pounds of SO_2 emissions per million Btu of heat content (lb/MBtu). Power plant coal typically has a sulfur content of between 0.2 and 5.5

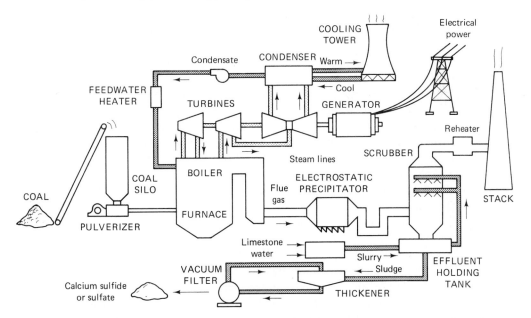

Figure 7.38 Typical modern coal-fired power plant using an electrostatic precipitator for particulate control and a limestone-based SO_2 scrubber. A cooling tower is shown for thermal pollution control.

percent (by weight) corresponding to an uncontrolled SO_2 emission rate of about 0.4 to 10 lb/MBtu. Recall that New Source Performance Standards for coal plants restrict emissions to no more than 1.2 lb/MBtu. Coal that can meet that emission rate without controls is sometimes referred to as "compliance" coal, or more loosely, as "low sulfur" coal.

In round numbers, just under half of U.S. coal reserves are in the eastern half of the United States (principally Illinois, West Virginia, Kentucky, Pennsylvania, and Ohio), and just over half are in the western United States (almost entirely Montana, Wyoming, Colorado, and Texas). While that distribution is nearly even, the location of low-sulfur coal reserves is very unevenly divided. Almost all of our low-sulfur coal, about 85 percent, is located in the western United States (the only notable exceptions are small reserves in West Virginia and Kentucky). But two-thirds of U.S. coal consumption and three-fourths of coal production occurs east of the Mississippi (OTA, 1984). Increasing reliance on fuel switching, then, would increase transportation costs to get the coal from mine to power plant. It would also significantly affect the economies of a number of states if eastern mines decrease production while western ones increase theirs. In addition, western coal often has different ash, moisture, and energy content, which may require modifications to existing power plants to allow fuel switching. Such costs, however, are relatively minor compared to the increased cost of low-sulfur fuel.

The other precombustion approach currently used to reduce sulfur emissions is coal cleaning. Sulfur in coal is either bound into organic coal molecules themselves, in which case precleaning would require chemical or biological treatment (at the research stage now), or it can be in the form of inorganic pyrite (FeS_2), which can readily be separated using physical treatment. Pyrite has a specific gravity that is 3.6 times greater than coal, and that difference allows various coal "washing" steps to separate the two. Such physical cleaning not only reduces the sulfur content of coal, but it also reduces the ash content, increases the energy per unit weight of fuel (which reduces coal transportation and pulverization costs), and it creates more uniform coal characteristics that can increase boiler efficiency. These benefits can offset much of the cost of coal cleaning. It has been estimated that an SO_2 reduction of about 1.5 million tons (out of 14 million tons released by coal plants in the United States) could be achieved by using simple, physical coal cleaning.

Fluidized-Bed Combustion

One of the most promising technologies emerging for cleaner and more efficient combustion of coal is fluidized-bed combustion (FBC). In an FBC boiler, crushed coal mixed with limestone is held in suspension (fluidized) by fast rising air injected from the bottom of the bed. Sulfur oxides formed during combustion react with the limestone to form solid calcium sulfate, which falls to the bottom of the furnace and is removed. Sulfur removal rates can be higher than 90 percent.

In an FBC boiler, the hot, fluidized particles are in direct contact with the boiler tubes. This enables much of the heat to be transferred by conduction rather than by the less efficient convection and radiation processes that conventional boilers rely on. The increase in heat transfer efficiency enables the boilers to operate at about half the temperature of conventional boilers (800 °C versus 1600 °C), greatly reducing the formation of NO_x. In addition to efficient SO_x and NO_x control, FBC boilers are less sensitive to variations in coal quality. Coal with higher ash content can be burned without fouling heat exchange surfaces since the lower combustion temperature is below the melting point of ash.

As of 1988, there were more than 70 industrial applications of the fluidized-bed combustion system in the United States alone, burning everything from cow manure and tree bark to coal and oil shale (Shepard, 1988). Utility-scale FBC applications are just beginning, the first unit having gone into operation in 1986. The future of this technology looks promising, however, and it is likely to become an important part of the utilities' approach to emissions control.

Low NO_x Combustion

Recall that nitrogen oxides are formed partly by the oxidation of nitrogen in the fuel itself (fuel NO_x) and partly as a result of the oxidation of nitrogen in the combustion air (thermal NO_x). Coal-fired power plants emit about 25 percent of U.S. NO_x, forming roughly twice as much fuel NO_x as thermal NO_x. Modifications to the combustion processes described below are designed to reduce both sources of NO_x. In one technique called *low excess air*, the amount of air made available for combustion is carefully controlled to restrict it to the minimum amount required for complete combustion. Low excess air technology can be retrofitted onto some boilers at a modest cost, yielding from 15 to 50 percent lower emissions.

A new technology that promises greater NO_x removal efficiencies and that can be retrofitted onto more existing furnaces, is the second-generation, *low NO_x burner*. Low NO_x burners employ a staged combustion process that delays mixing the fuel and air in the boiler. In the first stage of combustion, the fuel starts burning in an air-starved environment, causing the fuel-bound nitrogen to be released as nitrogen gas, N_2, rather than NO_x. The following stage introduces more air to allow complete combustion of the fuel to take place. Potential NO_x reductions of 45–60 percent seem likely.

Another combustion modification incorporates a staged burner for NO_x control combined with limestone injection for SO_2 control. This *limestone-injection multistage burner* (LIMB) technology is still under development, but looks promising.

Flue-Gas Desulfurization (Scrubbers)

Flue-gas desulfurization (FSD) technologies can be categorized as being either *wet* or *dry* depending on the phase in which the main reactions occur, and either *throwaway* or *regenerative,* depending on whether or not the sulfur from the flue gas is discarded or recovered in a usable form. Most scrubbers operating in the United States use wet, throwaway processes.

In most wet scrubbers, finely pulverized limestone ($CaCO_3$ plus inert siliceous compounds) is mixed with water to create a slurry that is sprayed into the flue gas. The SO_2 is absorbed by the slurry, producing a calcium sulfite or a calcium sulfate precipitate. The precipitate is removed from the scrubber as a sludge. Although the chemical reactions between SO_2 and limestone involve a number of steps, an overall relationship resulting in the production of inert calcium sulfite dihydrate is

$$CaCO_3 + SO_2 + 2H_2O \longrightarrow CaSO_3 \cdot 2H_2O + CO_2 \qquad (7.51)$$

About 90 percent of the SO_2 can be captured from the flue gas using limestone in wet scrubbers.

Wet scrubbers sometimes use lime (CaO) instead of limestone in the slurry, and the following overall reaction applies:

$$CaO + SO_2 + 2H_2O \longrightarrow CaSO_3 \cdot 2H_2O \qquad (7.52)$$

Lime slurries can achieve greater SO_2 removal efficiencies, up to 95 percent. However, lime is more expensive than limestone, so it is not widely used. Dry scrubbers must use lime, and that increased cost is one reason for their relative lack of use.

Although wet scrubbers can capture very high fractions of flue gas SO_2, they have been accepted by the utilities with some reluctance. They are expensive, costing on the order of $200 per kilowatt of generating capacity (10–20 percent of the total capital cost of the power plant). The total added revenue requirements for new plants using scrubbers is about 1–2 cents per kWhr. If they are installed on older plants, with less remaining lifetime, their capital costs must be amortized over a shorter period of time, and annual revenue requirements increase accordingly. Scrubbers also reduce the net energy delivered to the transmission lines. The energy to run scrubber pumps, fans, and flue gas reheat systems requires close to 5 percent of the total power produced by the plant. Scrubbers are also subject to corrosion, scaling, and plugging problems, which may reduce overall power plant reliability.

Scrubbers also use large amounts of water and create similarly large volumes of sludge that has the consistency of toothpaste. A large, 1000-MW plant uses up to 1000 gallons of water per minute; while burning 3 percent sulfur coal, for example, it produces enough sludge each year to cover a square mile of land to a depth of over one foot (Shepard, 1988). Sludge treatment often involves oxidation of calcium sulfite to calcium sulfate, which precipitates easier, thickening, and vacuum filtration. Calcium sulfate (gypsum) can be reused in the construction industry.

Particulate Control

There are a number of gas cleaning devices that can be used to remove particulates. The most appropriate device for a given source will depend on such factors as size, concentration, corrosivity, toxicity of particles, volumetric flow rate, required collection efficiency, allowable pressure drops, and costs.

For relatively large particles, the most commonly used control device is the centrifugal, or *cyclone,* collector. As shown in Figure 7.39, particle-laden gas enters tangentially near the top of the cyclone. As the gas spins in the cylindrical shell, the centrifugal force causes the particles to collide with the outer walls, and then gravity causes them to slide down into a hopper at the bottom. The spiraling gases then exit the collector from the top. Efficiencies of cyclones can be above 90 percent for particles larger than 5 μm, but that drops off rapidly for the small particle sizes that are of most concern for human health. While they are not efficient enough to meet emission standards, they are relatively inexpensive and maintenance free, which makes them ideal as precleaners for more expensive, and critical, final control devices, such as baghouses and electrostatic precipitators.

To collect really small particles, either *baghouses* or *electrostatic precipitators* are used, with the majority of utility power plants using precipitators. Figure 7.40 shows one configuration for an electrostatic precipitator consisting of vertical

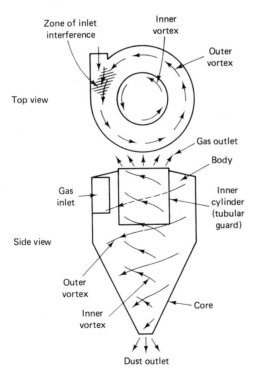

Figure 7.39 Conventional reverse-flow cyclone (USHEW, 1969).

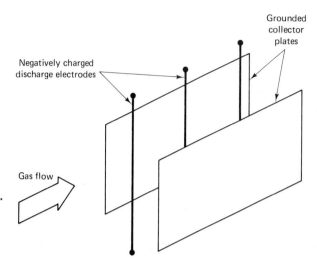

Figure 7.40 Schematic representation of a flat-surface-type electrostatic precipitator. Particles in the gas stream acquire negative charge as they pass through the corona, and are then attracted to the grounded collecting plates.

wires placed between parallel collector plates. A strong electric field is created between the wires and the grounded plates by impressing a very high negative voltage on the wires (as high as 100 000 V). The intense field created near the wires causes a corona discharge, ionizing gas molecules in the airstream. The negative ions and free electrons thus created move toward the grounded plates, and along the way, some attach themselves to passing particulate matter. The particles now carry a charge, which causes them to move under the influence of the electric field to a grounded collecting surface. They are removed from the collection electrode either by gravitational forces, by rapping, or by flushing the collecting plate with liquids. Actual electrostatic precipitators, such as the one shown in Figure 7.41, may have hundreds of parallel plates, with total collection areas measured in the tens of thousands of square meters.

Electrostatic precipitators can easily remove more than 98 percent of the particles passing through them, including particles of submicrometer size. Some have efficiencies even greater than 99.9 percent. They can handle large flue gas flow rates with little pressure drop, and they have relatively low operation and maintenance costs. They are quite versatile, operating on both solid and liquid particles. Precipitators are very efficient, but they are expensive and take a lot of space. Area requirements increases nonlinearly with collection efficiency, in accordance with the following relationship, known as the Deutsch–Anderson equation:

$$\eta = 1 - e^{-wA/Q} \tag{7.53}$$

where η is the fractional collection efficiency, A is the total area of collection plates, Q is the volumetric flow rate of gas through the precipitator, and w is a parameter known as the effective drift velocity. The effective drift velocity is the terminal speed at which the particle approaches the collection plate under the influence of the electric field. It is determined from either pilot studies or previous

Penthouse enclosing insulators and gas seals

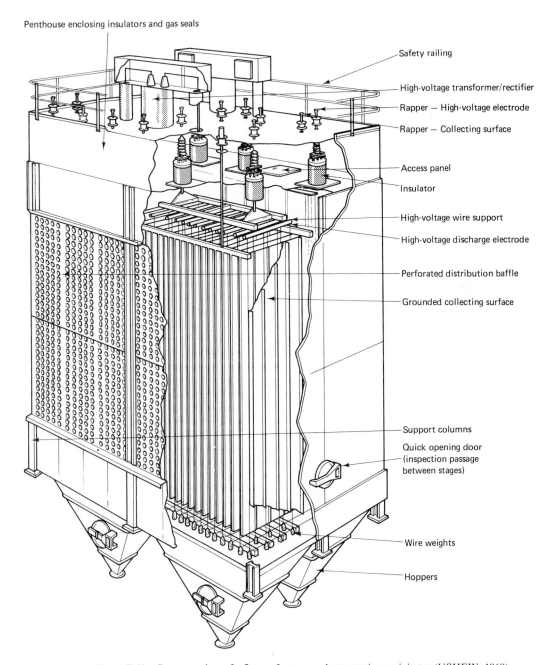

Safety railing

High-voltage transformer/rectifier

Rapper — High-voltage electrode

Rapper — Collecting surface

Access panel

Insulator

High-voltage wire support

High-voltage discharge electrode

Perforated distribution baffle

Grounded collecting surface

Support columns

Quick opening door
(inspection passage
between stages)

Wire weights

Hoppers

Figure 7.41   Cutaway view of a flat-surface-type electrostatic precipitator (USHEW, 1969).

experience with similar units. Any consistent set of units can be used for these quantities.

Example 7.15 Electrostatic Precipitator Area

An electrostatic precipitator with 6000 m^2 of collector plate area is 97 percent efficient in treating 200 m^3/s of flue gas from a 200-MW power plant. How large would the plate area have to be to increase the efficiency to 98 percent and to 99 percent?

Solution Rearranging (7.53) to solve for the drift velocity w, gives

$$w = -\frac{Q}{A} \ln(1 - \eta) = -\frac{200 \text{ m}^3/\text{s}}{6000 \text{ m}^2} \ln(1 - 0.97)$$

$$= 0.117 \text{ m/s}$$

To achieve 98 percent efficiency, the area required would be

$$A_{98} = -\frac{Q}{w} \ln(1 - \eta) = -\frac{200 \text{ m}^3/\text{s}}{0.117 \text{ m/s}} \ln(1 - 0.98) = 6690 \text{ m}^2$$

To achieve 99 percent, the area required would be

$$A_{99} = -\frac{200}{0.117} \ln(1 - 0.99) = 7880 \text{ m}^2$$

As these calculations suggest, the additional collector area required to achieve incremental improvements in collection efficiency goes up rapidly. To increase from 97 to 98 percent required 690 m^2 of added area, while the next 1-percent increment requires 1190 m^2.

The major competition that electrostatic precipitators have for efficient collection of small particles is *fabric filtration*. Dust-bearing gases are passed through fabric filter bags, which are suspended upside-down in a large chamber, called a baghouse, as shown in Figure 7.42. A baghouse may contain thousands of bags that are often distributed among several compartments. This allows individual compartments to be cleaned while others remain in operation.

Part of the filtration is accomplished by the fabric, but a more significant part of the filtration is caused by the dust that accumulates on the inside of the bags. Efficiencies approach 100 percent removal of particles as small as 1 μm; substantial quantities of particles as small as 0.01 μm are also removed. Baghouses have certain disadvantages, however. As is the case for precipitators, baghouses are large and expensive. They can be harmed by corrosive chemicals in the flue gases, and they cannot operate in moist environments. There is also some potential for fires or explosions if the dust is combustible. The popularity of baghouse filters is rising, however, and they now rival precipitators in total industrial sales.

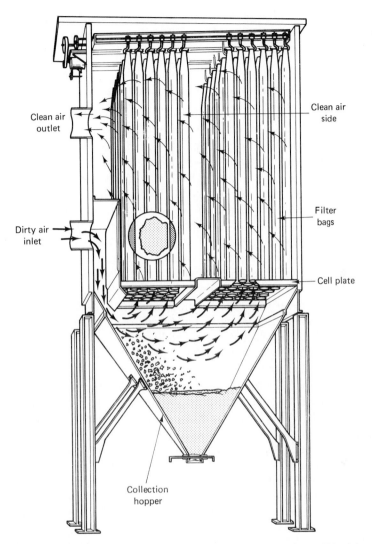

Figure 7.42 Typical simple fabric filter baghouse. (*Source:* Courtesy Wheelabrator Air Pollution Control.)

Emission Controls for Highway Vehicles

Whereas ships, locomotives, and aircraft are also included in the transportation sources category, it is highway vehicles that are by far the most important in terms of total emissions and location of the emissions relative to people. In 1986 there were 186 million registered highway vehicles in the United States (including 135 million passenger cars), consuming a total of 1.3 billion gallons of fuel. They emitted 58 percent of the nation's total CO, 38 percent of the lead, 34 percent of the NO, 27 percent of the VOCs, and 16 percent of the particulates (USEIA, 1988; USEPA, 1988a).

Legislation of Motor Vehicle Emissions

The legislative history of auto emission controls began in California in 1959 with the adoption of state standards to control exhaust hydrocarbons (HC) and carbon monoxide. It was supplemented in 1960 by standards to control emissions resulting from crankcase blowby. These standards, however, were not fully implemented until they were deemed technologically feasible in 1966. Federal standards for CO and HC began in 1968, and by 1970, the year in which major amendments to the Clean Air Act were enacted, the auto industry had reduced hydrocarbon emissions by almost three-fourths and carbon monoxide by about two-thirds (USEPA, 1971). At that time, no controls were required for NO_x, and to some extent, improvements in hydrocarbon and CO emissions were made at the expense of increased nitrogen oxide emissions.

The Clean Air Act amendments of 1970 required that emissions from automobiles be reduced by 90 percent compared to levels already achieved by the 1970 model year. These standards were to be reached by 1975 for CO and HC, and by 1976 for NO_x, with a 1-year delay allowed if industry could adequately prove that the technology was not available to meet them. The 1970 Amendments were unusual in that they were "technology forcing," that is, they mandated specific emission levels before technology was available to meet the standards. The automobile industry successfully argued the case for the 1-year delay, in spite of the fact that the 1973 Honda with a stratified charge engine, was able to meet the 1975 emission requirements while getting 40 mpg.

Subsequently, in response to the oil embargo of 1973 and concerns for the impact of emission standards on fuel economy, the Federal Energy Supply and Coordination Act of 1974 rolled the standards back again to 1978-79. Finally, the Clean Air Act amendments of 1977 delayed the standards once more, setting the date for HC and CO compliance at 1981. At the same time, the 1977 Amendments eased the standard for NO_x from 0.4 to 1 g/mile, calling the 0.4 g/mile a "research objective" and allowed the looser standard to be met as late as 1983. California, which is allowed to set its own auto emission standards as long as they are more stringent than federal requirements, has stayed with the original 0.4-g/mile NO_x standard since 1983. Emission standards as they existed in 1989 are summarized in Table 7.15.

Table 7.15 indicates that vehicle emission regulations are expressed in terms of grams of pollutant per mile of driving. Since emissions vary considerably as driving conditions change, it has been necessary to carefully define a "standard" driving cycle and to base emission regulations on that cycle. The standard driving cycle used is based on an elaborate study of Los Angeles traffic patterns. It consists of a simulated drive on a dynamometer of 7.5 miles, taking 22.8 min and

TABLE 7.15   HIGHWAY VEHICLE EMISSION STANDARDS (g/mile)

	Automobiles		Light/medium duty trucks, Federal
	Federal	California	
Hydrocarbons	0.41	0.41	0.9
Carbon monoxide	3.4	3.4	17
Nitrogen oxides	1.0	0.4	2.0

including 17 stops. Two tests are run, one that begins with a cold engine, and one on an engine that has been completely warmed up. Then the results are averaged.

Internal Combustion Engines

The most common internal combustion engine is a 4-stroke, spark-ignited, piston engine whose operation is described in Figure 7.43. On the first, or *intake* stroke, the descending piston draws in a mixture of fuel and air through the open intake valve. The *compression* stroke follows, in which the rising piston compresses the air/fuel mixture in the cylinder against the now closed intake and exhaust valves. As the piston approaches the top of the compression stroke, the spark plug fires, igniting the mixture. In the *power* stroke, the burning mixture expands and forces the piston down, which turns the crankshaft and delivers power to the drive train. In the fourth, or *exhaust* stroke, the exhaust valve opens and the rising piston forces combustion products out of the cylinder, through the exhaust system, and into the air. Diesel engines operate with a similar cycle, but during the compression stroke, the fuel injected into the cylinder reaches a high enough temperature due to compression that it ignites itself without needing a spark.

The single most important factor in determining emissions from an internal combustion engine is the ratio of air to fuel in the mixture as it enters the cylinders during the intake stroke. To analyze that mixture ratio and its impact on emissions, let us begin with the stoichiometry of gasoline combustion. While modern gasolines are blends of various hydrocarbons, an average formulation can be represented as C_7H_{13}. We can write the following to represent its complete combustion in oxygen:

$$C_7H_{13} + 10.25O_2 \longrightarrow 7CO_2 + 6.5H_2O \tag{7.54}$$

If we want to show complete combustion in air, we can modify this reaction to account for the fact that about 3.76 moles of N_2 accompany every mole of O_2 in air. Thus, $10.25 \times 3.76 = 38.54$ mol of N_2 can be placed on each side of the reaction, yielding

$$C_7H_{13} + 10.25O_2 + 38.54N_2 \longrightarrow 7CO_2 + 6.5H_2O + 38.54N_2 \tag{7.55}$$

where any oxidization of nitrogen to nitrogen oxides has been neglected.

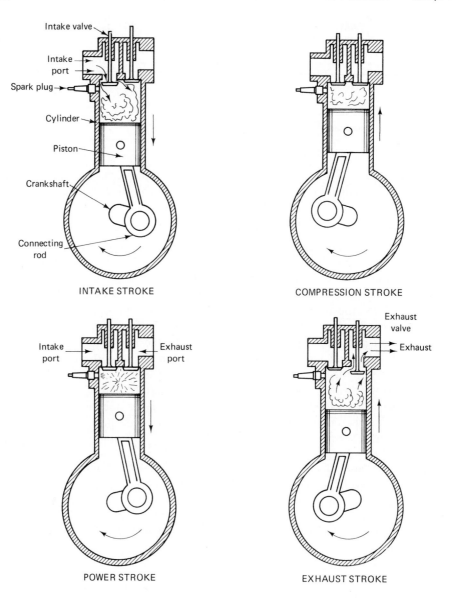

Figure 7.43   Schematic of a four-stroke, spark-ignited internal combustion engine. *Intake:* Intake valve open, piston motion sucks in fresh air/fuel charge. *Compression:* Both valves closed, air/fuel mixture is compressed by rising piston, spark ignites mixture near end of stroke. *Power:* Air fuel mixture burns, increasing temperature and pressure, expansion of combustion gases drives pistons down. *Exhaust:* Exhaust valve open, spent gases are pushed out of cylinder by rising piston. (Powell and Brennan, 1988.)

Example 7.16   Stoichiometric Air–Fuel Ratio

Determine the ratio of air to fuel required for complete combustion of gasoline.

Solution   For each mole of gasoline, 10.25 mol of O_2 and 38.54 mol of N_2 are required. Using (7.55), we can determine the masses of each constituent as

$$1 \text{ mol } C_7H_{13} = 7 \times 12 + 13 \times 1 = 97 \text{ g}$$

$$10.25 \text{ mol } O_2 = 10.25 \times 2 \times 16 = 328 \text{ g}$$

$$38.54 \text{ mol } N_2 = 38.54 \times 2 \times 14 = 1079 \text{ g}$$

Considering air to be made up of only O_2 and N_2, the air–fuel ratio needed for complete oxidation of gasoline is

$$\frac{\text{Air}}{\text{Fuel}} = \frac{328 + 1079}{97} = 14.5$$

This is known as the *stoichiometric ratio* for gasoline.

If the actual air–fuel mixture has less air than what the stoichiometric ratio indicates is necessary for complete combustion, the mixture is said to be *rich*. If more air is provided than is necessary, the mixture is *lean*. A rich mixture encourages production of CO and unburned hydrocarbons since there is not enough oxygen for complete combustion. On the other hand, a lean mixture helps reduce CO and HC emissions unless the mixture becomes so lean that misfiring occurs. Production of NO_x is also affected by the air–fuel ratio. For rich mixtures, the lack of oxygen lowers the combustion temperature, reducing NO_x emissions. In the other direction, beyond a certain point lean mixtures may have enough excess air that the dilution lowers flame temperatures and reduces NO_x production.

Figure 7.44 shows the relationship between CO, HC, and NO_x emissions, and the air–fuel ratio. Also shown is an indication of how the air–fuel ratio affects both power delivered and fuel economy. As can be seen, maximum power is obtained for a slightly rich mixture, while maximum fuel economy occurs with slightly lean mixtures. Before catalytic converters became the conventional method of emission control, most cars were designed to run slightly rich for better power and smoothness. When the first emission limitations were written into law, only CO and HC were controlled. Cars were then simply redesigned to run on a less rich mixture, which had the desired effect of reducing CO and HC; at the same time, however, NO_x emissions were increased. The difficulty automotive engineers have had in achieving control of all three pollutants, CO, HC, and NO_x, all at the same time is apparent from the figure.

The air–fuel ratio is, of course, not the only factor that influences the quantity of pollutants created during combustion. Other influencing factors include ignition timing, compression ratio, combustion chamber geometry, and, very importantly, whether the vehicle is idling, accelerating, cruising, or decelerating. Not all of the pollutants created in the combustion chamber pass directly into the exhaust system. Some find their way around the piston during the power and

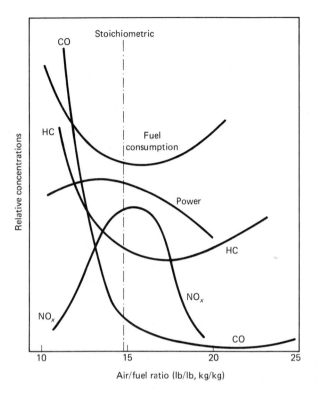

Figure 7.44 Effect of air–fuel ratio on emissions, power, and fuel economy.

compression strokes, into the crankcase. This *blowby,* as it is called, used to be vented from the crankcase to the atmosphere, but now it is recycled back into the engine air intake system to give it a second chance at being burned and released in the exhaust stream. The control method is called *positive crankcase ventilation* (PCV), and the main component is a crankcase ventilator valve, as shown in Figure 7.45. The PCV valve adjusts the rate of removal of blowby gases to match the changing air intake requirements of the engine, making sure that the desired air–fuel ratio is not upset by these added gases.

Fuel combustion is not the only process that creates pollutants in a motor vehicle. About 20 percent of the hydrocarbon emissions from an uncontrolled vehicle are the result of evaporative losses from the gas tank and carburetor. Every time the gas tank of a car is filled, for example, a volume of gasoline vapors equal to the volume of gasoline added is displaced and must go somewhere. Most gasoline stations in areas with air quality problems now have vapor recovery nozzles to keep those hydrocarbons from being emitted to the air. The nozzles are equipped with a rubberized gasket that effectively seals the entire filler pipe opening when the nozzle is in place. As gasoline is pumped into the automobile's tank, the displaced vapors are drawn back into the service station's fuel supply tanks.

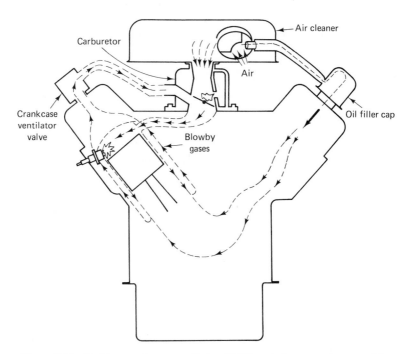

Figure 7.45 Positive crankcase ventilation (PCV) emission control system (Flagan and Seinfeld, 1988).

Within the vehicle itself, there are now vapor recovery systems that help control evaporative losses from both the fuel tank and the carburetor. Since vaporization rates are very dependent on temperature, evaporation in the vicinity of the engine is especially high just after the engine is turned off, when the fans, cooling systems, and airflow due to vehicle movement are no longer operative. Carburetor losses are especially high during this *hot soak* period. A vapor-recovery system that reduces these evaporative emissions is shown in Figure 7.46. In this approach, the crankcase air space is used to temporarily store fuel vapors before they are recycled back into the combustion air intake by the PCV system. Another way to control evaporative emissions is with a canister of activated carbon, which collects vapors from the fuel tank and carburetor. When the engine is running, air is drawn through the canister to collect accumulated gasoline vapors and return them to the intake manifold.

Evaporative emissions can also be reduced by using fuels with less tendency to vaporize. Gasoline itself can be modified to be less volatile, but in order to achieve substantial improvements in evaporative emissions and still have the vehicle start easily and run smoothly, extensive modifications in the fuel intake and carburetion system are required. Instead, the mechanical systems just mentioned have been the preferred approach. It is appropriate to note that one fuel

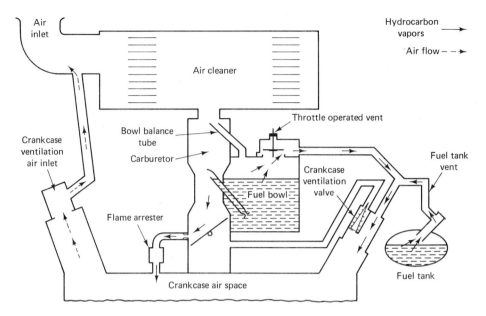

Figure 7.46 Use of crankcase air space and PCV system to reduce evaporative emissions (Flagan and Seinfeld, 1988).

modification that has become fairly readily available is gasohol, which is a mixture of 90 percent unleaded gasoline and 10 percent ethyl alcohol (ethanol). The interaction between alcohol and gasoline makes gasohol more volatile than either one alone, and evaporative emissions are correspondingly increased. While evaporative emissions of hydrocarbons increase using gasohol, it appears that CO emissions are reduced while NO_x and exhaust emissions of HC may increase or decrease depending on the emission control systems used (Gunther, 1980).

Exhaust System Controls

During the exhaust stroke, combustion gases are pushed through the exhaust manifold and out the tailpipe and it is in this exhaust system that most of the control of automobile emissions now occurs. The most commonly used systems for treatment of exhaust gases are *thermal reactors, exhaust gas recirculation* (EGR), and *catalytic converters.*

A thermal reactor is basically an afterburner that encourages the continued oxidation of CO and HC after these gases have left the combustion chamber. The reactor consists of a multipass, enlarged exhaust manifold with an external air source. Exhaust gases in the reactor are kept hot enough and enough oxygen is provided to allow combustion to continue outside of the engine itself, thus reducing CO and HC emissions. Usually, the carburetion system is designed to cause the engine to run rich in order to provide sufficient unburned fuel in the reactor to allow combustion to take place. This has the secondary effect of modestly reducing NO_x emissions, although it also increases fuel consumption since some fuel is not burned in the cylinders.

Some degree of control of NO_x can be achieved by recirculating a portion of the exhaust gas back into the incoming air–fuel mixture. This relatively inert gas that is added to the incoming mixture absorbs some of the heat generated during combustion without affecting the air–fuel ratio. The heat absorbed by the recirculated exhaust gas helps reduce the combustion temperature and, hence, helps decrease the production of NO_x. The coupling of exhaust gas recirculation with a thermal reactor, as shown in Figure 7.47, reduces emissions of all three pollutants, CO, HC, and NO_x, but at the expense of performance and fuel economy.

The approach most favored by American automobile manufacturers to achieve the emission standards dictated by the Clean Air Act has been the three-way catalytic converter. A three-way catalyst is able to oxidize hydrocarbons and carbon monoxide to carbon dioxide, while reducing NO_x to N_2. These catalysts are very effective in controlling emissions, and they have the advantage of allowing the engine to operate at near stoichiometric conditions where engine performance and efficiency are greatest. In fact, they *must* operate within a very narrow band of air–fuel ratios near the stoichiometric point or else their ability to reduce all three pollutants at once is severely compromised, as shown in Figure 7.48a. Maintaining that degree of control has required the development of precise elec-

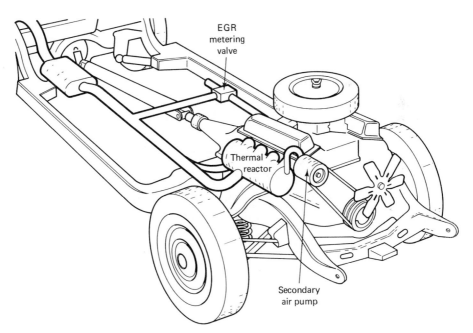

Figure 7.47 Exhaust gas recirculation reduces NO_x by lowering the combustion temperature, while the thermal reactor helps control CO and HC. (*Source:* Courtesy Gould Inc.)

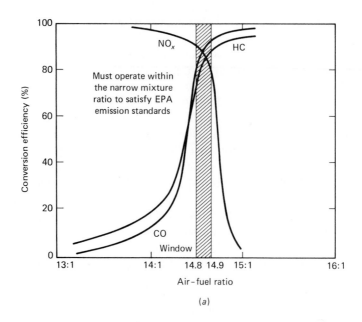

(a)

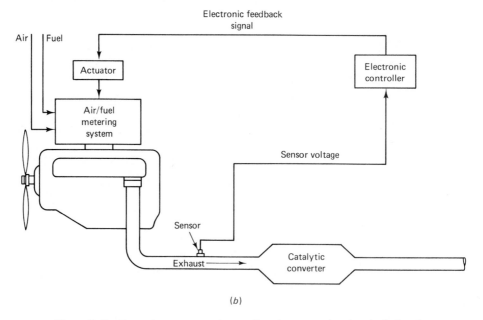

(b)

Figure 7.48 For a three-way catalyst to function correctly, the air–fuel ratio must be maintained within a very narrow band, as shown in (a). To maintain that ratio, a closed-loop control system monitors the composition of exhaust gases and sends corrective signals to the air–fuel metering system (b). (After Powell and Brennan (1988).)

tronic feedback control systems that monitor the composition of exhaust gas and feed that information to a microprocessor controlled carburetor or fuel-injection system as shown in Figure 7.48*b*.

Industry acceptance of the catalytic converter as its emission-control method of choice has another environmental advantage. Catalysts are quickly destroyed if leaded fuels are burned, so almost all new cars now must burn unleaded gasoline. As new cars gradually replace older ones, the fraction of vehicles that use unleaded fuel increases, and total atmospheric lead emissions drop. For example, in 1983, 59 percent of the gasoline sold in the United States was unleaded, while in 1985, only two years later, that fraction had already increased to 70 percent (USEIA, 1988). Unfortunately, leaded gasoline tends to be cheaper than unleaded, and some motorists have ruined their emission control systems by attempting to use the cheaper leaded fuel. In an effort to decrease the likelihood of such misfueling in cars equipped with catalysts, areas with air quality problems have been required to institute periodic vehicle inspection and maintenance programs. Vehicles must pass these annual or biannual inspections before their registrations can be renewed. Studies have shown considerably less misfueling catalyst damage and emission control tampering in states with these inspection programs.

Modifications to the Internal Combustion Engine

At this time, the only real competition to the basic spark-ignition, internal combustion engine, is the *diesel*. Diesel engines have no carburetor since fuel is injected directly into the cylinder, and there is no conventional ignition system with plugs, points, and condenser since the fuel ignites spontaneously during the compression stroke. Diesels have much higher compression ratios than conventional (Otto cycle) engines and, since they do not depend on spark ignition, they can run on very lean mixtures. Thus, they are inherently more fuel efficient. The increased fuel efficiency of diesels was responsible for their momentary surge in popularity just after the oil crises of the 1970s.

Since diesels run with very lean mixtures, emissions of hydrocarbons and carbon monoxide are inherently very low. However, because high compression ratios create high temperatures, NO_x emissions are relatively high. Moreover, since the fuel is burned with so much excess oxygen, catalysts that require a lack of oxygen to chemically reduce NO_x are ineffective. NO_x control, then, is achieved using less effective techniques such as exhaust gas recirculation coupled with modifications to the combustion process. These modifications tend to be at the expense of drivability and fuel economy. In addition, diesels emit significant quantities of carbonaceous particles known as soot, some of which are mutagenic and possibly carcinogenic. Filters have been proposed but complex schemes would be required to periodically clean the filter of its accumulated particulate load. The difficulty in controlling NO_x and soot, along with the lack of interest in fuel economy after the crash in oil prices in the 1980s, has led to the virtual demise of the diesel in new American automobiles.

Another engine design that has shown considerable promise in reducing air pollutants is the *stratified charge engine,* shown in Figure 7.49. As Figure 7.44 suggests, if an engine can be made to run on a very lean mixture, then CO, HC, and NO_x would all be controlled at once. In a conventional internal combustion engine, as the air–fuel ratio goes beyond about 17, the mixture does not ignite properly, leading to misfiring, poor performance, and increased emissions. In the late 1950s, Ralph Heintz patented an engine design that gets around those problems, allowing very lean mixtures to be burned for emission control without the associated misfiring problems (Gunn, 1973). The result is a straticharge engine, the basis for Honda's CVCC (compound vortex controlled combustion) engines that met the 1975 Clean Air Act emission standards in the early 1970s without catalytic converters. The key is the dual chamber configuration shown in Figure 7.49, in which an upper, precombustion chamber receives a fuel-rich charge that ignites easily and burns quickly. The flame front carries into the main chamber where the mixture is lean, causing the charge to continue burning in an oxygen-rich environment that holds down CO and HC emissions. The overall combustion process takes longer and operates at lower temperatures so NO_x emissions are also controlled.

There are other engine designs that have been proposed to help reduce emissions, such as gas turbines, Wankel (or rotary) engines, and even steam engines, but none seems likely to replace the conventional spark-ignition, internal combustion engine so common today. One old idea, however, that continues to intrigue many in the engineering community, is the electric vehicle. From an air quality perspective, a fleet of electric cars would shift pollution from the city

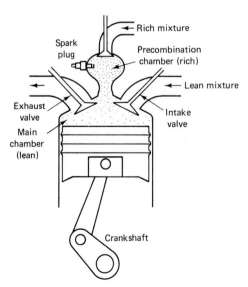

Figure 7.49 A stratified charge engine has a fuel-rich precombustion chamber where combustion begins. The flame spreads into the main chamber, which is lean, leading to reduced CO and HC emissions. Burning is slower and cooler, which reduces NO_x emissions.

where vehicles and people are concentrated, to the countryside where power plants that would generate electricity for the vehicles would tend to be located. Not only would the location of emissions shift, but the type of emissions would change as well. If the power plants burn fossil fuels, then HC and CO emissions would be reduced, NO_x could still be a problem, and particulate and SO_x emissions would greatly increase. Photochemical smog problems in the city could be eased but acid rain and other industrial smog problems could be exacerbated. There would be some trade-off between human health and the health of forests and other ecosystems. Of course, to the extent that electric power production might shift from fossil fuels, especially coal, to renewable energy sources or nuclear power, there would be an obvious air quality advantage. Before any significant shift to electric vehicles can occur, major breakthroughs in storage battery technology are required. At this time, the electric vehicle is a promising, but not yet viable, approach to pollution control.

Alternative Fuels

There are a number of alternatives to gasoline that are being investigated as possible fuels for the future. These include methanol, ethanol, compressed natural gas, propane, and hydrogen. Of these, methanol (mixed with gasoline) is the most likely to become common in the 1990s, and it will be described here.

Methanol (CH_3OH) can be produced from natural gas, coal, or biomass. Most commercial processes currently use natural gas as the feedstock, producing methanol in the following two steps. First, methane is reacted with steam to produce a "synthesis gas" consisting of carbon monoxide and hydrogen:

$$CH_4 + H_2O \longrightarrow CO + 3H_2 \qquad (7.56)$$

The second step utilizes a catalyst to promote the following reaction, producing liquid methanol:

$$CO + 2H_2 \longrightarrow CH_3OH \qquad (7.57)$$

Methanol has long been the fuel of choice in high-performance racing cars. It has a much higher octane rating than gasoline, which enables engines to be designed with higher compression ratios for increased power. It burns with a lower flame temperature than does either gasoline or diesel fuel, so NO_x emissions are reduced. It burns more completely, as well, so that hydrocarbon and CO emissions are also reduced. Moreover, the hydrocarbons that are released are less photochemically reactive. As a result of these improved emission characteristics, the ozone-producing potential of methanol is perhaps half that of gasoline (California Energy Commission, 1989).

Methanol is not without its problems. Its emissions are much higher in formaldehyde (HCHO), an eye irritant and a suspected carcinogen. The low volatility of pure methanol makes it difficult to start engines in cold conditions, and it is highly toxic. It burns without a visible flame, which can lead to especially dangerous conditions in an accident involving fire. Moreover, its energy content is only about half that of gasoline, so either much larger fuel tanks would be required or the driving range would be considerably reduced.

Some of the problems of pure methanol can be overcome by using a blend of methanol and gasoline. A mixture of 85 percent methanol and 15 percent gasoline, called M85, eliminates the cold-start problem and yields a visible flame as well. It also tastes much worse than pure methanol, which should discourage ingestion. The blend has an octane rating of 102 versus 87–92 for gasoline, and an energy content of 65 000 Btu per gallon compared with 116 000 Btu for gasoline. While the increase in combustion efficiency does slightly offset the low fuel energy content, the driving range for M85 is still only about 60 percent of that for gasoline. To help ease the transition to methanol, automobile manufacturers are introducing *flexible fuel vehicles* (FFVs) designed to run on M85, gasoline, or any combination of both. Eventually, if methanol becomes widely available, it would be possible to take advantage of the increased octane level with more efficient engines dedicated to M85 use.

With regard to emissions of the greenhouse gas, carbon dioxide, methanol derived from natural gas produces slightly less CO_2 than gasoline. On the other hand, if methanol is produced from coal, total CO_2 emissions released during coal processing and methanol combustion are double that of gasoline (Moyer, 1989).

PROBLEMS

7.1. Convert the following (8-hr) indoor air quality standards established by the U.S. Occupational Safety and Health Administration (OSHA) from ppm to mg/m^3 (at 25 °C and 1 atm), or vice versa:
  a. Carbon dioxide, 5000 ppm
  b. Formaldehyde, 3.6 mg/m^3
  c. Nitric oxide, 25 ppm

7.2. Consider a new 38-percent efficient, 600-MW power plant burning 9000 Btu/lb coal containing 1 percent sulfur. What would be the maximum allowable SO_2 emission rate (lb/day) and how efficient must the scrubber be?

7.3. In the above power plant, how many pounds per day of sulfur would be released from the fuel? If the scrubber were 90 percent efficient, how many pounds per day of sulfur would be released? How many lb/day of SO_2?

7.4. A new coal-fired power plant has been built using a sulfur emission control system that is 70 percent efficient. What maximum percent sulfur content can the fuel have if 15 000 Btu/lb coal is burned? If 9000 Btu/lb coal is burned, what maximum sulfur content can the fuel have?

7.5. What PSI and what air quality description should be reported for a day in which the following maximums occurred: 1-hr O_3 concentration, 230 $\mu g/m^3$; 8-hr CO, 12 mg/m^3; 24-hr TSP, 200 $\mu g/m^3$; 24-hr SO_2, 325 $\mu g/m^3$; and 1-hr NO_2, 100 $\mu g/m^3$?

7.6. The OSHA standard for worker exposure to 8 hr of CO is 50 ppm. What %COHb would be expected under those circumstances for a worker whose level of activity corresponds to $\alpha = 3$?

7.7. What hydrocarbon (RH) reacting with the OH· radical in (7.11) would produce formaldehyde (HCHO) in (7.14)?

7.8. Suppose propene (CH_2=CH—CH_3) is the hydrocarbon (RH) that reacts with the hydroxyl radical OH· in reaction (7.11). Write the set of chemical reactions that end with an aldehyde. What is the final aldehyde?

7.9. In 1986, 0.39×10^{12} g of particulates were released when 685 million (short) tons of coal were burned in power plants which produced 1400 billion kWhr of electricity. Assume the average heat content of the coal is 10 000 Btu/lb. What must have been the average efficiency (heat to electricity) of these coal plants? How much particulate matter would have been released if all of the plants met the New Source Performance Standards?

7.10. In the text, the following expression was used in the derivation of the dry adiabatic lapse rate: $dQ = C_p \, dT - V \, dP$ (Eq. 7.23). Derive that expression, starting with a statement of the first law of thermodynamics, $dQ = dU + dW$, where $dU = C_v \, dT$ is the change in internal energy when an amount of heat, dQ, is added to the gas, raising its temperature by dT, and causing it to expand and do work $dW = P \, dV$. C_v is the specific heat at constant volume. Then, use the ideal gas law, $PV = nRT$, where n is moles (a constant), and R is the gas constant, along with the definition of the derivative of a product: $d \, (PV) = P \, dV + V \, dP$, to find another expression for dQ. Finally, using the definition $C_p = (dQ/dT)$ with pressure held constant, show that $C_p = C_v + nR$, and you're about there.

7.11. Suppose the following atmospheric altitude versus temperature data have been collected.

Altitude (m)	Temperature (°C)
0	20
100	18
200	16
300	15
400	16
500	17
600	18

a. What would be the mixing depth?

b. How high would you expect a plume to rise if it is emitted at 21 °C from a 100-m stack if it rises at the dry adiabatic lapse rate? Would you expect the plume to be looping, coning, fanning, or fumigating?

7.12. For the temperature profile given in Problem 7.11, if the maximum daytime surface temperature is 22 °C, and a weather station anemometer at 10 m height shows winds averaging 4 m/s, what would be the ventilation coefficient? Assume stability class C and use the wind at the height halfway to the maximum mixing depth.

7.13. Suppose a bonfire emits CO at the rate of 20 g/s on a clear night when the wind is blowing at 2 m/s. If the effective stack height of the fire is 6 m, (a) what would you expect the ground-level CO concentration to be at 400 m downwind? (b) Estimate the maximum ground-level concentration.

7.14. A coal-fired power plant with effective stack height of 100 m emits 1.2 g/s of SO_2 per megawatt of power delivered. If winds are assumed to be 4 m/s at that height and just over 3 m/s at 10 m, how big could the plant be (MW) without having the ground level SO_2 exceed 365 $\mu g/m^3$? (First decide which stability classification leads to the worst conditions.)

7.15. A 35-percent efficient coal-fired power plant with effective height of 100 m emits SO_2 at the rate of 0.6 lb/10^6 Btu into the plant. If winds are assumed to be 4 m/s at the stack height and just over 3 m/s at 10 m, how large could the plant be (MW) without having the ground-level SO_2 exceed 365 $\mu g/m^3$?

7.16. A stack emitting 80 g/s of NO has an effective stack height of 100 m. The wind speed is 4 m/s at 10 m, and it is a clear summer day with the sun nearly overhead. Estimate the ground-level NO concentration:
  a. Directly downwind at a distance of 2 km.
  b. At the point downwind where NO is a maximum.
  c. At a point located 2 km downwind and 0.1 km off the downwind axis.

7.17. For stability class C, the ratio of σ_y/σ_z is essentially a constant, independent of distance x. Assuming that it is a constant, take the derivative of (7.34) and:
  a. Show that the distance downwind from a stack at which the maximum concentration occurs corresponds to the point where $\sigma_z = H/\sqrt{2} = 0.707H$.
  b. Show that the maximum concentration is

$$C_{max} = \frac{Q}{\pi \sigma_y \sigma_z u e} = \frac{0.117Q}{\sigma_y \sigma_z u}$$

  c. Show that C_{max} is inversely proportional to H^2.

7.18. The world's tallest stack is on a copper smelter in Sudbury, Ontario. It stands 380 m high, and has an inner diameter at the top of 15.2 m. If 130 °C gases exit the stack at 20 m/s while the ambient temperature is 10 °C and the winds at stack height are 8 m/s, use the Briggs model to estimate the effective stack height. Assume a slightly unstable atmosphere, class C.

7.19. Repeat Problem 7.18 for a stable, isothermal atmosphere (no temperature change with altitude).

7.20. A 200-MW power plant has a 100-m stack with radius 2.5 m, flue gas exit velocity 13.5 m/s, and gas exit temperature 145 °C. Ambient temperature is 15 °C, wind speed at the stack is 5 m/s, and the atmosphere is stable, class E, with a lapse rate of 5 °C/km. If the plant emits 300 g/s of SO_2, estimate the concentration at ground level at a distance 16 km directly downwind.

7.21. A source emits 20 g/s of some pollutant from a stack with effective height 50 m in winds that average 5 m/s. On a single graph, sketch the downwind concentration for stability classifications A, C, and F using Figure 7.30 to identify the peak concentration and distance.

7.22. A source emits 20 g/s of some pollutant in winds that average 5 m/s, on a class C day. Use Figure 7.30 to find the peak concentrations for effective stack heights of 50, 100, and 200 m. Note whether the concentration is roughly proportional to (1/H^2) as Problem 7.17 suggests.

7.23. A paper plant is being proposed for a location 1 km upwind from a town. It will emit 40 g/s of hydrogen sulfide, which has an odor threshold of about 0.1 mg/m³. Winds at the stack may vary from 4 to 10 m/s blowing toward the town. What minimum stack height should be used to assure concentrations are no more than 0.1 times the odor threshold at the near edge of town on a class B day. To be conservative, the stack will be designed assuming no plume rise. If the town extends beyond the 1 km distance, will any buildings experience higher concentrations than a residence at the boundary under these conditions?

7.24. A stack with effective height of 45 m emits SO_2 at the rate of 150 g/s. Winds are estimated at 5 m/s at the stack height, the stability class is C, and there is an inversion at 100 m. Estimate the ground-level concentration at the point where reflections begin to occur from the inversion, and at a point twice that distance downwind.

7.25. A long line of agricultural waste emits 0.3 g/m-s of particulate matter on a clear fall afternoon with winds blowing 3 m/s perpendicular to the line. Estimate the ground-level particulate concentration 400 m downwind from the line.

7.26. A freeway has 10 000 vehicles per hour passing a house 200 m away. Each car emits an average of 1.5 g/mi of NO_x, and winds are blowing at 2 m/s across the freeway toward the house. Estimate the NO_x concentration at the house on a clear summer day near noon (assuming NO_x is chemically stable).

7.27. Consider a box model for an air shed over a city 1×10^5 m on a side, with a mixing depth of 1200 m. Winds with no SO_2 blow at 4 m/s against one side of the box. SO_2 is emitted in the box at the rate of 20 kg/s. If SO_2 is considered to be conservative, estimate the steady-state concentration in the air shed.

7.28. With the same air shed and ambient conditions as given in Problem 7.27, assume the emissions occur only on weekdays. If emissions stop at 5 pm on Friday, estimate the SO_2 concentration at midnight. If they start up again on Monday at 8 am, what would the concentration be by 5 pm?

7.29. Assume that steady-state conditions have been reached for the city in Problem 7.27, and then the wind drops to 2 m/s. Estimate the concentration of SO_2 two hours later.

7.30. If the wind blowing into the air shed in Problem 7.27 has 5 $\mu g/m^3$ of SO_2 in it, and the SO_2 concentration in the air shed at 8 am Monday is 10 $\mu g/m^3$, estimate the concentration at noon, assuming that emissions are still 20 kg/s.

7.31. With the same air shed and ambient conditions as given in Problem 7.27, if SO_2 is not conservative, and, in fact, has a reactivity of 0.23/hr, estimate its steady-state concentration over the city.

7.32. A single-story home with infiltration rate of 0.5 ach has 200 m² of floor space and a total volume of 500 m³. If 0.6 pCi/m²-s of radon is emitted from the soil and enters the house, estimate the steady-state indoor radon concentration.

7.33. If the house in Problem 7.32 had been built as a two-story house with 100 m² on each floor and the same total volume, what would the estimated radon concentration be?

7.34. If an individual lived in the house given in Problem 7.32, estimate the lifetime risk of cancer from the resulting radon exposure. Use Nero's (1986) estimate that a house concentration of 1.5 pCi/L leads to a lifetime risk of 0.3 percent.

7.35. Consider a "tight" 300-m³ home with 0.2-ach infiltration rate. The only source of CO in the home is the gas range and the ambient concentration of CO is always zero. Suppose there is no CO in the home at 6 pm, but then the oven and two burners are on for 1 hr. Assume the air is well mixed in the house and estimate the CO concentration in the home at 7 pm and again at 10 pm.

7.36. A convective kerosene heater is tested in a well-mixed 27-m³ chamber having an air exchange rate of 0.39 ach. After 1-hr of operation, the NO concentration reached 4.7 ppm. Treating NO as a conservative pollutant:
   a. Estimate the NO source strength of the heater (mg/hr).
   b. Estimate the NO concentration that would be expected in the lab 1 hr after turning off the heater.
   c. If this heater were to be used in the home described in Problem 7.35, what steady-state concentration of NO would you expect to be caused by the heater?

7.37. Sidestream smoke from a cigarette contains roughly 100 μg of benzo-α-pyrene. How many cigarettes per day would yield enough sidestream smoke to increase the cancer risk for a nonsmoker by 1×10^{-6} if that individual spends all of his time indoors in a 150-m³ home having 0.5 ach? Assume that benzo-α-pyrene has a half-life of 3 days and assume that its ambient concentration is negligible.

REFERENCES

BRIGGS, G. A., 1972, Discussion of chimney plumes in neutral and stable surroundings, *Atmospheric Environment* 6(1).

California Energy Commission, 1989, *News and Comment, CEC Quarterly Newsletter,* No. 23, Fall.

COOPER, C. D., and F. C. ALLEY, 1986, *Air Pollution Control: A Design Approach,* PWS, Boston.

DIAMOND, R. C., and D. T. GRIMSRUD, 1984, *Manual on Indoor Air Quality,* prepared by Lawrence Berkeley Laboratory for the Electric Power Research Institute, Berkeley, CA.

DIMITRIADES, B., and M. WHISMAN, 1971, Carbon monoxide in lower atmosphere reactions, *Environmental Science and Technology* 5:213.

FLAGAN, R. C., and J. H. SEINFELD, 1988, *Fundamentals of Air Pollution Engineering,* Prentice Hall, Englewood Cliffs, NJ.

GIFFORD, F. A., 1961, Uses of routine meteorological observations for estimating atmospheric dispersion, *Nuclear Safety* 2(4).

GUNN, C. F., 1973, The amazing Ralph Heintz: Portrait of an inventor, *The Stanford Magazine,* Fall/Winter.

GUNTHER, A., 1980, *Gasohol, A Background Report,* Assembly Office of Research, California State Legislature, Sacramento, CA.

HINO, M., 1968, Maximum ground-level concentration and sampling time, *Atmospheric Environment* 2(3).

Interagency Task Force on Acid Precipitation, 1983, *Annual Report 1983 to the President and Congress,* Washington, DC.

MARTIN, D. O., 1976, The change of concentration standard deviation with distance, *Journal of the Air Pollution Control Association* 26(2).

MOLBURG, J., 1980, A graphical representation of the new NSPS for sulfur dioxide, *Journal of the Air Pollution Control Association* 30(2):172.

MOYER, C. B., 1989, *Global Warming Benefits of Alternative Motor Fuels,* Acurex Corporation, Mountain View, CA.

NAGDA, N. L., H. E. RECTOR, and M. D. KOONTZ, 1987, *Guidelines for Monitoring Indoor Air Quality,* Hemisphere, Washington, DC.

National Research Council, 1981, *Indoor Pollutants,* National Academy Press, Washington, DC.

NERO, A., 1986, The indoor radon story, *Technology Review,* January.

Office of Technology Assessment (OTA), 1984, *Acid Rain and Transported Air Pollutants, Implications for Public Policy,* Washington, DC.

PASQUILL, F., 1961, The estimation of the dispersion of windborne material, *Meterological Magazine* 90:1063.

PERKINS, H. C., 1974, *Air Pollution,* McGraw Hill, New York.

PETERSON, W. B., 1978, *User's Guide for PAL—A Gaussian Plume Algorithm for Point, Area, and Line Sources,* U.S. Environmental Protection Agency, Research Triangle Park, NC.

PORTELLI, R. V., and P. J. LEWIS, 1987, Meteorology, in *Atmospheric Pollution,* E. E. Pickett (ed.), Hemisphere, Washington, DC.

POWELL, J. D., and R. P. BRENNAN, 1988, *The Automobile, Technology and Society,* Prentice Hall, Englewood Cliffs, NJ.

REECE, N. S., 1988, Indoor air quality, *Homebuilding & Remodeling Resource,* November/December.

SEINFELD, J. H., 1986, *Atmospheric Chemistry and Physics of Air Pollution,* Wiley-Interscience, New York.

SHEPARD, M., 1988, Coal technologies for a new age, *EPRI Journal,* Electric Power Research Institute, Jan/Feb.

TRAYNOR, G. W., D. W. ANTHON, and C. D. HOLLOWELL, 1981, *Technique for Determining Pollutant Emissions from a Gas-Fired Range,* Lawrence Berkeley Laboratory, LBL-9522, December.

TRAYNOR, G. W., J. R. ALLEN, M. G. APTE, J. F. DILLWORTH, J. R. GIRMAN, C. D. HOLLOWELL, and J. F. KOONCE, JR., 1982, *Indoor Air Pollution from Portable Kerosene-Fired Space Heaters, Wood-Burning Stoves, and Wood-Burning Furnaces,* Lawrence Berkeley Laboratory, LBL-14027, March.

TURNER, D. B., 1970, *Workbook of Atmospheric Dispersion Estimates,* U.S. Environmental Protection Agency, Washington, DC.

USEIA, 1988, *Annual Energy Review 1987,* Energy Information Agency, Department of Energy, Washington, DC.

USEPA, 1971, *Annual Report of the Environmental Protection Agency to the Congress of the United States in Compliance with Section 202(b)(4), Public Law 90-148,* Washington, DC.

USEPA, 1977, *Air Quality Criteria for Lead,* Environmental Protection Agency, Washington, DC.

USEPA, 1986, *A Citizen's Guide to Radon,* Office of Air and Radiation, OPA-86-004, Washington, DC.

USEPA, 1987a, *Radon Reduction in New Construction, An Interim Guide,* Environmental Protection Agency, Washington, DC.

USEPA, 1987b, *Radon Reduction Methods, A Homeowner's Guide,* 2d ed., Environmental Protection Agency, Washington, DC.

USEPA, 1988a, *National Air Pollutant Emission Estimates 1940–1986,* Environmental Protection Agency, Washington, DC.

USEPA, 1988b, *National Air Quality and Emissions Trends Report, 1986,* Environmental Protection Agency, Washington, DC.

USHEW, 1967, *Air Quality Criteria for Sulfur Oxides,* National Air Pollution Control Administration, Washington, DC.

USHEW, 1969, *Control Techniques for Particulate Air Pollutants,* National Air Pollution Control Administration, Washington, DC.

USHEW, 1970a, *Air Quality Criteria for Carbon Monoxide,* AP-62, National Air Pollution Control Administration, Washington, DC.

USHEW, 1970b, *Control Techniques for Carbon Monoxide, Nitrogen Oxide, and Hydrocarbon Emissions from Mobile Sources,* National Air Pollution Control Administration, Washington, DC.

WADDEN, R. A., and P. A. SCHEFF, 1983, *Indoor Air Pollution, Characterization, Prediction, and Control,* Wiley-Interscience, New York.

WARK, K., and C. F. WARNER, 1981, *Air Pollution, Its Origin and Control,* Harper & Row, New York.

WILLIAMSON, S. J., 1973, *Fundamentals of Air Pollution,* Addison-Wesley, Reading, MA.

CHAPTER 8

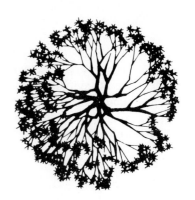

Global Atmospheric Change

THE GREENHOUSE EFFECT
AND STRATOSPHERIC OZONE DEPLETION

*Consider the reaction in the U.S. if the Soviet Union were
to threaten, as global climate change threatens, to invade
7000 square miles of U.S. coastal land, incapacitate a
significant fraction of U.S. agriculture, reduce
hydroelectric capacity and degrade water quality in many
regions, all in the next 50 years. What level of resources
would be committed to stopping this threat?*
Joel N. Swisher (1989)

8.1 INTRODUCTION

The quote above, which was made in reference to a USEPA report on global
warming (1989), provides a fitting introduction to this final chapter. Global warming and stratospheric ozone depletion threaten to become the dominant environmental issues of the 1990s.

Although the atmosphere is made up almost entirely of nitrogen and oxygen,
other gases and particles existing in very small concentrations are often of greater
interest in environmental science. In this chapter we will focus on several of these
other gases, including carbon dioxide (CO_2), nitrous oxide (N_2O), methane (CH_4),
and ozone (O_3), as well as a category of man-made gases called chlorofluorocarbons (CFCs). Minute quantities of these gases determine to a large extent the
habitability of our planet. However, they are all presently undergoing rapid
changes in their atmospheric concentrations as a result of human activities. The

two problems of greenhouse effect enhancement, leading to global climate change and stratospheric ozone depletion, which increases our exposure to life-threatening ultraviolet radiation, are linked to changes in these trace gases and are the subject of this chapter.

When the earth was formed, some four or five billion years ago, it probably had no atmosphere at all. Through volcanic activity, gases such as carbon dioxide, water vapor, and various compounds of nitrogen and sulfur were released over time, from which our current atmosphere was formed. It is thought that molecular oxygen in our air resulted largely from photosynthesis by plants that were evolving underwater, where life was protected from the sun's intense, biologically damaging, ultraviolet radiation. As photosynthesis gradually increased atmospheric oxygen levels, more and more ozone was formed in the atmosphere. The absorption of incoming ultraviolet radiation by that ozone provided the protection necessary for life to be able to emerge onto the land.

Table 8.1 shows the composition of the earth's atmosphere as it exists now, expressed in volumetric fractions (see Section 1.2 for a reminder of the difference between gaseous concentrations expressed by volume and by mass). The values given are for "clean," dry air and do not include the relatively small but extremely important amounts of water vapor and pollution. While most of the values in the table are essentially unchanging, the value for CO_2 is not—in fact, it is increasing by about $1\frac{1}{2}$ ppm per year.

In this chapter, not only is the composition of the atmosphere important, but so is its temperature. It is convenient to think of the atmosphere as being divided into various horizontal layers, each characterized by its temperature profile. Starting at the earth's surface, these layers are called the *troposphere, stratosphere, mesosphere,* and the *thermosphere.* Figure 8.1 shows an example temperature profile of the atmosphere, identifying these regions.

TABLE 8.1 COMPOSITION OF CLEAN, DRY AIR (fraction by volume)

Constituent	Concentration	
Nitrogen	0.7808	
Oxygen	0.2095	
Argon	0.0093	
Carbon dioxide	355[a]	ppm
Neon	18	ppm
Helium	5.2	ppm
Methane	1.8	ppm
Krypton	1.1	ppm
Nitrous oxide	0.3	ppm
Hydrogen	0.5	ppm
Ozone	0.01	ppm

[a] 1990.

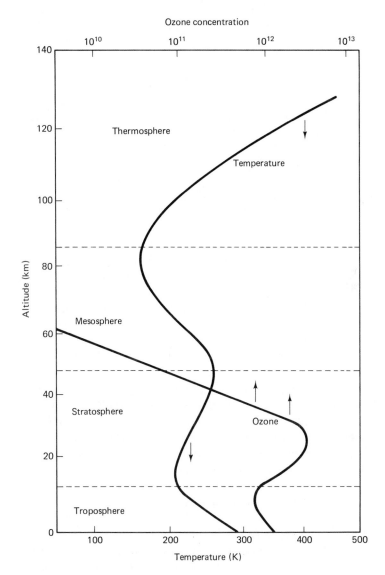

Figure 8.1 Example temperature profile of the atmosphere showing the four major layers (USEPA, 1986).

More than 80 percent of the mass of the atmosphere and virtually all of the water vapor, clouds, and precipitation occur in the troposphere. At mid-latitudes, the troposphere extends up to 10 or 12 km (about the altitude of a typical airline flight). At the poles it may be only about 5–6 km, while at the equator it is about 18 km. In the troposphere, temperatures typically decrease at 5–7 °C per km. That rate of change of temperature is essentially the wet adiabatic lapse rate (see

Section 7.4). The troposphere is usually a very turbulent place, that is, there are strong vertical air movements that lead to rapid and complete mixing. This mixing is good for air quality since it rapidly disperses pollutants.

Above the troposphere is a stable layer of very dry air called the stratosphere. Pollutants that find their way into the stratosphere may remain there for many years before they eventually drift back into the troposphere, where they can be more easily diluted and ultimately removed by such processes as rainfall or settling. In the stratosphere, shortwavelength ultraviolet energy is absorbed by ozone, causing the air to be heated. The resulting temperature inversion is what causes the stratosphere to be so stable. The troposphere and stratosphere combined account for about 99.9 percent of the mass of the atmosphere. Together they extend only about 50 km above the surface of the earth, a distance equal to less than 1 percent of the earth's radius.

Beyond the stratosphere lies the mesosphere, another layer where air mixes fairly readily, and above that the thermosphere. The heating of the thermosphere is due to the absorption of solar energy by atomic oxygen. Within the thermosphere is a relatively dense band of charged particles, called the *ionosphere*. (Before satellites, by the way, the ionosphere was especially important to worldwide communications because of its ability to reflect radio waves back to earth.)

8.2 GLOBAL TEMPERATURE

The history of the earth's climate is characterized by frequent, and sometimes rather sudden, temperature changes. In the past two million years, roughly the time that humans have inhabited the earth, there have been approximately 20 glacial and interglacial periods. While these temperature variations have been the result of nonanthropogenic forces, it now seems possible that we can change global climate through our own activities.

Climatologists have used a number of approaches to acquire data on past global temperatures, including gathering evidence from historical documents, tree rings, changes in ice volume and sea level, fossil pollen analysis, and geologic observations related to glacial movements. One of the most fruitful approaches involves an observed correlation between the world's ice cover and the concentration of the isotope ^{18}O in seafloor sediments. When water evaporates from the oceans it contains a mix of two isotopes of oxygen, ^{18}O and ^{16}O. Being slightly heavier, water vapor containing the isotope ^{18}O condenses and falls as precipitation somewhat sooner than water vapor containing ^{16}O. Thus, precipitation over the oceans tends to be slightly richer in ^{18}O than precipitation that must travel further to reach polar ice sheets. Precipitation that forms glaciers and ice sheets is relatively depleted of ^{18}O. As the world's ice increases, it selectively removes ^{16}O from the hydrologic cycle and concentrates the remaining ^{18}O in the oceans. Hence, marine organisms that build their shells out of calcium carbonate in seawater will have a higher ratio of ^{18}O to ^{16}O in their shells when it is cold and more of the world's water is locked up in glaciers and ice. By dating marine sediments and observing the ratio of the two oxygen isotopes in their carbonates, a historic thermometer can be created. A similar record can be obtained by noting ratios of ordinary hydrogen and the heavier isotope, deuterium, in snow.

In one of the most significant scientific studies of past climates, an ice core 2083 m long was recovered by the Soviets at Vostok in East Antarctica (Jouzel et al., 1987). By careful analysis of the isotope ratios in this Vostok ice core, a continuous 160 000-year temperature record has been obtained. Moreover, the composition of the air bubbles sealed off in the ice gives a corresponding record of the concentrations of various gases, especially carbon dioxide, in the atmosphere. Figure 8.2 shows the remarkable correlation between atmospheric CO_2 concentrations and Antarctic surface temperature. During glacial periods, CO_2 is low. During the warmer interglacial period, it is high.

By combining many types of climate data, a record of mid-latitude air temperatures has been reconstructed, some of which is summarized in Figure 8.3. A striking feature of this figure is how modest the temperature differences have been in the past 850 000 years, between the warmest interglacials such as we are in now, and the coldest glacial periods when ice covered much of North America. A rise in temperature of only a few degrees Celsius would make the earth warmer than it has been in the last million years.

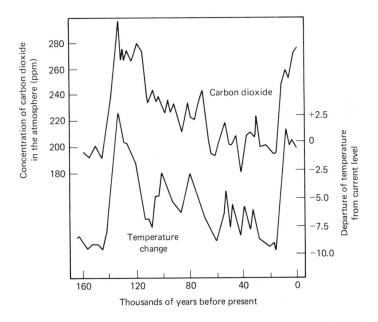

Figure 8.2 CO_2 concentrations (ppm) and Antarctic temperatures (°C) plotted against age in the Vostok record. Temperatures are referenced to current Vostok surface temperature. (*Source:* Barnola et al. Reprinted by permission from *Nature*, vol. 329. Copyright © 1987 Macmillan Magazines Ltd.)

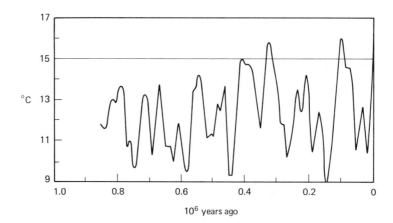

Figure 8.3 Northern Hemisphere, mid-latitude air temperatures over the past one million years. (*Source:* Clarke W. C. (ed.). *Carbon Dioxide Review 1982*. Copyright © 1982 by Oxford University Press. Reprinted by permission of Oxford University Press.)

The historical record of global temperatures shows a primary cycle between glacial episodes of about 100 000 years, mixed with secondary oscillations with periods of 23 000 years and 41 000 years. Although there is still not complete agreement on what has caused these historical temperature variations, evidence suggests that changes in the earth's orbit, which affects the amount of sunlight striking the atmosphere, can account for a significant portion of the changes. The shape of the earth's orbit oscillates from elliptical to more nearly circular with a period of 100 000 years. The earth's tilt fluctuates from 21.5° to 24.5° with a period of 41 000 years. Finally, there is a 23 000-year period associated with the precession, or wobble, of the earth's spin axis. This precession determines where in the earth's orbit a given hemisphere's summer occurs.

While variations in the earth's orbit correlate well with changes in global temperature, they do not provide a complete explanation. There is growing evidence that rather dramatic and sudden shifts in the circulation patterns of the oceans and atmosphere may play a key role (Broecker and Denton, 1990). Phytoplankton near the ocean's surface remove CO_2 during photosynthesis. As the oceans circulate, and as organic matter sinks into the deep sea, carbon is transferred from the surface into the abyss. During glacial periods, this biological pump seems to work better, and more CO_2 is removed from the atmosphere. As will be described in this chapter, carbon dioxide is a greenhouse gas that helps control global temperatures, so that fluctuations in its concentration can account for some of the observed temperature variation. Apparently, the ocean–atmosphere circulation system is subject to abrupt jumps between two stable patterns. These jumps can lead to rapid changes in the way that heat is distributed around the planet as well as the concentration of carbon dioxide in the atmosphere. Coupled with variations in the amount of sunlight hitting the earth caused by orbital changes, these shifts in ocean circulation patterns provide a consistent explanation for the cycles of glaciation that have been observed.

Simple Global Temperature Models

Modeling global climate and estimating the effect of increasing concentrations of influential gases such as carbon dioxide is an extremely important, but difficult, task. Such models range from very simple back-of-the-envelope calculations, to complex, three-dimensional atmospheric *general circulation models* (GCMs) that attempt to predict climate on a regional and seasonal basis. The most sophisticated of these models take weeks to run on a supercomputer, yet they must still be considered primitive. In comparison, our treatment here is just the briefest of introductions.

The starting point in predicting changes in *climate* begins with models that focus on factors influencing global *temperature*. Besides temperature, aspects of climate that are crucial include variables such as winds, ocean currents, precipitation, soil moisture, runoff, snow cover, polar sea ice, and glaciers. The impacts of any changes in these climate indicators on how major portions of the human population meet their basic needs for food and security cannot be underestimated.

The Earth as a Blackbody

Recall from Section 1.4 that a blackbody is a useful theoretical abstraction defining an object that absorbs all radiation impinging upon it, as well as one that radiates at the maximum possible rate for any object with the same dimensions and temperature.

Figure 8.4 shows a simple model that treats the earth as a blackbody. Radiation from the sun arrives just outside the earth's atmosphere with an average

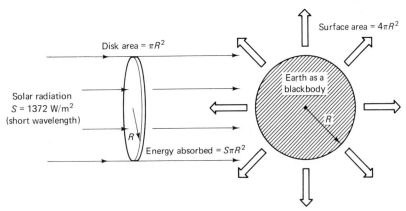

Figure 8.4   In this simplest model, the earth is treated as a blackbody, absorbing all radiation impinging upon it, and radiating an equal amount.

annual intensity, called the *solar constant, S,* equal to 1372 W/m². A simple way to calculate the total rate at which energy is absorbed by the earth is to note that all of the flux passing through a disk having radius equal to that of the earth, and placed normal to the incoming radiation, strikes the earth's surface. Since we are considering the earth to be a blackbody, all of that incoming radiation is absorbed, so we can write

$$E_a = S\pi R^2 \tag{8.1}$$

where E_a is the rate of energy absorption (watts, averaged over all latitudes for a year's time) and R is the radius of the earth (and disk) in meters.

If we assume that the earth is isothermal, with temperature T (K) everywhere, the energy radiated, E_r, from this "blackbody" earth with surface area $4\pi R^2$ is given by the Stefan–Boltzmann equation as

$$E_r = \sigma 4\pi R^2 T^4 \tag{8.2}$$

where σ = the Stefan–Boltzmann constant = 5.67×10^{-8} W/m²K⁴.

If we go on to assume steady-state conditions, that the earth's temperature is not changing with time, we can equate the rate at which energy from the sun is absorbed with the rate at which energy is radiated back to space. Since space is essentially a vacuum, there is no heat transfer to it by conduction or convection. And, since the earth's temperature is assumed constant, we assume that there is no change in internal energy.

All that is left is to equate energy absorbed with energy radiated

$$S\pi R^2 = \sigma 4\pi R^2 T^4 \tag{8.3}$$

and solve for the earth's temperature that would result from that equality. Notice that the radius of the earth conveniently drops out of the equation:

$$T = \left(\frac{S}{4\sigma}\right)^{1/4}$$

$$T = \left(\frac{1372 \text{ W/m}^2}{4 \times 5.67 \times 10^{-8} \text{ W/m}^2\text{K}^4}\right)^{1/4} = 279 \text{ K} \tag{8.4}$$

This answer is remarkably close to the actual global average surface air temperature of about 288 K (15 °C). Unfortunately, the seeming accuracy is to a large extent coincidental, rather than the result of a particularly good model.

A Simple Radiation Balance Model That Includes the Earth's Albedo

One simple modification to the blackbody model just described includes the reflection of incoming solar energy off of the atmosphere and earth's surface, back into space. Such reflected energy is not absorbed by the earth or its atmosphere and does not contribute to their heating.

The fraction of incoming solar radiation that is reflected is known as the *albedo*, and for the earth, the global annual mean value is usually estimated to be about 30 percent. Figure 8.5 shows the revised model, again under the assumption that the earth is a blackbody absorbing all of the nonreflected, incoming solar radiation. Setting the absorbed energy equal to the reradiated energy, assuming equilibrium conditions and a uniform temperature for the earth, yields

$$S(1 - \alpha)\pi R^2 = \sigma 4\pi R^2 T_e^4 \tag{8.5}$$

where α is the albedo and T_e is usually called the *effective temperature*, or the *equivalent* (blackbody) temperature of the earth. Solving for T_e yields

$$T_e = \left[\frac{S(1 - \alpha)}{4\sigma}\right]^{1/4} \tag{8.6}$$

Substituting appropriate values into (8.6) yields

$$T_e = \left[\frac{1372 \text{ W/m}^2(1 - 0.3)}{4 \times 5.67 \times 10^{-8} \text{ W/m}^2\text{K}^4}\right]^{1/4} = 255 \text{ K}$$

The correction that we have applied to account for albedo has unfortunately made our estimate worse. While we are only 11 percent off of the correct value, which might seem a modest error, in terms of life on earth the 255 K (-18 °C) estimate for T_e is extreme. We need to find an explanation for why the earth is (fortunately) not that cold. The key factor that makes our model differ so much from reality is that it does not account for interactions between the atmosphere and radiation that is emitted from the earth's surface. That is, it does not include the greenhouse effect.

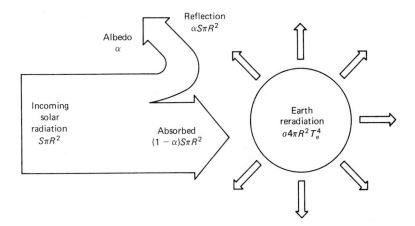

Figure 8.5 A more realistic model that includes the earth's albedo.

8.3 THE GREENHOUSE EFFECT

The surface of the earth is 33 °C higher than what is predicted by (8.6). To understand the reason for the higher temperature, it is helpful to begin by recalling the discussion in Chapter 1 concerning the relationship between the spectrum of wavelengths radiated by an object and its temperature. Wien's displacement rule, repeated here, gives the wavelength at which a blackbody spectrum peaks, as a function of absolute temperature:

$$\lambda_{max}(\mu m) = \frac{2898}{T (K)} \tag{8.7}$$

The sun can be represented as a blackbody with temperature around 6000 K, so its spectrum peaks at around 0.5 μm. The earth, at 288 K, has a peak wavelength of about 10.1 μm. Figure 8.6a shows the two spectra normalized to have the same areas. Notice that nearly all the incoming solar energy arrives extrater-

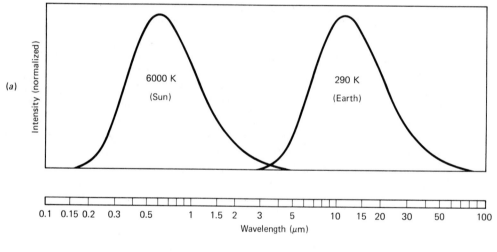

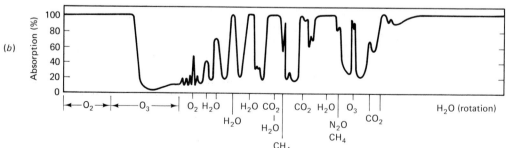

Figure 8.6 (a) Normalized blackbody radiation curves for the sun and earth. (b) Atmospheric absorption on a clear day. (Adapted from Wallace and Hobbs, 1977.)

restrially, with wavelengths less than 4 μm, while the outgoing energy radiated by the earth has essentially all of its energy in wavelengths greater than 4 μm. With so little overlap, it is convenient to speak of solar energy as being *short-wave-length* radiation, while energy radiated from the earth's surface is *long-wave-length,* or *thermal* radiation.

In Figure 8.6*b* the ability of various gases in the atmosphere to absorb radiation is shown as a function of wavelength. Notice that essentially all of the incoming solar radiation with wavelengths less than 0.3 μm (ultraviolet) is absorbed by oxygen and ozone. This absorption of ultraviolet occurs in the stratosphere, shielding the earth's surface from harmful ultraviolet radiation. Later in this chapter we will return to this important phenomenon in our discussion of the stratospheric ozone reduction being caused by chlorofluorocarbons (CFCs).

Figure 8.6*b* shows that most of the long-wavelength energy radiated by the earth is affected by a combination of radiatively active gases, most importantly water vapor (H_2O), carbon dioxide (CO_2), nitrous oxide (N_2O), and methane (CH_4). Water vapor strongly absorbs thermal radiation with wavelengths less than 8 μm and greater than 18 μm. Carbon dioxide shows a strong absorption band centered at 15 μm and extending from 13 to 18 μm, as well as bands centered at 2.7 and 4.3 μm. Between 7 and 12 μm there is a relatively clear sky for outgoing thermal radiation, referred to as the *atmospheric radiative window*. Radiation in those wavelengths easily passes through the atmosphere, except for a small, but important, absorption band between 9.5 and 10.6 μm associated with ozone (O_3).

Radiatively active gases that absorb wavelengths longer than 4 μm are called greenhouse gases. As Figure 8.6 suggests, these gases trap most of the outgoing thermal radiation attempting to leave the earth's surface. This absorption heats the atmosphere, which, in turn, radiates energy back to the earth as well as out to space, as shown in Figure 8.7. These greenhouse gases act as a thermal blanket around the globe, raising the earth's surface temperature beyond the equivalent temperature calculated earlier. As an aside, it is interesting to note that the term

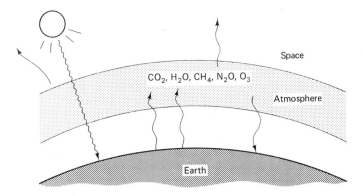

Figure 8.7   Greenhouse gases trap long-wavelength energy from the earth's surface, heating the atmosphere, which, in turn, heats the earth.

"greenhouse effect" is based on the concept of a conventional greenhouse with glass acting much like the above gases. Glass, which easily transmits short-wavelength solar energy into the greenhouse, is nearly opaque to the longer wavelengths radiated by the greenhouse interior. This radiation trapping is partly responsible for the elevated temperatures inside the greenhouse (the reduction in convective cooling of the interior space caused by the enclosure is thought to be even more important).

If the earth did not already have a greenhouse effect, its temperature would be 255 K as predicted by (8.6). That is, the planet would have an average temperature of -18 °C, or about 0 °F. In fact, one way to quantify the magnitude of the greenhouse effect is to compare the effective temperature T_e, given in (8.6) with the actual surface temperature T_s:

$$\text{Greenhouse effect (°C)} = T_s - T_e \tag{8.8}$$

Thus, since the actual temperature of the earth is 288 K and its effective temperature is 255 K, we can say that the greenhouse effect adds 33 °C of warming to the surface of the earth.

In Table 8.2, this notion is applied to Venus and Mars. Mars has almost no greenhouse effect. Even though the atmosphere of Mars is almost entirely carbon dioxide, there is so little atmosphere that the greenhouse effect is barely apparent. The atmospheric pressure on Venus, on the other hand, is nearly 100 times that of earth, and its atmosphere is 97 percent CO_2. The greenhouse effect on Venus is, correspondingly, very pronounced. It is interesting to note that without the greenhouse effect, the greater albedo of Venus would make it cooler than the earth, in spite of its closer proximity to the sun.

Let us add some quantitative information to the simple greenhouse diagram presented in Figure 8.7. As suggested there, it is convenient to consider the earth, its atmosphere, and outer space as three separate regions. We will normalize energy flows between these regions by expressing them in terms of rates per unit of surface area of the earth. For example, (8.1) indicates that the total amount of

TABLE 8.2 APPLICATION OF THE SIMPLE MODEL (8.6) TO COMPUTE EQUIVALENT TEMPERATURES, COMPARED WITH ACTUAL SURFACE TEMPERATURES. MARS SHOWS ALMOST NO GREENHOUSE EFFECT WHILE ON VENUS IT IS QUITE PRONOUNCED.

Planet	Insolation S (W/m²)	Albedo α	Equivalent temperature T_e (K)	Actual temperature T_s (K)
Earth	1372	0.30	255	288
Venus	2613	0.75	232	700
Mars	589	0.15	217	220

Source: USEPA (1983).

solar radiation striking the earth is $S\pi R^2$. Distributed over the entire surface of the earth, the incoming solar radiation is equal to

$$\frac{\text{Incoming solar radiation}}{\text{Surface area of earth}} = \frac{S\pi R^2}{4\pi R^2} = \frac{S}{4} = \frac{1372 \text{ W/m}^2}{4} = 343 \text{ W/m}^2 \quad (8.9)$$

Since the albedo is 30 percent, the amount of incoming radiation reflected back into space is

$$\text{Incoming radiation reflected} = 0.30 \times 343 \text{ W/m}^2 = 103 \text{ W/m}^2$$

Of this 103 W/m², it is estimated that 89 W/m² are reflected off of the atmosphere itself while the remaining 14 W/m² are reflected off of the earth's surface. (Most of the data in this section are taken from Harte (1985).)

The amount of incoming radiation absorbed by the atmosphere and earth is

$$\text{Incoming radiation absorbed} = 0.70 \times 343 \text{ W/m}^2 = 240 \text{ W/m}^2$$

Of that 240 W/m², 86 W/m² are absorbed by the atmosphere and the remaining 154 W/m² are absorbed by the surface of the earth.

If we assume that global temperatures are unchanging with time, then the rate at which the earth and its atmosphere absorb energy from space must equal the rate at which energy is being returned to space. That is, the earth and its atmosphere must radiate 240 W/m² back into space. If the earth's surface were at 255 K, it would radiate 240 W/m², which is just enough to balance the incoming energy. We know, however, that greenhouse gases would absorb most of that outgoing 240 W/m² so the required energy balance would not be realized. Therefore, to force enough energy through the atmosphere to create the necessary balance, the temperature of the earth's surface must be higher than 255 K.

If we treat the earth as a blackbody, we can use (8.2) to estimate the rate at which energy is radiated from the earth's surface toward the atmosphere. We could use the 288 K temperature that is usually considered to be the global average, but this is the temperature of the air just above the surface. The actual temperature of the solid and liquid surface of the earth itself is considered to be about 2 °C higher. Using 290 K in (8.2) gives

$$\frac{\text{Energy radiated by surface}}{\text{Surface area}} = \frac{\sigma 4\pi R^2 T_s^4}{4\pi R^2} = \sigma T_s^4$$

$$= 5.67 \times 10^{-8} \text{ W/m}^2\text{K}^4 \times (290 \text{ K})^4 = 401 \text{ W/m}^2$$

$$(8.10)$$

Of that 401 W/m², only 20 W/m² passes directly through the atmosphere, mostly through the atmospheric window. The remaining 381 W/m² is absorbed by greenhouse gases in the atmosphere. The atmosphere then radiates 344 W/m² back to the surface.

All of these energy flows are shown in Figure 8.8. If this model is internally self-consistent, the rate of energy gain should equal the rate of energy loss in each

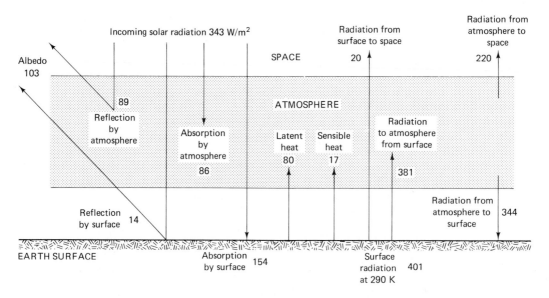

Figure 8.8 Global average energy flows between space, the atmosphere, and the earth's surface. Units are watts per square meter of surface area. (Based on Harte, 1985.)

of the three regions: space, the atmosphere, and the earth's surface. Consider the following checks:

$$\text{Rate of energy gain} = \text{Rate of energy loss?}$$

Earth's surface: $154 + 344 = 80 + 17 + 20 + 381$ (checks)

Atmosphere: $86 + 80 + 17 + 381 = 220 + 344$ (checks)

Space: $103 + 20 + 220 = 343$ (checks)

So, the model shows the necessary balances.

Greenhouse Effect Enhancement

Thus far, the greenhouse effect has been described as a natural phenomenon that is responsible for earth having an average near-surface air temperature 33 °C warmer (288 vs 255 K) than it would have if it did not have radiatively active gases in the troposphere. As is now well known, anthropogenic sources of a number of gases are enhancing the greenhouse effect, leading us into a future of uncertain global climate.

Figure 8.9 provides an overview for our discussion of global climate change. Driving factors in the problem are population, economic activity, and the technology used to produce the goods and services used by society. In the process of meeting our wants and needs, greenhouse gases are added to the atmosphere. Carbon dioxide from the combustion of fossil fuels and deforestation has

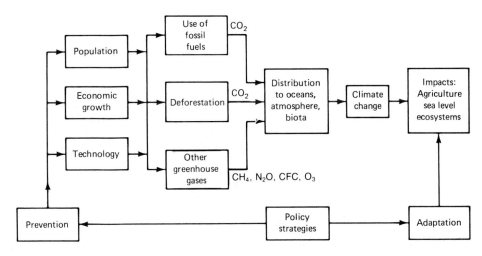

Figure 8.9   An overview of the global warming problem.

received the most attention, but other greenhouse gases, including nitrous oxide, methane, and chlorofluorocarbons, are rapidly becoming as important as CO_2 in changing our climate. These gases all absorb within the 7- to 13-μm atmospheric window, so they are especially effective at influencing the earth's radiative balance.

Table 8.3 summarizes some of the characteristics of these gases, including the relative greenhouse efficiency of each compared with carbon dioxide. A molecule of CFC, for example, exerts approximately 15 000 times the greenhouse warming effect as a molecule of CO_2. Table 8.3 also points out the relative global

TABLE 8.3   MAJOR GREENHOUSE GASES AND THEIR CHARACTERISTICS

Gas	Atmospheric concentration (ppm)	Annual concentration increase (%)	Relative greenhouse efficiency ($CO_2 = 1$)	Current greenhouse contribution (%)	Principal sources of gas
Carbon dioxide	351	0.4	1	57	Fossil fuels, deforestation
CFCs	0.00225	5	15 000	25	Foams, aerosols, refrigerants, solvents
Methane	1.675	1	25	12	Wetlands, rice, livestock, fossil fuels
Nitrous oxide	0.31	0.2	230	6	Fuels, fertilizer, deforestation

Source: Flavin (1989).

warming importance of each of these greenhouse gases. Carbon dioxide is the dominant contributor, with 57 percent. The next biggest source of global warming in the late 1980s were chlorofluorocarbons (CFCs). Both carbon dioxide and chlorofluorcarbons will receive considerable attention in later sections of this chapter.

While the easiest measure of global climate is average temperature, of greater importance is how temperature affects such factors as the timing and length of agricultural growing seasons and the amount, location, and timing of precipitation. Climate changes obviously have important implications for agriculture. Some areas of the world may become better suited to agriculture, while others will be adversely affected. Many models, for example, project that the grain belt in North America will shift northward from the midwestern states in the United States up into Canada. Unfortunately, shifts in agricultural land use patterns will not respect political boundaries, and although the developed countries may well be able to adapt to the changes, the potential for international conflict in the overcrowded developing world is extreme. Climate changes can also have disastrous effects on ecosystems since they cannot very easily move to keep up with shifting climate patterns.

Increasing temperatures will also cause thermal expansion of the oceans. This could raise sea level by significant amounts, possibly flooding large areas of coastlines. Calculations indicate a rise in sea level of about 1/4 m per °C, with most projections indicating a total rise of on the order of 1–3 m by the end of the next century (Hoffman and Keyes, 1983). If the West Antarctic ice sheet ultimately slides into the sea, levels could rise another 5 or 6 m, though this might take several hundred years.

Figure 8.9 suggests two complementary approaches to dealing with potential climate change. One approach is labeled *adaptation,* where attention is focused on steps that can be taken to adjust to the changing climate. Adaptive strategies might include construction of conveyance systems to redistribute water, fortification of sensitive coastal areas to avoid damage from rising sea levels, and migration of populations toward regions that are more conducive to continued survival. The second strategy is *prevention,* where steps are taken to delay or limit the emission of greenhouse gases in the first place. Energy efficiency improvements, shifts from fossil fuels to nuclear or solar energy, reductions in CFC production, and reforestation projects are examples.

8.4 CARBON DIOXIDE

Carbon dioxide has been recognized for its importance as a greenhouse gas for almost a century. Arrhenius (1896) is usually credited with the first calculations on global temperature as a function of atmospheric CO_2 content, and his results are not that far from those obtained today. It is the gas that has received the most attention in discussions on the greenhouse effect, but its impact is quickly being approached by other trace gases that will be covered later.

The Carbon Cycle

The carbon and oxygen cycles are closely linked through the processes of photosynthesis and respiration. During photosynthesis, green plants use solar energy to combine carbon dioxide and water, producing energy-rich carbohydrates and, in the process, releasing oxygen to the atmosphere. The reaction is summarized by the following:

$$6CO_2 + 6H_2O \xrightarrow{\text{sunlight, chlorophyll}} C_6H_{12}O_6 + 6O_2 \qquad (8.11)$$

where the carbohydrate suggested above is glucose. Three absolutely crucial services are thus performed by green plants. Complex organic molecules, which are used as fundamental building blocks for plants, are created from simple inorganic carbon dioxide and water. The sunlight captured and stored provides the energy needed by plants themselves or by whatever consumes the plants. Finally, the process liberates molecular oxygen.

Reversing the above reaction yields the equation describing respiration whereby energy is derived from food. During respiration, living organisms break down complex organic molecules, returning carbon dioxide and water.

$$C_6H_{12}O_6 + 6O_2 \longrightarrow 6CO_2 + 6H_2O + \text{energy} \qquad (8.12)$$

Carbon thus moves continually from the atmosphere into the food chain during photosynthesis and returns to the atmosphere during respiration. From the atmosphere it can be assimilated by plants on the land or in the oceans, or it can dissolve into the seawater, as shown in Figure 8.10.

Respiration by living things, including decomposers that are feeding on dead organic matter, return carbon dioxide either to the oceans or the atmosphere. A very small portion of the dead organic matter each year ends up being buried in sediments. The slow, historical accumulation of buried organic matter is the source of our fossil fuels—oil, gas, and coal. When these are burned, carbon in the form of carbon dioxide is returned to the atmosphere. The rather rapid accumulation of carbon dioxide in the atmosphere that we are now experiencing, due in part to combustion of fossil fuels and in part to the return of carbon during deforestation, is the cause of most of our concern for future global warming.

Historical Emissions of Carbon Dioxide

The first reliable, continuous measurements of atmospheric carbon dioxide were begun by C. D. Keeling at Mauna Loa, Hawaii, in 1957. By 1990, the concentration had reached 355 ppm and was climbing at about 1.5 ppm per year. The oscillations shown in the Mauna Loa data plotted in Figure 8.11 are caused by

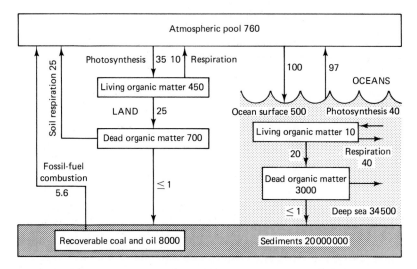

Figure 8.10 Simple biogeochemical cycle for carbon. Quantities are either in billions of tons of carbon (Gt C), for standing crops, or billions of tons per year (Gt C/yr) for flow rates. Numbers based on Bolin (1970), adjusted to 1990.

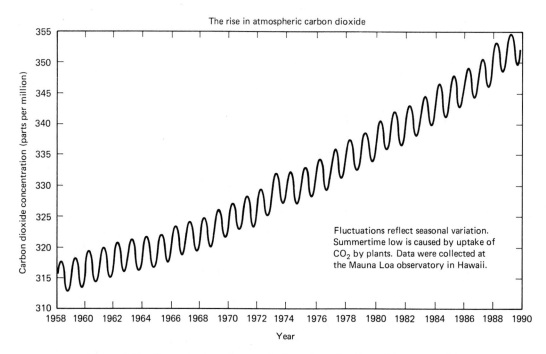

Figure 8.11 Concentration of atmospheric carbon dioxide at Mauna Loa Observatory, Hawaii. (*Source:* Courtesy C. D. Keeling et al., 1990.)

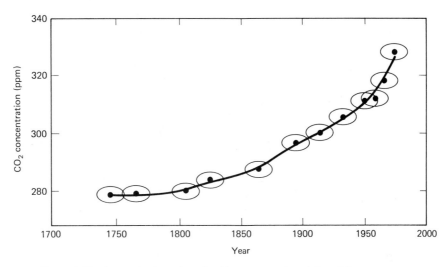

Figure 8.12 Atmospheric carbon dioxide concentrations inferred from measurements of glacial ice formed during the last 200 years. (*Source:* Neftel et al. Reprinted by permission from *Nature*, vol. 315. Copyright © 1985 Macmillan Magazines Ltd.)

seasonal variations in the rates of photosynthesis and respiration, with maximum concentrations occurring in May and minimums in October. Measurements of CO_2 have been extended tens of thousands of years back in time by analyzing air trapped in glacial ice, as was shown in Figure 8.2. Similar measurements of glacial ice over the past 200 years suggest that the concentration of carbon dioxide at the beginning of the nineteenth century was around 280 ppm (Figure 8.12). This 280-ppm value is often used as a reference point for comparison with the modern values. Carbon dioxide concentrations in 1990 were more than 25 percent higher than those just before the industrial revolution.

Example 8.1   Carbon Content of the Atmosphere

Estimate the tons of carbon in the atmosphere corresponding to a concentration of 360 ppm of CO_2. Assume the total mass of air equals 5.1×10^{18} kg. The density of air at standard temperature and pressure (STP, 0 °C, and 1 atm) is 1.29 kg/m^3.

Solution   One g-mole of CO_2 contains 44 g ($12 + 2 \times 16$), and each mole at STP occupies a volume of 22.4×10^{-3} m^3 (see Section 1.2). At 360 ppm, the concentration of CO_2 (by weight at STP) is given by

$$CO_2 = \frac{360 \text{ m}^3 \text{ CO}_2}{1 \times 10^6 \text{ m}^3 \text{ air}} \times \frac{\text{mol}}{22.4 \times 10^{-3} \text{ m}^3 \text{ CO}_2} \times \frac{44 \text{ g}}{\text{mol}} = 0.707 \text{ g/m}^3$$

Since 44 g of CO_2 contains 12 g of C, the total amount of carbon in the atmosphere is

$$C = \frac{0.707 \text{ g CO}_2}{\text{m}^3 \text{ air}} \times \frac{12 \text{ g C}}{44 \text{ g CO}_2} \times \frac{5.1 \times 10^8 \text{ kg air}}{1.29 \text{ kg/m}^3 \text{ air}} = 7.62 \times 10^{17} \text{ g}$$

which, at 10^6 g/ton, is equivalent to 762×10^9 tons or 762 gigatons (Gt).

Example 8.1 provides a very convenient conversion factor between ppm CO_2 concentration and tons of atmospheric carbon:

$$1 \text{ ppm } CO_2 = 1 \text{ ppm } CO_2 \times \frac{762 \text{ Gt C}}{360 \text{ ppm } CO_2} = 2.12 \text{ Gt C} \qquad (8.13)$$

Equation 8.13 is a useful link that will help us estimate the relationship between future carbon emissions and expected changes in atmospheric CO_2 concentration. By itself, however, it is not enough. Since some carbon added to the atmosphere will be absorbed by the oceans or taken up by plants during photosynthesis, not all emissions will result in increased carbon dioxide concentrations. It is convenient to represent the fraction of emissions that remain in the atmosphere with a quantity called the *airborne fraction:*

$$\text{Airborne fraction} = \frac{\Delta \text{ C atmosphere}}{\Delta \text{ C emissions}} \qquad (8.14)$$

where (Δ C atmosphere) is the change in carbon content of the atmosphere, and (Δ C emissions) is the total amount of carbon added to the atmosphere. Carbon is being added to the atmosphere primarily as the result of fossil fuel combustion. At 1987 emission rates, fossil fuel combustion was adding 5.6 billion tons of carbon to the atmosphere per year (Flavin, 1990). Carbon is also added to the atmosphere when the amount of carbon locked in the earth's biomass is reduced. When forests are cleared and soil degrades, carbon that had been locked into vegetation or humus is released, adding to the CO_2 emissions from fossil fuels. Siegenthaler and Oeschger (1987) estimate such net biospheric inputs in the period 1959–1983 at between 0 and 0.9 Gt C/year. Higher estimates for the biomass contribution are not uncommon. Bolin (1986), for example, estimates mid-1980s reductions in biomass to be contributing 1.6 ± 0.8 Gt C/year.

If we estimate the contribution from biomass to be 0.9 Gt C/year and add that to the 5.6 Gt/year from fossil fuels, the total 1987 carbon emission rate would be 6.5 Gt C/year. If all of that stayed in the atmosphere, the increase in CO_2 would be

$$6.5 \text{ Gt C/yr} \times \frac{1 \text{ ppm}}{2.12 \text{ Gt C}} = 3.1 \text{ ppm } CO_2/\text{yr}$$

Since carbon dioxide is actually increasing at about 1.5 ppm/year, we can estimate the airborne fraction to be

$$\text{Airborne fraction} = \frac{1.5 \text{ ppm/yr}}{3.1 \text{ ppm/yr}} = 0.48 \cong 0.5$$

Values of airborne fraction have been estimated at anywhere from 0.4 to 0.7, with 0.5 being a commonly used ratio. The airborne fraction would probably be higher for rapid increases in carbon emissions due to the slow uptake rate of the oceans and biota. Similarly, for slow increases in CO_2, the airborne fraction would likely be less.

Rotty and Masters (1984) have estimated that between 1860 and 1980, a total of 162 Gt of carbon were released due to combustion of fossil fuels and cement manufacturing. On the biomass side, Siegenthaler and Oeschger (1987) estimate the net cumulative biospheric carbon release from 1800 to 1980 to have been between 90 and 150 Gt. More than half of that, they calculate, was released before 1900. Using a middle estimate of 120 Gt C for the biomass contribution suggests that total emissions of carbon since the industrial revolution have been about 280 Gt. During that time atmospheric carbon dioxide rose 58 ppm, from 280 to 338 ppm, which corresponds to an increased atmospheric carbon content of 123 Gt C. This corresponds to an atmospheric fraction of 44 percent. The discrepancy between this and the higher value of 48 percent obtained earlier is only modest, although it does suggest that over longer periods of time the airborne fraction decreases somewhat.

One way to illustrate the removal of carbon from the atmosphere is to imagine adding a pulse of carbon dioxide to the atmosphere, and then plot the gradual decline in its concentration. The results from one such modeling experiment are shown in Figure 8.13. The initial concentration is arbitrarily set at 100 units so that the vertical axis becomes the percentage of the initial amount of carbon that remains in the atmosphere. The model suggests that there is an initial, rapid decline in concentration, followed by a much slower removal rate. The time required to remove half of the carbon is on the order of 50 years.

Using measured concentrations of carbon dioxide in ice bubbles, Siegenthaler (1986) has produced a historical estimate of carbon emissions over the past 180 years. By subtracting known historical fossil-fuel carbon emissions from the total, the net carbon additions by deforestation over this period of time can be estimated. The results presented in Figure 8.14 show the individual contributions of fossil fuels and non-fossil-fuel inputs, as well as the total.

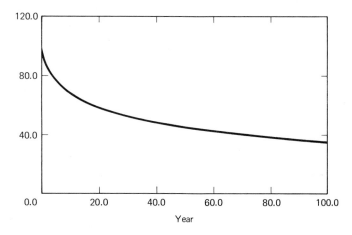

Figure 8.13 Atmospheric response to an initial pulse input of CO_2. The initial value is arbitrarily set to 100 units. (*Source:* Siegenthaler and Oeschger, 1987.)

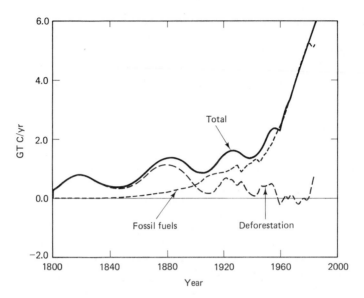

Figure 8.14   Total CO_2 production rate (solid line) calculated from ice cores.   The short dashed line indicates fossil-fuel contribution; the difference between the two curves is non-fossil inputs, shown with long dashes.   (*Source:* Siegenthaler, 1986.)

The following example utilizes the simple exponential growth model developed in Chapter 1 to make our first-cut estimate of future carbon dioxide concentrations in the atmosphere.   In the next section we will refine our forecasting techniques.

Example 8.2   Exponential Growth of CO_2 Concentration

The concentration of CO_2 in 1965 was 320 ppm, while in 1990 it was 355 ppm.   If we model that growth with a simple exponential function, what growth rate would that correspond to?   At that exponential rate of growth, in what year would atmospheric concentrations be twice the 280 ppm preindustrialization value?

Solution   Using (3.3) to express the concentrations in an exponential form,

$$C = C_0 e^{rt}$$

we have

$$355 = 320 \, e^{r(1990-1965)} = 320 \, e^{25r}$$

or

$$r = \frac{1}{25} \ln \frac{355}{320} = 0.0041 = 0.41 \text{ percent/yr}$$

If growth continues at 0.41 percent/year, using (3.3) we can compute the time required to raise the concentration from the 1990 level of 355 to 560 ppm (double the preindustrialization value):

$$2 \times 280 = 355 \, e^{0.0041t}$$

$$t = \frac{1}{0.0041} \ln \frac{560}{355} = 110 \text{ yr}$$

That is, it would be reached in the year 1990 + 110 = 2100.

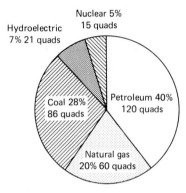

Figure 8.15 World primary energy production by fuel source (omitting biomass), 1985 (Energy Information Administration, 1986).

Carbon Emissions from Fossil Fuels

Fossil fuels supply about 88 percent of (nonbiomass) world energy needs, with the remainder coming almost entirely from hydroelectric facilities and nuclear power plants. Figure 8.15 shows the energy delivered by each of these sources, with values given in the most common American unit, *quads* (1 quad = 1 quadrillion BTU = 1 × 10^{15} BTU = 1.055 × 10^{18} J). Table 8.4 shows the distribution of

TABLE 8.4 CARBON EMISSIONS FROM FOSSIL FUELS, SELECTED COUNTRIES, 1960 AND 1987

Country	Carbon (million tons)		Carbon per dollar GNP (g)		Carbon per capita (tons)	
	1960	1987	1960	1987	1960	1987
United States	791	1224	420	276	4.38	5.03
Soviet Union	396	1035	416	436	1.85	3.68
China	215	594	n.a.	2024	0.33	0.56
Japan	64	251	219	156	0.69	2.12
West Germany	149	182	410	223	2.68	2.98
Poland	55	128	470	492	1.86	3.38
India	33	151	388	655	0.08	0.19
Nigeria	1	9	78	359	0.04	0.03
World total	2547	5599	411	327	0.82	1.08

Source: Flavin (1990).

carbon emissions from fossil fuels for selected countries. Emissions are also broken down into amounts of carbon per dollar of gross national product (GNP), and
carbon per capita. The United States, with less than 5 percent of the world's
population, emits 22 percent of the total carbon. On a per capita basis, the United
States emits nearly five times as much carbon per person as the world average,
and nearly double the amount emitted by other developed countries.

Table 8.5 summarizes carbon emission factors for fossil fuels, the amounts
of each fuel used in 1985, and the corresponding amounts of carbon released. The
differences in carbon emission factors for these fuels is signficant. About 75 percent more carbon is released per unit of energy when coal is burned instead of
natural gas. And the average synthetic oil or gas made from coal or shale oil,
including the emissions released during preparation of the fuel, produces about 80
percent more carbon than coal. In the 1970s, there was strong U.S. government
push toward development of synthetic fuels as a way to improve energy self-
sufficiency. The program was severely criticized, however, as being one of the
least cost-effective and most environmentally damaging ways to achieve that goal,
and the program was eventually abandoned.

The emission factors given in Table 8.5 suggest that one way to reduce
carbon emissions is to switch from coal to oil or natural gas. For example, if a
power plant could burn gas instead of coal, carbon emissions would decrease by
42 percent. The following example suggests that even more dramatic reductions
can be realized if heating systems that now use electricity derived from coal were
to switch to gas at the point of use.

TABLE 8.5   CARBON EMISSION FACTORS FOR VARIOUS FUELS, AND 1985 WORLD FUEL AND
EMISSIONS DATA

				Synthetic fuels			
Factor	Natural gas	Conventional oil	Coal	Oil from shale	Oil from coal	Gas from coal	Non-fossil fuel
10^6 ton C/quad	14.5	20.8	25.2	50.2	40.7	42.9	0
Quads/yr	60	120	86	0	0	0	36
Carbon emissions (Gt C/yr)	0.9	2.5	2.2	0	0	0	0

Source: Data from Seidel and Keyes (1983) and EIA (1986).

Example 8.3   Replacing Coal-Generated Electricity with Direct Consumption of Gas

Suppose utilities generate electricity using 33-percent efficient coal-fired power
plants. As a carbon-reducing measure, suppose electric water heaters that convert
electricity into hot water with 100 percent efficiency are replaced with gas water
heaters with a 70 percent conversion efficiency. By what fraction would carbon
emissions be reduced?

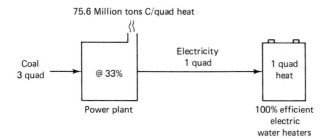

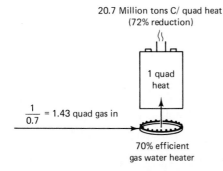

Figure 8.16 Carbon emissions can be reduced by 73 percent when coal-fired electricity is replaced with gas used on site. A gas water heater is more expensive to install, but cheaper to operate. Numbers correspond to Example 8.3.

Solution   Let us approach this problem by imagining 1 quad of electricity being delivered to those electric water heaters, as suggested in Figure 8.16. Since the power plants are 33 percent efficient, 3 quads of heat would need to be delivered to the power plant. Using the emission factor for coal given in Table 8.5, the total carbon emissions would be

$$\text{Coal emissions} = 3 \text{ quads} \times 25.2 \times 10^6 \text{ ton C/quad} = 75.6 \times 10^6 \text{ ton C}$$

To get 1 quad of heat using 70-percent efficient gas-fired water heaters would require

$$\text{Heat input} = 1 \text{ quad}/0.70 = 1.43 \text{ quads}$$

Corresponding carbon emissions would be

$$\text{Gas emissions} = 1.43 \text{ quads} \times 14.5 \times 10^6 \text{ ton C/quad} = 20.7 \times 10^6 \text{ ton C}$$

That is, there would be a 73 percent reduction in emissions.

While Example 8.3 suggests that sizeable reductions in carbon emissions are possible by switching from coal to natural gas, it is unfortunate that most of the world's fossil fuel reserves and resources are in the form of coal. Recall the distinction between reserves and resources introduced in Chapter 3. Reserves are quantities that can reasonably be assumed to exist and that are producible with existing technology under present economic conditions; resources are amounts

TABLE 8.6 ENERGY AND CARBON
CONTENT OF THE WORLD'S FOSSIL FUEL
RESERVES AND RESOURCES

Fossil fuel	Gas	Oil	Coal
Reserves			
Quads	3600	4100	22 000
Gt C	52	85	550
Resources			
Quads	10 000	11 000	280 000
Gt C	145	230	7100

Sources: Reserve data, EIA (1986); resource
data, IIASA (1981).

thought to be ultimately recoverable. Table 8.6 presents data on the world's supplies of fossil fuels expressed both by energy content and carbon content. Unless we can quickly develop our solar or nuclear options, it appears carbon emissions will jump as we grow evermore dependent on coal.

Example 8.4 Potential CO_2 Increases

For each of the fossil fuels, estimate the increase in atmospheric carbon dioxide that would result from complete combustion of the entire resource. Assume an airborne fraction of 50 percent.

Solution For natural gas, the total carbon content of the resource is given in Table 8.6 as 145 Gt C. If all of it is released and half of that remains in the atmosphere, then using the conversion given in (8.13) the increase in CO_2 would be

$$\text{Natural gas} = 145 \text{ Gt C} \times 0.50 \times \frac{1 \text{ ppm } CO_2}{2.12 \text{ Gt C}} = 34 \text{ ppm } CO_2$$

Similarly, for oil,

$$\text{Oil} = 230 \text{ Gt C} \times 0.50 \times \frac{1 \text{ ppm } CO_2}{2.12 \text{ Gt C}} = 54 \text{ ppm } CO_2$$

and coal,

$$\text{Coal} = 7100 \text{ Gt C} \times 0.50 \times \frac{1 \text{ ppm } CO_2}{2.12 \text{ Gt C}} = 1675 \text{ ppm } CO_2$$

Example 8.4 very clearly shows that it is the carbon currently locked up in coal that holds the greatest threat to global climate. Burning all of the world's remaining oil and gas would increase the atmospheric concentration of carbon dioxide by 25 percent over 1990 levels. However, under the above assumptions, burning all of the world's coal could more than quadruple CO_2. It is worth noting that two-thirds of the world's coal reserves exist in just three countries: the United States, the Soviet Union, and China. If South Africa and West Germany are added to the list, 80 percent of the world's reserves are accounted for. As a result, the magnitude of the global warming problem will be controlled to a large extent by policies adopted by just a handful of governments.

Estimating Future Fossil-Fuel Emissions

Table 8.4 gives the growth in carbon emissions over a very turbulent period of time that included the two oil shocks of 1973 and 1979. In the decades before the first oil shock, carbon emissions from fossil fuels had been growing at about 4.5 percent per year. During the 1970s the rate dropped to about 2.3 percent per year, and through the first half of the 1980s, emissions were nearly constant. Predicting the future based on extrapolations of past records would seem, therefore, to be a questionable approach. Some argue that we will return to the high growth rates of the 1950s and 1960s, while others contend that the more recent past is evidence that through conservation and greater emphasis on non-carbon-emitting fuels, emission rates can be controlled. To give a sense of the extreme uncertainty that still exists in forecasting future carbon emissions, consider Figure 8.17, which shows four realistic scenarios developed by the World Resources Institute (Mintzer, 1987).

To gain some insight of our own into carbon emission scenarios, consider a disaggregation of emissions into the product of a number of individual factors (Section 3.2), such as the following:

$$\text{C emissions} = \text{Population} \times \frac{\text{GNP}}{\text{Person}} \times \frac{\text{Energy}}{\text{GNP}} \times \frac{\text{C emissions}}{\text{Energy}} \qquad (8.15)$$

Of course, (8.15) is not unique; any number of other disaggregations of emissions are possible. This expression does, however, let us deal with a number of the most important driving forces individually so we can then combine the results into an overall scenario. Studies using disaggregations of this sort usually are done separately for different regions of the world and then the results are combined (for example, Perry, 1982).

Recall from Chapter 3, that if we can characterize each of the factors in a product as a quantity that is growing (or decreasing) exponentially, then the overall growth rate is merely the sum of the growth rates of each factor. That is, assuming each of the individual factors in (8.15) is growing exponentially, then the overall growth rate of carbon emissions, r, is given by

$$r = r_{\text{pop}} + r_{\text{GNP/cap}} + r_{\text{energy/GNP}} + r_{\text{emissions/energy}} \qquad (8.16)$$

By adding the individual rates of growth, as has been done in (8.16), an overall growth rate can be obtained for use in the following equation.

$$\text{Emission rate } (t) = \text{Emission rate } (0) \ e^{rt} \qquad (8.17)$$

The first factor in (8.16), population growth, is self-explanatory. The world's population in 1990 was growing at about 1.8 percent per year. It is likely

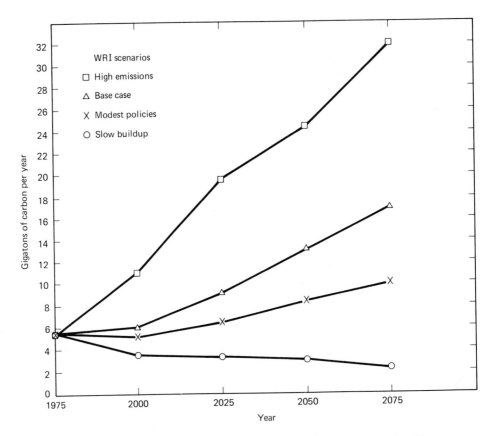

Figure 8.17 Total emissions of carbon in four plausible scenarios developed by the World Resources Institute. (*Source:* Mintzer, 1987.)

to slow down some in the future, due especially to progress being made in China. However, as we saw in Chapter 3, due to the inherent momentum associated with population growth, changes will come slowly. If our projections are to be made for the next half century or so, it is unlikely, barring catastrophe, that the average population growth rate will be less than about 1 percent.

The per capita GNP is included as an indicator of world economic growth, with many midrange projections using values on the order of 1–2 percent per year. Since almost all future population growth will be in developing countries, the growth rate in per capita GNP in our scenarios should be heavily biased toward economic growth that can be expected in the less developed countries.

The third term in (8.16), growth in energy per unit of GNP, is the key factor in this equation that allows us to include improvements in energy efficiency. One of the most encouraging changes since the oil shocks of the 1970s has been the decoupling of economic growth from energy growth. For example, in the United States in 1973, the energy to GNP ratio was 27.1 thousand Btu per 1982 dollar. By

1986, it had fallen to 20.1 (EIA, 1986). That drop is equivalent to an exponential decay rate of -2 percent per year. Proponents of energy efficiency hold that cost-effective conservation improvements have only begun to be incorporated into our society and that this decoupling can continue for the foreseeable future. More-over, developing countries have the opportunity to incorporate energy efficiency at the outset, to do it right the first time. Goldemberg et al. (1985), for example, argue that by incorporating the most energy-efficient technologies currently avail-able, developing countries could approach a standard of living equivalent to that of Western Europe with little or no increase in per-capita energy consumption.

The last term in (8.16), emissions per unit of energy, allows us to account for changes in the fuel mix in the future. Obviously some energy sources, such as hydroelectric, solar, and nuclear power, produce no carbon emissions. To the extent that they are used to meet future energy needs, overall emissions can decrease accordingly. Biomass fuels, where a crop is grown, harvested, and "burned," can also yield a near zero net release of carbon to the atmosphere. Carbon is removed from the atmosphere during plant growth, and that carbon is returned when the biomass is consumed.

As Table 8.5 indicates, carbon emissions per unit of energy consumed is very dependent on the fuel mix. Moreover, Example 8.3 illustrates the impor-tance of carefully analyzing the end uses of energy as well. To project the rate of change of emissions per unit of energy, the relative fuel mix as well as the amount of energy that is derived from fossil-fuel fired electric power plants must be carefully estimated.

Example 8.5   Future Fuel Mix Estimate

Suppose primary energy demand is doubled in 50 years, with the fuel mix at that point being 50 percent coal, 13 percent oil, 11 percent gas, and the remaining 26 percent hydro, nuclear, and solar. Assume initial carbon emissions equal to 5.5 Gt/year and total initial energy supply equal to 300 quads. Modeling the change with an exponential function, estimate the annual rate of growth in carbon emissions per unit of energy.

Solution   Initially,

$$\frac{\text{Emissions}}{\text{Energy}} = \frac{5500 \times 10^6 \text{ ton C/yr}}{300 \text{ quad/yr}} = 18.3 \times 10^6 \text{ ton C/quad}$$

We can compute carbon emissions for each fuel using emission factors from Table 8.5, coupled with the indicated fuel mix in 50 years. For example, with oil supplying 13 percent of 600 quads,

C from oil $= 0.13 \times 600 \text{ quad} \times 20.8 \times 10^6 \text{ ton C/quad} = 1622 \times 10^6 \text{ ton C/yr}$

Natural gas would emit

C from gas $= 0.11 \times 600 \times 14.5 \times 10^6 = 957 \times 10^6 \text{ ton C/yr}$

and coal,

C from coal $= 0.50 \times 600 \times 25.2 \times 10^6 = 7560 \times 10^6 \text{ ton C/yr}$

Total emissions per unit of energy would be

$$\frac{\text{Emissions}}{\text{Energy}} = \frac{(1622 + 957 + 7560) \times 10^6 \text{ ton C/yr}}{600 \text{ quad/yr}}$$

$$= 16.9 \times 10^6 \text{ mt/quad} \quad \text{(in 50 years)}$$

If that change occurred at an exponential rate, we can write

$$16.9 \times 10^6 = 18.3 \times 10^6 \, e^{50r}$$

or

$$r = \frac{1}{50} \ln \frac{16.9}{18.3} = -0.0016 = -0.16 \text{ percent/yr}$$

By combining estimates for key parameters, such as airborne fraction, population and economic growth rates, energy efficiency improvements, fuel mixes, and emission factors, we can generate scenarios of our own to compare with those done with more precise models. Consider the following example.

Example 8.6   A "Conventional Wisdom" Carbon Dioxide Scenario

Let us assume the following annual growth rates, and let their values remain constant throughout the scenario:

1. Population, 1.0 percent
2. Per capita GNP, 1.2 percent
3. Energy per unit of GNP, −0.8 percent
4. Carbon emissions per unit of energy, −0.2 percent

Also, use the following data and estimates:

5. Airborne fraction = 0.58
6. 1985 carbon emission rate = 5.5 Gt/yr
7. 1985 CO_2 concentration = 345 ppm
8. Preindustrial atmospheric CO_2 concentration = 280 ppm
9. 1985 primary energy supply = 300 quads

Using exponential growth models:

a. Determine energy consumption and carbon emissions versus time.
b. Find the concentration of carbon dioxide versus time.
c. Find the year in which atmospheric carbon dioxide is double the preindustrial level.

Solution

a. To graph energy supply, note

$$\text{Energy} = \text{Population} \times \frac{\text{GNP}}{\text{Person}} \times \frac{\text{Energy}}{\text{GNP}}$$

so the energy growth rate would be

$$r_{energy} = r_{population} + r_{GNP/cap} + r_{energy/GNP}$$

$$= 1.0 + 1.2 - 0.8 = 1.4 \text{ percent/yr}$$

Modeling energy consumption as an exponential function that begins in 1985 at 300 quads and grows at 1.4 percent/yr gives

$$\text{Energy}(t) = 300 \, e^{0.014t} \text{ quads/yr}$$

Using (8.15) for carbon emissions,

$$\text{C emissions} = \text{Population} \times \frac{\text{GNP}}{\text{Person}} \times \frac{\text{Energy}}{\text{GNP}} \times \frac{\text{C emissions}}{\text{Energy}}$$

$$r_{emissions} = r_{pop} + r_{GNP/cap} + r_{energy/GNP} + r_{emissions/energy}$$

$$= 1.0 + 1.2 - 0.8 - 0.2 = 1.2 \text{ percent/yr}$$

The initial carbon emission rate is 5.5 Gt C/yr and it is projected to grow at 1.2 percent, so

$$\text{Carbon emission rate } (t) = 5.5 \, e^{0.012t} \text{ Gt C/yr} \qquad (8.18)$$

These are plotted in Figure 8.18.

b. To find the additional carbon in the atmosphere at any given time, we need to find the total amount of carbon released up until that time, and multiply that by the airborne fraction. We can find the total amount of carbon released by integrating the exponential function (8.18). We can use (3.12), derived in Section 3.3:

$$Q = \frac{P_0}{r} \, (e^{rt} - 1) \qquad (3.12)$$

where Q = the total carbon added between up until time t
P_0 = the initial emission rate at time $t = 0$
r = the rate of growth of emissions

For this example,

$$\text{Accumulated emissions} = \frac{5.5}{0.012} \, (e^{0.012t} - 1) \text{ Gt C}$$

$$= 458 \, (e^{0.012t} - 1) \text{ Gt C}$$

To find the added carbon remaining in the atmosphere, we need to multiply emissions by the airborne fraction, 0.58. To convert that to carbon dioxide, we can use the conversion given in (8.13) of 1 ppm = 2.12 Gt C:

$$\text{Added CO}_2 = 458 \text{ Gt C} \times 0.58 \times \frac{1 \text{ ppm}}{2.12 \text{ Gt C}} \, (e^{0.012t} - 1)$$

$$= 125(e^{0.012t} - 1) \text{ ppm}$$

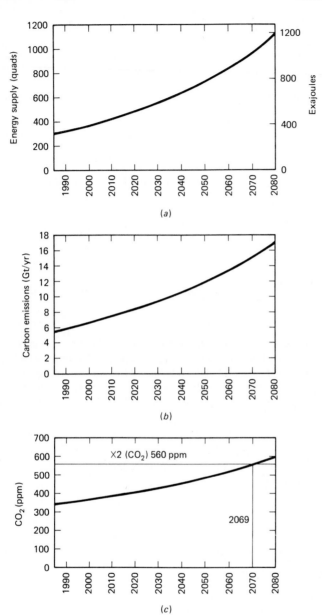

Figure 8.18 Plots of energy consumption (*a*), carbon emissions (*b*), and carbon dioxide concentration (*c*), for the scenario in Example 8.6.

This needs to be added to the initial concentration of 345 ppm to get the total CO_2 concentration,

$$CO_2 = 345 + 125(e^{0.012t} - 1) \text{ ppm}$$

$$= 220 + 125e^{0.012t} \text{ ppm}$$

which is plotted in Figure 8.18*c*.

c. We can find the year in which the concentration is double the preindustrial level, by setting the CO_2 level at 2×280 ppm and solving for time:

$$560 = 220 + 125e^{0.012t}$$

$$t = \frac{1}{0.012} \ln \frac{340}{125} = 84 \text{ yr}$$

That is, in the year $1985 + 84 = 2069$.

Example 8.6 was generated to approximate the results obtained in the much more carefully done World Resources Institute "Base Case Scenario" shown in Figure 8.17 (Mintzer, 1987). In this scenario energy growth inertia prevails, no new policies to slow CO_2 emissions are enacted, there is minimal stimulus to improve end use efficiency and solar energy, and there is no policy to limit tropical deforestation or encourage reforestation. Certainly we can imagine society doing better.

Lovins et al. (1981) have been leading advocates of approaching energy scenarios from a different perspective than what we have used here thus far. Their approach, which is sometimes referred to as *end-use,* or *demand-side,* modeling, is based on estimates of the specific tasks for which energy is required. Then they apply the most efficient, cost-effective, technologies available for meeting those tasks. This methodology has been applied to a number of countries, and the results consistently affirm the possibility of avoiding the carbon dioxide-driven greenhouse effect through renewable energy and energy efficiency.

Extending models beyond the next century is an exercise that does provide some insight, but it must be treated as highly speculative. We can apply the techniques developed in Section 3.3 for ultimate production of a resource to the carbon emission problem. Recall that one way to represent the total production of a resource, or in this case, the total release of carbon during fossil fuel combustion, is with a Gaussian distribution function of the sort

$$P = P_m \exp \left[-\frac{1}{2} \left(\frac{t - t_m}{\sigma} \right)^2 \right] \tag{3.14}$$

where, in this context, P = the rate of release of carbon (Gt/yr)
P_0 = the rate of release of carbon at time $t = 0$
Q_∞ = the total carbon ever released (Gt)
t_m = the time at which the maximum rate of release is reached (yr)
P_m = the maximum rate of release of carbon (Gt/yr)
σ = the standard deviation of the Gaussian distribution (yr)

where σ can be found from

$$Q_\infty = \sqrt{2\pi}\sigma P_m \tag{3.15}$$

and

$$t_m = \sigma \sqrt{2 \ln \frac{P_m}{P_0}} \qquad (3.17)$$

Example 8.7   Ultimate Production of Fossil Fuels and Resulting Carbon Emissions

Suppose the total ultimate worldwide release of carbon due to fossil fuel combustion is equivalent to burning 200 000 quads of coal at 25.2×10^6 t C/quad. The 1980 rate of release of carbon was 5.2 Gt/year. If the peak emission rate is eventually 33 Gt C/year, find the year when peak emissions would result.

Solution   The total tons of carbon ever to be released would be 200 000 quads $\times$ 25.2×10^6 t C/quad, which is 5.0×10^3 Gt of carbon. From Eq. 3.15, the standard deviation of the Gaussian emission function would be

$$\sigma = 5.0 \times 10^3 / (33 \sqrt{2\pi}) = 60.4 \text{ yr}$$

and from (3.17)

$$t_m = 60.4 \left(2 \ln \frac{33}{5.2} \right)^{1/2} = 116 \text{ yr}$$

which is in the year 1980 + 116 = 2096. We can use (3.14) to plot the total emission rate curve for the conditions given, and this has been done in Figure 8.19.

Keeling and Bacastow (1977) derived curves similar to the one shown in Figure 8.19 using the same estimate for ultimate carbon release. These are shown in Figure 8.20a along with resulting carbon dioxide concentrations in the atmosphere. As can be seen, these production curves produce peak concentrations of CO_2 of between 2000 and 2300 ppm. Concentrations gradually decay thereafter as the deep oceans slowly take up the atmospheric carbon.

Equilibrium Temperature Increase Caused by CO_2

Considerable effort has gone into attempting to quantity the relationship between expected global temperature change and carbon dioxide concentration. A sort of

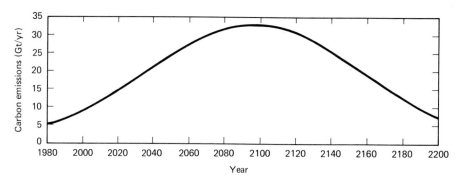

Figure 8.19   A symmetrical production curve for ultimate carbon emissions utilizing data given in Example 8.7.

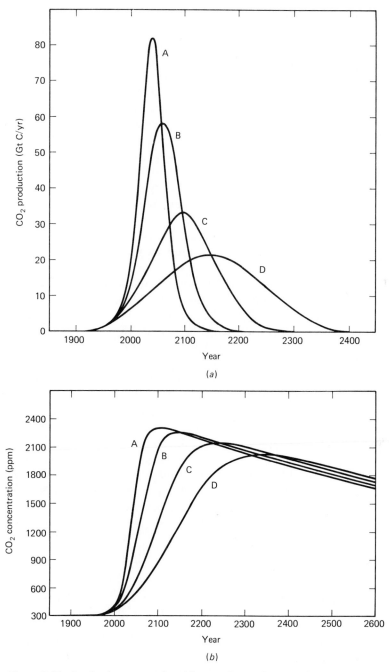

Figure 8.20   Production curves for ultimate release of carbon from fossil fuels and resulting carbon dioxide concentrations. (*Source:* Reprinted from Keeling and Bacastow, *Energy and Climate*, 1977, with permission from the National Academy of Sciences, National Academy Press, Washington, DC.)

"benchmark" result that is reported for almost every climate model being tested is the expected change in mean global surface air temperature induced by a doubling of the atmospheric concentration of CO_2. It should be noted that these values are usually given for equilibrium conditions that, due to thermal inertia, will not be reached until some time after the doubling of carbon dioxide concentration has actually been reached. Typical of current understanding is that a doubling of CO_2 will likely result in an eventual global warming of approximately 1.5–4.5 °C. The probability of a temperature change of less than 1 °C or more than 7 °C is considered to be extremely small (Dickinson, 1986). For perspective, an increase of only 1.5 °C over the preindustrial temperature would make the earth warmer than it has been in the last 10 000 years. An increase of 4.5 °C would create a climate last experienced during the Mesozoic era—the age of the dinosaurs—several hundred million years ago.

The large range of uncertainty in the equilibrium temperature change is due in large part to climate feedback loops that are difficult to model. One such complication is the ice–albedo feedback loop that links the albedo to changes in temperature. As temperature increases and ice and snow areas retreat, less sunlight is reflected, the albedo decreases, and temperatures increase even more. A more complicated, much less understood, feedback loop is the one dealing with changes in clouds as global temperature increases. Elevated temperatures increase evaporation, increasing the amount of water vapor in the air. Since water vapor is a greenhouse gas, it might cause even more warming. On the other hand, increased cloudiness may increase the albedo. Increasing the albedo would lead to global cooling. The trade-off between greenhouse and albedo effects is dependent on latitude and season, as well as on the type of clouds formed. This is understandably one of the most difficult and controversial aspects of current modeling efforts.

The effect of carbon dioxide on the equilibrium, average global surface temperature is often modeled using a logarithmic function since each increment in CO_2 is less and less able to affect temperature as the absorption within the CO_2 band becomes more and more saturated. Using the temperature change associated with a doubling of CO_2 as a scaling parameter, the equilibrium surface temperature change is conveniently described as (National Research Council, 1983):

$$\Delta T = \frac{\Delta T_d}{\ln 2} \times \ln \left[\frac{(CO_2)}{(CO_2)_0} \right] \qquad (8.19)$$

where ΔT = the equilibrium, global average change in surface air temperature
ΔT_d = the equilibrium temperature change predicted for a doubling of carbon dioxide
CO_2 = the concentration of carbon dioxide at some future time
$(CO_2)_0$ = the initial concentration of carbon dioxide

Since Eq. 8.19 is logarithmic, it suggests that for every doubling of CO_2 the temperature goes up by the same amount. That is, if ΔT_d is 3 °C, then the first doubling raises temperatures by 3 °C, the second doubling another 3 °C, and so on.

Example 8.8   Predicted Temperature Change for 1984

What would be the equilibrium temperature change corresponding to a CO_2 increase from 280 ppm in 1850 to 345 ppm in 1984, if a doubling of CO_2 is predicted to produce a 3.0 °C increase?

Solution   Using (8.19), the expected change would be

$$\Delta T = \frac{3.0}{\ln 2} \times \ln \frac{345}{280} = 0.9 \,°C$$

Given the rise in carbon dioxide concentration in the atmosphere that has already been experienced, it is interesting to compare the historic average global temperature with the results calculated in Example 8.8. Figure 8.21 shows the record of global temperature as reported by the Goddard Institute for Space Sciences (Hansen and Lebedeff, 1988). The 1980s were especially warm, with the six hottest years on record being 1988, 1987, 1983, 1981, 1989, and 1980 (in decreasing order). It is tempting to suggest that this warming has been caused by the greenhouse effect, but that conclusion is still controversial.

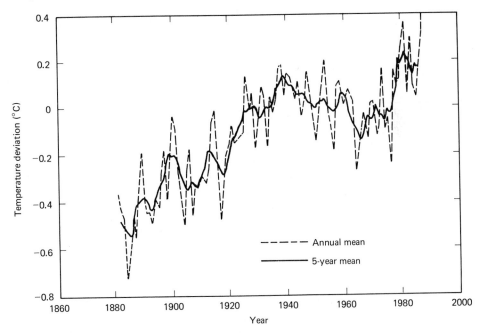

Figure 8.21 Global average surface temperature changes as reported by the Goddard Institute for Space Sciences. (*Source:* Hansen and Lebedeff, *Geophysical Research Letters*, Vol. 15, pp. 323–326. Copyright © 1988 by the American Geophysical Union.)

Analyzing historic temperature records is an exercise that has been subject to considerable criticism. As Schneider (1989) points out, cities have often grown up around temperature recording stations, altering the local climate by creating urban heat islands. In many circumstances, thermometers have been moved from time to time, especially from city centers to airports, or up and down mountains. Also, thermometers have not been uniformly distributed around the globe, so that remote land and ocean locations have not been equally represented in the record versus major metropolitan centers.

Figure 8.21, which is quite controversial, shows an increase of perhaps 0.7 °C over the past hundred years or so. After correcting for urban heat island effects, it has been argued that an increase of approximately 0.5 °C has actually occurred (Schneider, 1989). It is tempting to suggest that this warming has been caused by the greenhouse effect, but that conclusion is still uncertain. The changes have not been at all regular and the rather poorly understood magnitude of the fluctuations and patterns of change makes it difficult to conclude unequivocally that the net increase is due to the greenhouse effect.

Transient Response

If we accept that there has been a global temperature increase of approximately 0.5 °C over the past century, and if we are willing to attribute that temperature rise to accumulating carbon dioxide, then the estimate of 0.9 °C calculated in Example 8.8 seems to be much too high. One explanation could be that the 3.0 °C predicted for a doubling of CO_2 is wrong by nearly a factor of two. More likely, the discrepancy is due to the thermal inertia of the oceans, which causes a delay between the time that the climate system is perturbed and the time that the earth finally moves to a new temperature equilibrium. While it would take only a few years for the well-mixed top 50–100 m of the oceans to equilibrate with the atmosphere, Dickinson (1986) estimates that an ocean reservoir almost 1 km in depth would eventually be affected. This would take considerably longer, with perhaps 50 years being required to approach equilibrium.

If the calculation in Example 8.8 is correct, then we have experienced only about half of the total temperature change that existing levels of CO_2 eventually will cause. This suggests that we are already committed to roughly another 0.4 °C temperature increase even if there is no further increase in CO_2.

8.5 CHLOROFLUOROCARBONS

Chlorofluorocarbons (CFCs) are molecules that contain chlorine, fluorine, and carbon. As opposed to the other greenhouse gases, CFCs do not occur naturally and their presence in the atmosphere is due entirely to human activities. CFCs absorb strongly in the atmospheric window and tend to have long atmospheric residence times; hence, they are potent greenhouse gases. The two CFCs that have received the most attention, in both the ozone and climate change contexts, are trichlorofluoromethane, $CFCl_3$ (CFC-11), and dichlorodifluoromethane, CF_2Cl_2 (CFC-12). These molecules are inert and non-water soluble, so they are not destroyed through chemical reactions or removed with precipitation. The only known removal mechanism is photolysis by short-wavelength solar radiation, which occurs after the molecules drift into the stratosphere. It is the chlorine freed during this process that can go on to destroy stratospheric ozone.

CFCs are mainly used as refrigerants, solvents, foaming agents in the production of rigid and flexible foams, and (in most parts of the world) as aerosol propellants for such products as deodorants, hairspray, and spray paint. Table 8.7 shows the fraction of global CFC use in each of these categories. While the largest single use for CFCs globally is for aerosols, in the United States such applications were essentially banned in 1979. As a result, aerosol propellants account for only 4 percent of U.S. consumption (USEPA, 1987).

CFCs are referred to in the literature by a variety of names. Since some of the most important CFCs are based on a one-carbon methane structure, such as trichlorofluoromethane ($CFCl_3$) and dichlorodifluoromethane (CF_2Cl_2), they were often referred to in earlier literature as *chlorofluoromethanes,* or CFMs. The DuPont tradename Freon has also been used. When the molecules contain only fluorine, chlorine, and carbon, they are called *fully halogenated* CFCs. Some CFCs contain hydrogen as well as chlorine, fluorine, and carbon, and they are called *hydrochlorofluorocarbons,* or HCFCs. HCFCs have the environmental advantage that, due to the hydrogen bond, they are less stable in the atmosphere and, hence, are less likely to reach the stratosphere to affect the ozone layer. In fact, their ozone-depletion potential is typically only 2–5 percent of that for the most commonly used CFCs. As a result, HCFCs are often advocated, and used, as replacements for fully halogenated CFCs.

Finally, when no chlorine is present in the molecule they are called *hydrofluorocarbons,* or HFCs. HFCs are especially important replacements for CFCs since their lack of chlorine means they do not threaten the ozone layer.

CFCs are referred to using a number system developed years ago by DuPont. For example, trichlorofluoromethane ($CFCl_3$) is CFC-11, and dichlorodiflu-

TABLE 8.7   GLOBAL CFC USE BY CATEGORY, 1985

Use	Share of total (%)
Aerosols	25
Rigid foam insulation	19
Solvents	19
Air conditioning	12
Refrigerants	8
Flexible foam	7
Other	10

Source: Shea (1989).

oromethane (CF_2Cl_2) is CFC-12. To determine the chemical formula from a CFC number, simply add 90 to the number and interpret the three digit result as follows: the rightmost digit is the number of fluorine atoms, the middle digit is the number of hydrogen atoms, and the left digit is the number of carbons. For example, to figure out what CFC-12 is, add 90 to 12:

$$\text{CFC-12} \quad 12 + 90 = 102 \quad \text{implies} \quad \begin{array}{l} \text{1 carbon} \\ \text{0 hydrogen} \\ \text{2 fluorine} \end{array}$$

To determine the number of chlorine atoms, start by imagining a methane or ethane building block (depending on whether there are 1 or 2 carbon atoms):

$$\begin{array}{cc} \overset{|}{\underset{|}{-C-}} & \overset{|}{\underset{|}{-C}}\,\overset{|}{\underset{|}{C-}} \\ \text{Methane} & \text{Ethane} \end{array}$$

Each vacant bonding site not taken up by fluorine or hydrogen is occupied by chlorine. Methane has one carbon atom and 4 bonding sites; ethane has two carbons with 6 bonding sites. For example, CFC-12 has 2 of its 4 sites occupied by fluorine, leaving room for 2 chlorines. Thus, CFC-12 is CF_2Cl_2.

Example 8.9   CFC Numbering

   a. What is the chemical composition of CFC-115?
   b. What is the CFC number for CCl_2FCClF_2?

Solution

   a. CFC-115: Adding 90 to 115 gives 205. Thus, a molecule contains 2 carbons, no hydrogen, and 5 fluorines. Two carbons have 6 bonding sites, 5 of which are taken by fluorine. The remaining site is taken by chlorine. The chemical formula would therefore be: C_2ClF_5 (or CF_3CF_2Cl).
   b. CCl_2FCClF_2 has 2 carbons, no hydrogen, and 3 fluorine atoms so its number is 203. Subtracting 90 from 203 gives 113. This is CFC-113, a very commonly used solvent.

The most widely used CFCs are CFC-11, CFC-12, and CFC-113. Table 8.8 indicates that worldwide production rates for these three amounts to almost 1 million metric tons per year, with 30 percent of that being used in the United States.

Major Uses for CFCs

Aerosols

When CFCs were first hypothesized as a danger to the ozone layer by Molina and Rowland (1974), over half of worldwide emissions were from aerosol propellants. At that time the United States alone was using over 200 000 tons per

TABLE 8.8 ESTIMATED 1985 USE OF MAJOR CFCs (thousands of tons)

Chemical	World	United States	Other CMA[a] reporting countries	Centrally planned countries
CFC-11	341.5	75.0	225.0	41.5
CFC-12	443.7	135.0	230.0	78.7
CFC-113	163.2	73.2	85.0	5.0
Total	948.4	283.2	540.0	125.2
	(100%)	(30%)	(57%)	(13%)

[a] Chemical Manufacturer's Association.

Source: Hammitt et al. (1986).

year of CFC-11 and CFC-12 in aerosols. The U.S. Environmental Protection Agency (EPA) responded rather quickly to this new environmental threat and, acting under the Toxic Substances and Control Act, banned the use of CFCs in nonessential aerosol propellant applications beginning in 1979. Norway, Sweden, and Canada adopted similar restrictions. While use of CFCs in aerosols in the United States has dropped to less than 10 000 tons per year, worldwide use in 1985 was still about 250 000 tons per year. Replacements for CFCs in aerosols include isobutane, propane, and carbon dioxide. In some applications, simple pumps or "roll on" systems have replaced the propellants. Future emissions in this important application can be reduced dramatically if worldwide cooperation can be attained. Figure 8.22 shows the impact of aerosol reductions on production of CFC-11 and CFC-12 as reported by those companies that supply data to the Chemical Manufacturers Association (CMA), (essentially all non-centrally planned countries). As can be seen, the drop in aerosol production has been offset by increases in nonaerosol applications.

Foamed Plastics

The second most common use for CFCs in general, and the most common use for CFC-11 in particular, is in the manufacture of various rigid and flexible plastic foams, where they are used as blowing agents to form the foam holes. Sheets of rigid "closed-cell" urethane or isocyanurate foams are used primarily as thermal insulation in buildings and refrigeration equipment. In such applications, the CFCs trapped in foam cells reduce the heat transfer capabilities of the product. CFCs are also used to manufacture nonurethane, rigid foams such as extruded polystyrene (Dow's trade name is Styrofoam), used extensively for egg cartons and food service trays, and expanded polystyrene foam, which is used to make drinking cups. Since CFCs are trapped in the holes in closed-cell foams, they are only slowly released into the atmosphere as the material ages or is eventually crushed.

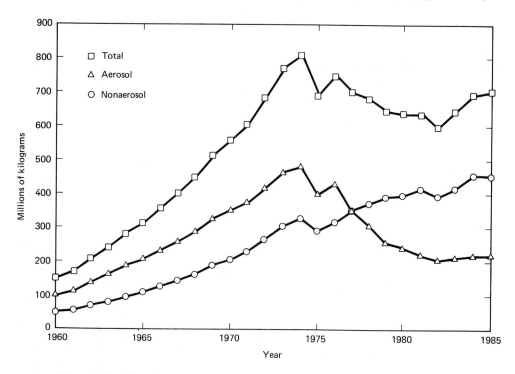

Figure 8.22 Historical production of CFC-11 and CFC-12 as reported to the Chemical Manufacturer's Association. Decreases in aerosol production have occurred mostly in the United States (USEPA, 1987).

Replacements for these foams are possible. Fiberglass insulation, though thermally not as effective per unit of thickness, contains no CFCs and can be used to replace rigid foams used in the construction industry. Various cardboards and other paper products can be used to replace many of the polystyrene applications in the food industry. Finally, the CFC foaming agent itself can be modified or replaced. HCFC-22, which is less damaging to the ozone layer, is sometimes used, as is pentane or methylene chloride (CH_2Cl_2).

Flexible foams, which are used in furniture, automobile seats, and packaging, have cells that are open to the atmosphere ("open-cell"). Hence, CFC release is almost immediate. These foams are made using carbon dioxide as the primary blowing agent, but the CO_2 is often augmented with CFC-11 or methylene chloride. It should be noted that some flexible foams are manufactured without any chlorofluorocarbons.

Refrigerants

CFCs were originally developed to satisfy the need for a nontoxic, nonflammable, efficient refrigerant for home refrigerators. Before they were introduced in the early 1930s, the most common refrigerants were ammonia, carbon dioxide, isobutane, methyl chloride, methylene chloride, and sulfur dioxide. All had significant disadvantages. They were either toxic, noxious, highly flammable, or required high operating pressures that necessitated heavy, bulky equipment. From those perspectives, CFCs are far superior to any of the gases that they replaced.

Most refrigerators, freezers, and automobile air conditioners use CFC-12 as the refrigerant. Large building air conditioning systems tend to use CFC-11. CFC refrigerants do not wear out, so as long as they are sealed into equipment the total usage rate can be small. Automobile air conditioners, however, tend to develop leaks that necessitate periodic recharging of the system. When car air conditioners are serviced, the old refrigerant is usually vented to the atmosphere rather than being captured and recycled, which compounds the loss rate. As a result, the use of CFCs in automobile air conditioners is a major source of emissions, accounting for nearly 20 percent of total U.S. consumption. Given the importance of this application, finding a suitable replacement will be crucial. HFC-134a, which contains no chlorine, is currently being considered.

When CFCs are used as refrigerants, they are frequently referred to with an "R" number. Thus, for example, CFC-12 is designated R-12. And, in a break with the numbering scheme described earlier, a popular refrigerant is made from a mix of CFC-22 and CFC-115 and it is called CFC-502.

Solvents

The third most heavily used chlorofluorocarbon (after CFC-11 and CFC-12) is CFC-113. CFC-113 is used primarily as a solvent in the electronics industry. It is used in various critical cleaning and degreasing operations including the defluxing of printed circuit boards. Although its role as a greenhouse gas is somewhat uncertain, there is considerable concern for its potential role as a catalyst in the destruction of stratospheric ozone. Due in large part to the rapid growth of the electronics industry, the growth in atmospheric concentration of CFC-113 has outpaced that of CFC-11 or CFC-12. In the 1980s, production of CFC-113 increased by about 10 percent per year, while annual production increases of CFC-11 and CFC-12 were closer to 5 percent (USEPA, 1987).

Global Warming and Ozone-Depletion Impacts of CFCs

Since CFCs are of concern both for their ozone-depletion potential and for their impact on global temperature, it is important to pay attention to both problems when proposing replacements. Figure 8.23 shows the relationship between ozone depletion and global-warming potentials for a number of CFCs, HCFCs, and HFCs. CFC-11 is given an arbitrary value of 1 on each axis, so the impacts of other compounds are measured relative to it. The areas of the circles are proportional to the atmospheric lifetimes of each substance.

Fully halogenated CFCs have long atmospheric lifetimes, contain relatively large amounts of chlorine, and absorb strongly within the 7- to 13-μm atmospheric

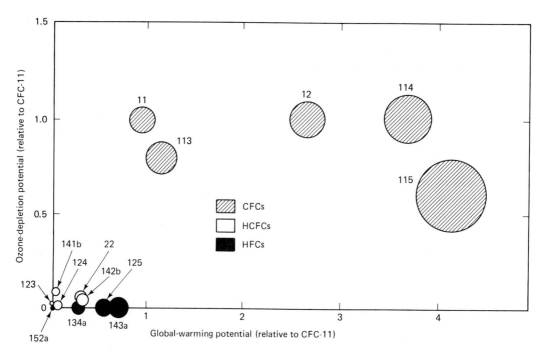

Figure 8.23 Global warming potential and ozone-depletion potential for fully halogenated CFCs, hydrochlorofluorocarbons (HCFCs), and hydrofluorocarbons (HFCs) measured relative to CFC-11. The size of the circles is proportional to atmospheric lifetimes. (*Source:* DuPont.)

window. As a result, they have considerable potential to affect both global warming and stratospheric ozone depletion. For example, CFC-11 is 77 percent chlorine, has strong absorption bands at 9.22 and 11.82 μm, and its atmospheric residence time is estimated at between 60 and 110 years. CFC-12 is 59 percent chlorine, absorbs at 8.68, 9.13, and 19.93 μm, and its atmospheric residence time is between 55 and 400 years. As can be seen by the smaller size of the circles in Figure 8.23, the HCFCs have short atmospheric lifetimes. They break down relatively quickly and thus have only modest potential to affect either ozone or global warming. The HFCs contain no chlorine to threaten the ozone layer, but they do have some potential to affect global warming.

Estimating Future Atmospheric Concentrations of CFCs

Because of their importance in the stratospheric ozone problem, CFCs have come under increasing regulatory attention and, hopefully, emissions will be reduced rapidly. The 1987 Montreal Protocol on Substances That Deplete the Ozone Layer, for example, requires a 20-percent reduction in CFC emissions below the 1986 level by 1994, and a total reduction of 50 percent by 1998. The gases specifically listed in the Protocol are CFC-11, CFC-12, CFC-113, CFC-114, and CFC-

115. The Protocol was signed by more than 35 countries, including the United States.

Unfortunately, even if the Montreal Protocol is completely successful, the long atmospheric lifetimes of CFCs will mean their atmospheric concentrations will continue to increase. To demonstrate this important phenomenon, let us develop a simple model to quantity the relationship between emissions and atmospheric concentrations. We can begin by postulating that if emissions were totally eliminated, then the mass of a given CFC in the atmosphere would follow a simple exponential decay process:

$$m(t) = m(0) \, e^{-t/\tau} \tag{8.20}$$

where $m(t)$ = the mass of CFC at time t after elimination of all emissions
$\quad m(0)$ = the initial mass
$\quad\quad \tau$ = the average atmospheric residence time

Notice the *residence time* is just the time constant τ in (8.20). When $t = \tau$, the amount of CFC drops to $1/e = 36.8$ percent of its initial value. We can use (8.20) to find the rate of removal of CFC

$$\text{Rate of CFC removal} = -\frac{dm}{dt} = -m(0)e^{-t/\tau}\left(-\frac{1}{\tau}\right)$$
$$= \frac{m}{\tau} \tag{8.21}$$

This gives us another interpretation of residence time τ. That is, residence time is the amount present divided by the rate of removal (or addition) of the quantity in question.

We can extend this simple model using the mass balance concepts developed in Section 1.3. If CFCs are being emitted into the atmosphere at the same time that they are being removed by various natural processes, we can write

Rate of increase in CFC mass = Rate of CFC addition − rate of CFC removal

$$\tag{8.22}$$

Letting P (g/yr) represent the rate of addition of CFC mass and substituting (8.21) into (8.22) gives

$$\frac{dm}{dt} = P - \frac{m}{\tau} \tag{8.23}$$

If the emission rate P is constant, then we can find the ultimate steady-state mass of CFC in the atmosphere by setting $dm/dt = 0$, giving

$$m(\infty) = P\tau \tag{8.24}$$

For constant P, the solution to (8.23) is

$$m(t) = m(0)e^{-t/\tau} + m(\infty)(1 - e^{-t/\tau}) \tag{8.25}$$

Since it is concentrations that are usually desired, (8.25) can be converted to give

$$C(t) = C(0)e^{-t/\tau} + C(\infty)(1 - e^{-t/\tau}) \tag{8.26}$$

Concentrations of trace gases are usually expressed on a volumetric basis (ppm or ppb). Since volumetric units are independent of temperature or pressure, we can work with standard temperature and pressure conditions (STP) to help us find $C(\infty)$. We will need to use some data on the earth's atmosphere coupled with the fact that 1 mole of a gas at STP occupies 22.4×10^{-3} m^3. The mass of the atmosphere is 5.14×10^{18} kg and its density at STP is 1.293 kg/m^3 (Harte, 1985). We can write the ultimate steady-state concentration as the ratio of the volume of CFC to the volume of air:

$$C(\infty) = \frac{m(\infty) \text{ g} \times (\text{mol/MW g}) \times 22.4 \times 10^{-3} \text{ m}^3 \text{ CFC/mol}}{(5.14 \times 10^{18} \text{ kg})/(1.293 \text{ kg/m}^3 \text{ air})}$$

Where MW is the molecular weight of the CFC in question. Substituting (8.24) into this equation and simplifying, gives

$$C(\infty) = \frac{P(\text{g/yr}) \ \tau(\text{yr})}{\text{MW}} \times 5.63 \times 10^{-21} \tag{8.27}$$

Example 8.10   Future CFC Concentrations with Reduced Emission Rates

Assume the following statistics for CFC-12 (CF$_2$Cl$_2$):

Atmospheric residence time = 150 yr

1985 emission rate = 0.44×10^{12} g/yr

1985 atmospheric concentration = 0.40 ppb

Suppose the emission rate of CFC-12 is instantaneously reduced to 50 percent of its 1985 value and held constant thereafter:

a. What would be the final, steady-state atmospheric concentration of CFC-12?

b. What cut in the emission rate would be required for CFC-12 concentrations to remain constant at the 1985 level of 0.40 ppb?

c. Plot the transient response of CFC concentrations versus time for various (constant) emission rates corresponding to (1) 1985 rates, (2) rates cut by 15 percent, and (3) rates cut by 50 percent.

Solution

a. To find the ultimate steady-state concentration, we first need to find the molecular weight of CFC-12 (CF$_2$Cl$_2$). The atomic weights of C, F, and Cl are 12, 19, and 35.5, respectively (Table 2.1), so the molecular weight is $12 + 2 \times 19 + 2 \times 35.5 = 121$ g/mol. Using (8.27) with emissions reduced to $0.50 \times 0.44 \times 10^{12}$ g/yr gives

$$C(\infty) = \frac{P\tau}{\text{MW}} \times 5.63 \times 10^{-21}$$

$$C(\infty) = \frac{0.5 \times 0.44 \times 10^{12} \times 150}{121} \times 5.63 \times 10^{-21}$$

$$= 1.54 \times 10^{-9} = 1.54 \text{ ppb}$$

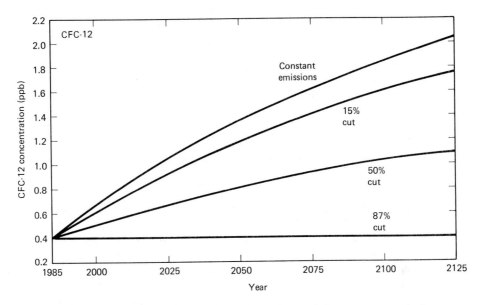

Figure 8.24 Even with significant cuts in CFC-12 emission rates, atmospheric concentrations rise substantially for many decades (see Example 8.10).

Thus, even with a 50 percent reduction in emissions, the concentration of CFC-12 would eventually reach a value almost 4 times the 1985 (0.40 ppb) level. This example suggests that the Montreal Protocol of 1987, while setting the significant goal of a 50 percent reduction in CFC emissions, still falls far short of what would be required to hold or reduce CFC concentrations in the future.

b. To find the rate of emissions that would maintain the current concentration of 0.40 ppb, we simply rearrange (8.27):

$$P = \frac{C(\infty) \times MW}{\tau \times 5.63 \times 10^{-21}}$$

$$= \frac{0.40 \times 10^{-9} \times 121}{150 \times 5.63 \times 10^{-21}} = 0.057 \times 10^{12} \text{ g/yr}$$

which, compared to the 1985 emission rate, represents a decrease of 87 percent.

c. To find the transient response, we can simply plot (8.26) for each emission rate P and corresponding value of $C(\infty)$ from (8.27). This has been done in Figure 8.24. Notice that concentrations increase for several centuries.

8.6 OTHER GREENHOUSE GASES

Methane is a naturally occurring gas that is increasing in concentration as a result of human activities. It is produced by bacterial fermentation under anaerobic conditions, such as occur in swamps, marshes, rice paddies, as well as in the digestive systems of ruminants and termites. Significant contributions of methane to the atmosphere are the result of human food-growing activities, such as increased cattle production and increases in areas planted in rice paddies. It is also released during the production, transportation, and consumption of fossil fuels as well as when biomass fuels are burned.

After its release, methane is thought to have an atmospheric residence time of around 8–11 years. It eventually is removed through oxidation with various OH radicals. The best evidence of long-term trends in methane concentration has been obtained from the analysis of air bubbles trapped in ice. As shown in Figure 8.25, concentrations have increased rapidly in the past 200 years, and in fact, they correlate quite well with human population size. In the most recent years, methane concentration has been increasing in the range of 1–2 percent per year. Some models suggesting that a doubling of methane would lead to a global temperature increase of between 0.2 and 0.3 °C (Dickinson, 1986). It absorbs on the edge of the atmospheric window at 7.66 μm, yielding a global warming impact of about 25 times that of carbon dioxide.

Nitrous oxide is another naturally occurring greenhouse gas that has been increasing in concentration due to human activities. It is released into the atmosphere mostly during the nitrification portion of the nitrogen cycle:

$$NH_4^+ \longrightarrow N_2 \longrightarrow N_2O \longrightarrow NO_2^- \longrightarrow NO_3^-$$

The reverse reaction of denitrification, is no longer considered a significant source of atmospheric N_2O. Combustion of fossil fuels and nitrogen fertilizer consumption are thought to be the two most important human activities leading to increases in nitrous oxide levels. It apparently has no significant tropospheric sinks

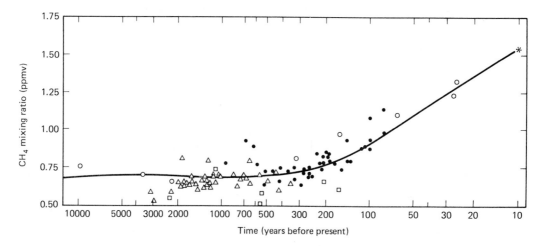

Figure 8.25 History of methane concentrations as determined by measurements of air trapped in ice cores. (*Source:* Bolle et al. How much carbon will remain in the atmosphere? In B. Bolin et al. (eds.), *The Greenhouse Effect, Climatic Change, and Ecosystems.* Copyright © 1986 by John Wiley & Sons, Inc.)

and is only slowly degraded in the stratosphere by photolysis. The destruction process in the stratosphere involves a reaction with atomic oxygen leading to formation of nitric oxide, which in turn reacts catalytically with ozone. The desired removal of nitrous oxide, then, has the undesired side effect of reducing stratospheric ozone. Nitrous oxide has an extremely long residence time in the atmosphere, on the order of 150 years, which means perturbations in the natural cycle will have long-lasting repercussions. Its concentration is growing at about 0.2 percent per year. It has strong absorption bands at 4.5, 7.8, and 17 μm, and it is thought to be about 230 times as potent as carbon dioxide in causing global warming (Table 8.3).

Ozone plays important roles in both the troposphere and stratosphere, and in each of those regions of the atmosphere it impacts both life and climate. About 90 percent of atmospheric ozone resides in the stratosphere, and it is there that it protects life by absorbing short-wavelength ultraviolet radiation. Stratospheric ozone also affects climate, but in a very complex way. Incoming solar energy is absorbed, which heats the stratosphere. This, however, reduces the radiation arriving at the earth's surface, thereby cooling the surface. On the other hand, the warmed stratosphere radiates energy back to the earth's surface, thereby heating it. The net effect is uncertain.

In the troposphere, ozone is a component of photochemical smog, and it poses a serious health problem in many cities. As a factor in climate, tropospheric ozone absorbs strongly at around 9.6 μm right in the middle of the atmospheric window. Increasing concentrations could contribute to raising global temperatures. Ozone, however, has a rather short residence time in the troposphere, measured in days. It is irregularly distributed by time of day, geographic location, and altitude, so it has been difficult to assess its overall change with time, leaving us uncertain as to its impact on climate.

Combined Effect of Greenhouse Gases

A number of studies have attempted to model the combined effects on global temperature of carbon dioxide along with other greenhouse gases (Lacis et al., 1981; Ramanathan et al., 1985; Mintzer, 1987). Using an analysis similar to that of Mintzer, but with the impact of CO_2 modified slightly to match (8.19), the equilibrium temperature increase ΔT (°C) associated with a combination of carbon dioxide, nitrous oxide, methane, CFC-11, and CFC-12 is given by

$$\Delta T = \frac{\Delta T_d}{\ln 2} \times \ln \left[\frac{(CO_2)}{(CO_2)_0} \right] + 0.057[(N_2O)^{0.5} - (N_2O)_0^{0.5}]$$

$$+ 0.019[(CH_4)^{0.5} - (CH_4)_0^{0.5}]$$

$$+ 0.14[(CFC\text{-}11) - (CFC\text{-}11)_0] + 0.16[(CFC\text{-}12) - (CFC\text{-}12)_0] \quad (8.28)$$

where concentrations are expressed in ppb (though, of course, carbon dioxide can be expressed in any concentration units since it is used in a ratio). Since (8.28) is given as a temperature change, each compound must have some reference concentration, which is what the subscript 0 designates.

Example 8.11   Combined Greenhouse Gases

Using data from the following table for approximate preindustrial concentrations and 1985 concentrations, estimate the combined equilibrium temperature change for 1985. Using the assumed growth rates, estimate the equilibrium temperature increase (compared to preindustrial times) in the year 2075. Assume ΔT_d is 3 °C.

Gas	1850	1985	Assumed growth rate 1985–2075
CO_2	280 ppm	345 ppm	0.57%
CH_4	1150 ppb	1790 ppb	1%
N_2O	285 ppb	305 ppb	0.5%
CFC-11	0 ppb	0.24 ppb	2.5%
CFC-12	0 ppb	0.40 ppb	2.5%

Solution   The 1985 equilibrium temperature increase would be

$$\Delta T(CO_2) = \frac{3.0}{\ln 2} \times \ln \frac{345}{280} = 0.904 \text{ °C}$$

$$\Delta T(N_2O) = 0.057[(305)^{0.5} - (285)^{0.5}] = 0.033 \text{ °C}$$

$$\Delta T(CH_4) = 0.019[(1790)^{0.5} - (1150)^{0.5}] = 0.159 \text{ °C}$$

$$\Delta T(CFC\text{-}11) = 0.14(0.24 - 0) = 0.034 \text{ °C}$$

$$\Delta T(CFC\text{-}12) = 0.16(0.40 - 0) = 0.064 \text{ °C}$$

Thus, the total equilibrium temperature increase between 1850 and 1985 would be

$$\Delta T(1985) = 0.904 + 0.033 + 0.159 + 0.034 + 0.064 = 1.2 \text{ °C}$$

To determine the equilibrium temperature in 2075, each of the concentrations must be extrapolated using their exponential growth rates. Let us work out one component and then summarize the rest. The concentration of methane in 2075 would be

$$CH_4(2075) = CH_4(1985)e^{r(2075-1985)} = 1790e^{0.01 \times 90} = 4402 \text{ ppb}$$

So, the equilibrium temperature increase between 1850 and 2075 caused by methane would be

$$\Delta T(CH_4) = 0.019[(4402)^{0.5} - (1150)^{0.5}] = 0.62 \text{ °C}$$

Table 8.9 summarizes the temperature increase in 2075.

A graph of this equilibrium temperature increase, including the contribution of each greenhouse gas, is shown in Figure 8.26. Recall that the earth's thermal inertia

TABLE 8.9 TEMPERATURE CHANGES CAUSED BY
INDIVIDUAL GREENHOUSE GASES IN EXAMPLE 8.11

Gas	2075 concentration	Equilibrium temperature change (°C) (1850–2075)	% of total temperature change
CO_2	576 ppm	3.12	63
CH_4	4402 ppb	0.62	12
N_2O	478 ppb	0.28	6
CFC-11	2.28 ppb	0.32	7
CFC-12	3.80 ppb	0.61	12
Total		4.95	100%

would cause the actual global temperature rise to lag behind this plot by several decades.

While the above example suggests that approximately two-thirds of the 5 °C equilibrium temperature increase in 2075 (over preindustrial times) would be caused by CO_2, a more dramatic way to indicate the importance of the other greenhouse gases is to note that in the later years of this scenario, the other greenhouse gases actually contributed more to each decade's temperature increment than carbon dioxide did. This perspective emphasizes the important conclusion that controlling future temperature increases is going to require stringent controls on not only carbon dioxide, but the other greenhouse gases as well— in particular, the chlorofluorocarbons.

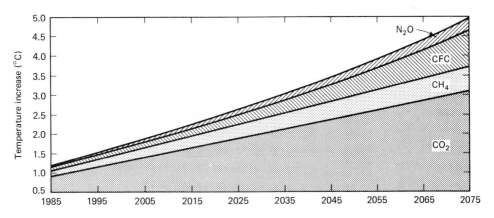

Figure 8.26 Equilibrium temperature increase (over preindustrial value) based on data given in Example 8.11.

8.7 REGIONAL EFFECTS OF TEMPERATURE INCREASES

The relatively simple models introduced in this chapter are useful for estimating average global temperature increases, but the regional implications of a global warming are of more importance and are much harder to predict. A starting point in assessing regional changes is provided by three-dimensional general circulation models (GCMs) that can give an indication of how parameters such as temperature, cloud cover, soil moisture, and ice cover will vary by geographical location and time of year. These models are based on three-dimensional cells in which the atmosphere is divided into a series of layers, each several hundred kilometers on a side. Each cell is a node that interacts with adjacent nodes in the model. The grid from an early GCM shown in Figure 8.27 puts the size of individual cells into some perspective. GCMs are rapidly being improved, but unfortunately these models are still in the early stage of development and it is not at all uncommon for different models to disagree with each other.

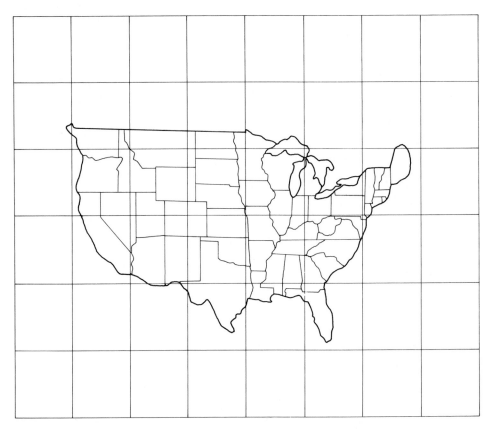

Figure 8.27 Grid covering the United States from an early GCM by the Goddard Institute for Space Studies (California Energy Commission, 1989).

One way to illustrate predicted temperature changes is to look at specific cities in the United States, as was done by Hansen et al. (1986). In Table 8.10, they compare current average temperatures with changes expected for a doubling of CO_2 (or its equivalent in all greenhouse gases). The number of days with temperature over 90 °F are suggestive of increased demands for air conditioning, photochemical smog potential, and generally uncomfortable conditions. On the other hand, temperature increases in the winter suggest decreased heating demands and longer growing seasons. For example, in Washington, DC, the average temperature in July is expected to increase by nearly 7 °F, the number of days with temperature over 90 °F increases from 36 per year to 87, but the growing season is increased by a month and a half. These changes are significant and would be easily noticeable by the general public. They are also extremely uncertain. The model uses a 4 °C scaling factor for a doubling of CO_2 that is itself uncertain by several degrees either way. Also, it is assumed that no major changes in ocean currents would take place, which could in themselves have major effects on any of the coastal cities listed. Of course, significant climate differences for regions located only a few miles apart are common, but these models are far too crude to capture such subtleties.

How greenhouse changes would affect agriculture is, of course, crucial, but it is also largely unpredictable at this point. Increasing the concentration of CO_2 might stimulate some plant growth, and global warming would extend growing seasons. Net precipitation is usually predicted to increase, but most climate models show rainfall patterns shifting northward. The shift in rainfall might improve growing conditions in Canada, Siberia, and perhaps sub-Saharan Africa, but could recreate the Dust Bowl in the American plains. Increasing temperatures also increase the rate of evaporation. The net effect of changes in precipitation and evaporation is crucial to a critical agricultural parameter, *soil moisture*. Soil

TABLE 8.10   TEMPERATURE INCREASES DUE TO EFFECTIVE DOUBLED CO_2 FOR VARIOUS CITIES

City	ΔT (°F) Jan	ΔT (°F) July	Days $T_{max} > 90$ °F Today	Days $T_{max} > 90$ °F $2CO_2$	Growing season (days) Today	Growing season (days) $2CO_2$
Washington	7.9	6.6	36	87	233	278
New York	7.9	6.7	15	48	236	271
Chicago	10.0	5.2	16	56	178	232
Omaha	10.4	6.2	37	86	179	231
Denver	11.7	6.1	33	86	162	206
Los Angeles	9.3	7.5	5	27	365	365
Dallas	7.8	7.4	100	162	253	313
Memphis	8.5	6.7	65	145	234	311

Source: Hansen et al. (1986).

moisture is related to the difference between precipitation and evaporation, but it also depends on soil saturation and runoff. If the soil is already saturated, then precipitation (and snow melt) in excess of evaporation causes runoff rather than an increase in soil moisture. Summer soil moisture is often predicted to be reduced over extensive middle and high latitude areas, including the North American Great Plains, western Europe, northern Canada, and Siberia (Manabe and Wetherald, 1986).

Increasing temperatures and decreasing rainfall in the western portion of the United States could have devastating impact on water supplies. Stockton and Boggess (1979) and Revelle and Waggoner (1983) studied the effects of a 2 °C warming and a 10-percent reduction in precipitation in the United States. They found that the impact would not be serious in regions east of the 100th meridian, but the drainage basins of the Missouri, Arkansas–White–Red, Rio Grande, and Colorado rivers, and the river basins draining into the Gulf of Mexico from the northern portion of Texas and the California river system, would experience an average decline in surface water supply of as much as 50 percent. For California, which is critically dependent on snow melt for summer water supplies, it has been predicted that a 3 °C temperature increase would raise the average snowline by 1500 ft and decrease the April 1 snowpack area by 54 percent (California Energy Commission, 1989). With a greater fraction of precipitation falling as rain rather than snow, winter flooding and reduced summer runoff would be expected.

Warmer temperatures would accelerate the current rate of sea level rise. Sea levels that have risen 10–15 cm over the last century could rise another 10–20 cm by the year 2025, and 50–200 cm by 2100. Disintegration of the West Antarctic ice sheet might raise sea level by an additional 6 m over the next few centuries (Titus, 1986). A rise of a meter or two could permanently inundate a large fraction of the world's wetlands, threatening fisheries, accelerating coastal erosion, and increasing the salinity of estuaries and coastal aquifers. Some areas of the world have substantial populations living in areas that would be flooded. Bangladesh, for example, is one of the most vulnerable nations. It has been estimated that as much as 20 percent of its land area would be flooded with a 2-m rise in sea level unless massive coastal protection facilities were to be constructed. While rich nations might be able to adapt to rising sea levels, the poor countries of the world could be devastated.

8.8 CHANGES IN STRATOSPHERIC OZONE

The changes occurring in the stratosphere's protective layer of ozone are closely linked to the greenhouse problem just discussed. Many of the same gases are involved, including CFCs, methane, and nitrous oxide, as well as ozone itself. There are some new gases, too, including methyl chloroform, carbon tetrachloride, and halons (fluorocarbons that contain bromine atoms, used in some fire extinguishers). Some of the gases that enhance the greenhouse effect actually reduce the problem of ozone depletion, so the two problems really must be considered together.

Ozone is continuously being created in the stratosphere by the absorption of short-wavelength ultraviolet (UV) radiation, while at the same time it is continuously being removed by various chemical reactions that convert it back to molecular oxygen. The rates of creation and removal at any given time and location dictate the concentrations of ozone present. The balance between creation and removal is being affected by increasing stratospheric concentrations of chlorine, nitrogen, and bromine, which act as catalysts speeding up the removal process. The most prominent type of ozone-destructive gases are the chlorofluorocarbons, which played such an important role in our description of the greenhouse effect. CFCs tend to be very stable compounds that are relatively unaffected by the usual pollutant removal processes in the troposphere. When they drift to the stratosphere, CFC molecules can be broken by ultraviolet radiation, freeing the chlorine that is then available to destroy ozone. The reaction involving CFC-12, for example, is

$$CCl_2F_2 + h\nu \longrightarrow Cl + CClF_2 \qquad (8.29)$$

where $h\nu$ represents solar radiation. The chlorine freed by reactions such as (8.29) acts as a catalyst in the ozone removal process, which means it contributes to the reaction but is unaffected by it. A single chlorine atom may break down tens of thousands of ozone molecules before it returns to the troposphere, where it is rained out as hydrochloric acid.

Concern over possible destruction of stratospheric ozone was first expressed in the early 1970s (Molina and Rowland, 1974), but it was not until 1985, with the dramatic announcement of the discovery of a "hole" in the ozone layer over Antarctica the size of the continental United States, that the world began to acknowledge the seriousness of the problem. The hole, which begins to open up with the first sunlight of the Antarctic spring in late August and continues until roughly the end of November, was not predicted by any of the then-existing stratospheric ozone depletion models. In fact, even though its presence had been measured since 1977, the reductions in ozone were so unexpected and severe that the research groups collecting the data did not believe their findings for many years, and, hence, did not report them until 1985.

Since ozone absorbs biologically damaging ultraviolet radiation before it can reach the earth's surface, ozone destruction increases the risks associated with UV exposure. UV radiation is linked with human skin cancer, cataracts, and suppression of immune system response. Moreover, many plants and aquatic organisms have been shown to be adversely affected by increases in UV exposure. And finally, increases in terrestrial UV flux could increase urban air pollution through the photolysis of formaldehyde, a common component of photochemical smog.

The Ozone Layer as a Protective Shield

Ozone formation in the stratosphere can be described by the following pair of reactions. In the first, atomic oxygen (O) is formed by the photolytic decomposition of molecular oxygen (O_2)

$$O_2 + h\nu \longrightarrow O + O \tag{8.30}$$

where the ultraviolet radiation in this case has wavelengths less than 242 nanometers (nm). The atomic oxygen, in turn, reacts rapidly with molecular oxygen to form ozone,

$$O + O_2 + M \longrightarrow O_3 + M \tag{8.31}$$

where M represents a third body (N_2 or O_2) necessary to carry away the energy released in the reaction.

Opposing the above formation process is ozone removal by photodissociation

$$O_3 + h\nu \longrightarrow O_2 + O \tag{8.32}$$

wherein ozone absorbs UV radiation in the 200- to 320-nm wavelength region. The above combination of reactions form a long chain in which oxygen atoms are constantly being shuttled back and forth between the various molecular forms. The net effect of the above reactions is the creation of layer of ozone in the stratosphere that absorbs biologically damaging short-wavelength ultraviolet radiation. In addition, the heating that results from this UV absorption is what creates the temperature inversion that characterizes the stratosphere in the first place. That temperature inversion produces stable atmospheric conditions that lead to long residence times for stratospheric pollutants.

The effectiveness of these reactions in removing short-wavelength UV is demonstrated in Figure 8.28, in which the extraterrestrial solar flux and the flux actually reaching the earth's surface are shown. The radiation reaching the surface has been drawn for a clear day with the sun assumed to be 60° from the zenith (overhead), corresponding roughly to a typical mid-latitude site in the afternoon. As can be seen, the radiation reaching the earth's surface is rapidly reduced for wavelengths less than about 320 nm. In fact, virtually none reaches the surface with wavelengths shorter than about 295 nm. In Figure 8.28, the UV wavelengths have been divided into two bands designated as UV-A and UV-B, where the UV-A waveband extends from 320 to 400 nm and the UV-B corresponds to wavelengths less than 320 nm. The UV-A portion of the spectrum is not carcinogenic at usual exposure levels on the earth's surface. The UV-B wavelengths, on the other hand, are the ones that biological organisms are most sensitive to; they are, for example, usually responsible for human sunburn and skin cancer. The crucial role played by ozone in reducing UV-B is apparent.

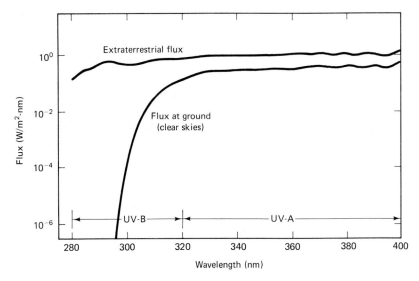

Figure 8.28   Extraterrestrial solar UV flux and expected flux at the earth's surface on a clear day with the sun 60° from the zenith (Frederick, 1986).

Destruction of Stratospheric Ozone

Ozone can be removed from the stratosphere by catalytic reactions involving nitrogen, chlorine, bromine, or hydrogen oxides. It is the enhancement of these reactions by anthropogenic activities that are of concern. One way to describe the catalytic destruction of ozone is with the following set of reactions:

$$X + O_3 \longrightarrow XO + O_2$$

and

$$XO + O \longrightarrow X + O_2$$

for a net

$$O + O_3 \longrightarrow O_2 + O_2 \tag{8.33}$$

where X may be Cl, Br, OH, or NO. Notice in (8.33) how the X radical that enters the first reaction is released in the second, freeing it to go on and participate in another catalytic cycle. The net result of this pair of reactions is the destruction of one ozone molecule. The original catalyst that started the reactions, however, may go on to destroy thousands of ozone molecules before it eventually leaves the stratosphere.

More recently, the following set of reactions has been proposed to describe the role of chlorine and CFCs in the creation of the Antarctic ozone hole. After chlorine is released from a CFC (for example, Eq. 8.29), it reacts with ozone to form chlorine monoxide (ClO):

$$Cl + O_3 \longrightarrow ClO + O_2 \tag{8.34}$$

which reacts with nitrogen dioxide (NO_2) to form a relatively inert molecule of chlorine nitrate ($ClONO_2$):

$$ClO + NO_2 \longrightarrow ClONO_2 \tag{8.35}$$

At this point, the chlorine is effectively stored in an inactive form, unable to destroy more ozone. In the Antarctic winter, however, a unique atmospheric condition known as the *polar vortex* traps air above the pole and creates conditions that eventually allow the chlorine to be released. The polar vortex blocks warmer mid-latitude air from mixing with the air above the pole, creating extremely cold polar air temperatures. Stratospheric temperatures may fall below $-90\,°C$, which is cold enough to form polar stratospheric clouds even though the air is very dry. The ice crystals that make up polar clouds play a key role in the Antarctic phenomenon. By providing reaction surfaces, these ice crystals allow chlorine nitrate to react with water to form hypochlorous acid (HOCl):

$$H_2O + ClONO_2 \longrightarrow HOCl + HNO_3 \tag{8.36}$$

As soon as the sun rises in the Antarctic in August, the chlorine stored in HOCl is freed by photolysis

$$HOCl + h\nu \longrightarrow Cl + OH \tag{8.37}$$

A number of possible catalytic reactions have been proposed whereby the freed chlorine can proceed to destroy ozone. As described by Rowland (1989), the chlorine from (8.37) can destroy an ozone molecule, creating chlorine monoxide:

$$Cl + O_3 \longrightarrow ClO + O_2 \tag{8.38}$$

and the OH radical can destroy another ozone:

$$OH + O_3 \longrightarrow HO_2 + O_2 \tag{8.39}$$

The ClO and HO_2 formed in (8.38) and (8.39) can react with each other to form HOCl,

$$ClO + HO_2 \longrightarrow HOCl + O_2 \tag{8.40}$$

which can be photolyzed, releasing chlorine once again

$$HOCl + h\nu \longrightarrow Cl + OH \tag{8.37}$$

The net result of reactions (8.37) to (8.40) is

$$2O_3 \longrightarrow 3O_2 \tag{8.41}$$

The destruction of ozone as the sun first appears in the Antarctic spring, proceeds as described until the nitric acid formed in (8.36) photolyzes, forming NO_2. Photolysis of nitric acid does not begin with the first rays of sunlight, however. It requires wavelengths that are not readily available when the sun is near the horizon in the first weeks of Antarctic spring. Eventually, however, the photolysis proceeds, creating NO_2 that ties up chlorine (8.35), stopping the ozone destruction.

Impacts of Other Greenhouse Gases

The destruction of stratospheric ozone due to increases in man-made gases, such as the CFCs, is partially offset by other human activities. For example, greenhouse gases cause heating of the troposphere, but that, in turn, causes cooling in the stratosphere. This cooling slows down the ozone destroying reactions (8.33), shifting the balance toward greater ozone concentrations.

The role of the greenhouse gas, methane, is much more complex. Methane helps remove ozone destroying chlorine as the following reaction indicates:

$$Cl + CH_4 \longrightarrow HCl + CH_3 \tag{8.42}$$

The CH_3 thus formed oxidizes to the peroxyradical CH_3O_2, which after a number of further oxidations eventually stabilizes as CO_2. Along the way, various oxides of nitrogen, which are often present in the troposphere due to combustion of fossil fuels, get involved:

$$CH_3O_2 + NO \longrightarrow CH_3O + NO_2 \tag{8.43}$$

Sunlight can break apart NO_2, liberating atomic oxygen

$$NO_2 + h\nu \longrightarrow NO + O \tag{8.44}$$

which can combine with molecular oxygen to produce ozone

$$O + O_2 + M \longrightarrow O_3 + M \tag{8.45}$$

Methane thus not only helps cleanse the atmosphere of ozone destroying chlorine, it also encourages the production of new ozone, largely in the troposphere. To some extent, then, methane offsets some fraction of the ozone destruction caused by other trace gases.

Models that attempt to predict changes in the ozone column (total ozone in both the stratosphere and troposphere) must include the interaction between these various ozone-affecting gases, with some gases causing increases and some causing decreases in total ozone. A summary of the important gases and a qualitative assessment of their impact on both stratospheric ozone and global temperature is given in Table 8.11.

TABLE 8.11 SUMMARY OF QUALITATIVE IMPACT OF GREENHOUSE GASES ON STRATOSPHERIC OZONE AND GLOBAL TEMPERATURE

Greenhouse gas	Stratospheric ozone	Global temperature
Chlorofluorocarbons	Decreases	Increases
Carbon dioxide	Increases	Increases
Methane	Increases	Increases
Nitrous oxide	Increases (at high chlorine levels)	Increases

Figure 8.29 shows the results produced by one such model (Hoffman, 1987). Average global ozone depletion is calculated for six different scenarios ranging from one in which global emissions of ozone depleting substances (CFCs) are reduced by 80 percent by 2010, to one in which growth in such emissions is at 5 percent per year. Emissions for all scenarios are assumed constant after 2050. Concentrations of greenhouse gases that counter ozone depletion are assumed to grow at historically extrapolated rates (CO_2 at 0.7 percent, N_2O at 0.2 percent, CH_4 at 0.017 ppm/year). The 80-percent reduction scenario shows a small decline in ozone levels until 2010, after which growth in the other greenhouse gases causes ozone to eventually *increase* beyond today's levels. On the other hand, the 5-percent growth scenario shows extremely rapid ozone depletion with drastic reductions occurring in the next 50 years.

Effects of Ozone Modification

As Figure 8.28 indicates, the ozone shield drastically reduces the amount of UV-B reaching the earth's surface; indeed, without it, life as we know it could not exist

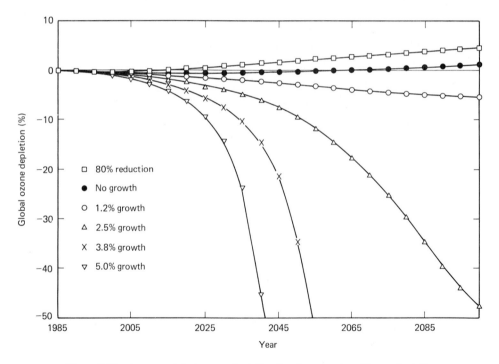

Figure 8.29 Average global ozone depletion for six scenarios. Ozone-depleting gases (CFCs) are assumed to grow at the indicated rates until 2050, and are constant thereafter. Greenhouse gases that counter ozone depletion are assumed to grow at historically extrapolated rates (Hoffman, 1987).

on earth. Increases in UV-B can be related to adverse effects on human health, terrestrial and aquatic plants, and a degradation of materials.

The biological response to increased ultraviolet radiation is often represented by an *action spectrum*. The action spectrum provides a quantitative way to express the relative ability of various wavelengths to cause biological harm. For example, the action spectrum for radiation damage to the DNA molecule is shown in Figure 8.30. Comparing Figure 8.28, which shows the amount of UV radiation reaching the earth's surface, with Figure 8.30, which shows the sensitivity of DNA to UV, clearly shows the importance of the UV shielding provided by ozone. For short wavelengths, where DNA is most sensitive to UV, ozone is fortunately most effective in blocking UV. For longer wavelengths in the UV-B portion of the spectrum, the ozone shield is less effective, but DNA is correspondingly less sensitive. It is the intermediate range of wavelengths, from about 300 to 315 nm, that cause greatest damage to DNA.

There is good evidence that human skin cancer arises from UV-induced changes in DNA, and much of the focus on human health effects caused by ozone depletion has been directed toward understanding that relationship. Skin cancer is usually designated as being either the more common nonmelanoma (basal cell or squamous cell carcinomas), or the much more life-threatening malignant melanoma. Nonmelanoma is unequivocably associated with exposure to sunlight. Nonmelanoma skin cancer is more prevalent among people with fair complexion, is most likely to occur on areas of the skin habitually receiving greatest exposure to the sun, and is associated with areas of the world having the greatest amount of sunshine. While it is rarely fatal, it often causes disfigurement. In the United

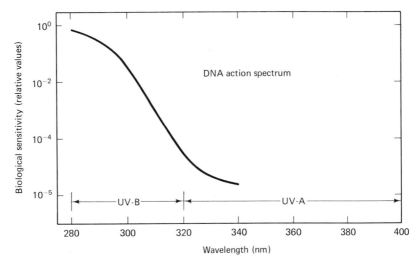

Figure 8.30 Action spectrum for damage to the DNA molecule. The vertical scale is a relative measure of biological sensitivity to radiation at each wavelength (Frederick, 1986).

States, there are about 400 000 new cases of nonmelanoma each year, with a fatality rate of about 1 percent.

Exposure to UV-B is directly correlated with location. At higher latitudes there is less surface-level UV-B, and nonmelanoma rates decrease by roughly a factor of 2 for each 10° increase in latitude. Using such data, it has been estimated that for every 1-percent decrease in ozone, the added UV-B exposure would cause the incidence rate of nonmelanoma to increase by approximately 2–3 percent (Hoffman, 1987). For perspective, it has also been pointed out that by simply moving 100 miles closer to the equator, UV-B exposure is increased by an amount comparable to what would be expected by a 5 percent thinning of the ozone layer (Singer, 1989).

The link between malignant melanoma and sunlight is strong, though less well established than is the case for nonmelanoma. Although melanomas are more frequent on individuals who have had severe and repeated sunburns in childhood, they are often not located on exposed parts of the body. On the other hand, there has been a rather alarming increase in melanoma in the last few decades among fair-skinned populations around the world that has not been observed among U.S. nonwhites. In fact, malignant melanoma death rates among whites in the United States have been increasing by about 2–3 percent annually for the past two decades in spite of improved survival rates among those who receive treatment (Emmett, 1986). There are about 25 000 new cases of melanoma in the United States each year and about 6000 deaths per year. For each 1-percent decrease in ozone, it has been estimated that the melanoma incidence rate will increase by 1–2 percent.

Other human health problems associated with UV exposure include ocular damage (cataracts and retinal degeneration) and immune system suppression. The effects of increasing UV-B radiation would not be felt by humans alone, of course. Although large uncertainties exist in the data, there is evidence that crops and terrestrial ecosystems would be adversely affected by increasing UV exposure. In a recent summary of the status of research on plant yields, it was noted that the effects of enhanced levels of UV-B radiation have been studied in only four of the ten major terrestrial plant ecosystems, but that in nearly half of the plant species that were examined, UV-B radiation deleteriously affected crop yield and quality (Hoffman, 1987). Aquatic ecosystems would be affected as well. Various experiments, for example, have shown that UV-B radiation damages fish, shrimp, and crab larvae, as well as copepods, and plants essential to the marine food web (summarized in Hoffman, 1987).

8.9 PERSPECTIVES ON GLOBAL ATMOSPHERIC CHANGE

A 1986 National Aeronautics and Space Administration report to Congress warns that "we are conducting a giant experiment on a global scale by increasing the concentration of trace gases in the atmosphere without knowing the environmental consequences.'' Another assessment made at a scientific meeting in Villach, Austria, in 1985 concludes, ''While some warming of climate now appears inevitable due to past actions, the rate and degree of future warming could be profoundly affected by governmental policies on energy conservation, use of fossil fuels, and the emission of some greenhouse gases.'' Finally, the EPA Science Advisory Board in 1987 argued that ''by the time it is possible to detect decreases in ozone concentrations with a high degree of confidence, it may be too late to institute corrective measures that would reverse this trend.''

These are global problems requiring international cooperation for their solution, but the United States, as a major contributor, bears a major responsibility for leading the way. A key question is whether to act now, while major scientific uncertainties abound, or to wait until more is known. Some of each may be called for. Fortunately, in this context anyway, the future is largely under our control and policy decisions will ultimately be more important than technical discoveries. The discovery of the Antarctic ozone hole in 1985 and the record temperatures and droughts of the 1980s have heightened public awareness of the potential seriousness of these problems and created a political atmosphere that is open to action. Policy areas that deserve immediate attention include the following.

1. *Adopting an energy policy that encourages reduction in carbon emissions.* The cornerstone of this policy would be increasing energy efficiency. Numerous studies have shown opportunities for large reductions in the energy required to meet end use demands, at costs that are substantially below current energy prices. Not only would energy efficiency improvements contribute to the reduction of CO_2 emissions, but they would also extend the lifetime of limited fossil-fuel resources, ease international tensions created by the need for access to those resources, and they would contribute to the solution of other environmental problems, including acid rain and urban air pollution. To encourage efficiency improvements, current subsidies for energy supply systems should be removed, and the environmental impacts of fuel use should be built into energy prices. Goals should be established to continue the improvement of energy efficiency at a rate of at least 2 percent annually.

   Renewable energy technologies such as solar-thermal systems for heat and electricity, wind-electric systems and photovoltaics for electricity generation, and biomass for portable fuels would be generously supported and encouraged. In the short term, natural gas used in efficient cogeneration systems that provide both heat and electricity could help reduce consumption of coal. Problems with nuclear power should be addressed, and assuming they can be adequately dealt with, the nuclear option should be reevaluated.

2. *Developing international agreements that control production and emissions of damaging trace gases.* The Montreal Protocol, which requires a 50-percent reduction in CFC use by 1998, is an important first step, but it is not

enough. The complete elimination of ozone-depleting substances should be established as a goal. Substitutes for CFCs in many applications are known, and in those cases where CFCs can be replaced without increasing energy consumption, substitutions should begin immediately. Development of new refrigerants and improved thermal insulations that do not use CFCs should be given priority.

A similar global warming agreement that would establish carbon reduction goals should be adopted.

3. *Supporting efforts to halt deforestation and encourage reforestation.* Reducing CO_2 emissions is only one of many reasons for encouraging better management of the world's forest resources, which are rapidly being decimated. Large-scale replanting and forest management projects in developing countries will require the financial support of the industrialized world, and international lending institutions will need to be encouraged in their efforts to include consideration of the ecological impacts of the projects that they fund. Financial mechanisms that will make it economically attractive for developing nations to protect their own forests, including debt-for-nature swaps, carbon taxes, and carbon emission offsets, should be developed (Swisher and Masters, 1989).

Tree planting goals should be established, especially in urban areas. Urban reforestation can be an especially effective way to reduce atmospheric carbon. Not only would carbon be sequestered in the trees themselves, but the resulting shade can reduce urban temperatures and decrease air conditioning loads.

We have arrived at a turning point in human history. The future habitability of our planet will be determined by decisions made and actions taken by this generation, by the people who are alive today. Our problems have solutions, and we all have our parts to play,

> Never doubt that a small group of thoughtful, committed citizens can change the world; indeed, it's the only thing that ever has.
>
> ——*Margaret Mead*

PROBLEMS

8.1. Suppose that the earth really is flat! Imagine an earth that is shaped like a penny, with one side that faces the sun at all times (see Figure P8.1). Also suppose that this flat earth is the same temperature everywhere (including the side that faces away from the sun). Neglect any radiation losses off of the rim of the earth and assume there is no albedo or greenhouse effect. Treating it as a perfect blackbody, estimate the temperature of this new, flat earth.

Figure P8.1

8.2. a. Use the data for insolation and albedo given in Table 8.2 to confirm the values given there for the equivalent temperature T_e, of Mars and Venus.

  b. If Mars radiates as a blackbody at 217 K, what would be the peak wavelength in its spectrum?

8.3. The solar flux S striking a planet will be inversely proportional to the square of the distance from the planet to the sun. That is, $S = k/d^2$, where k is some constant, and d is the distance.

  a. Use the data given in Table 8.2 to estimate k, given the distance from the sun to earth is 150×10^9 m.

  b. Use your result from (a) to estimate the distance from the sun to Venus.

  c. Mercury is 58×10^9 m from the sun, and has an albedo equal to 0.06. Estimate its equivalent temperature.

8.4. The solar flux S arriving at the outer edge of the atmosphere, varies by ± 3.3 percent as the earth moves in its orbit (reaching its greatest value in early January). By how many degrees would the effective temperature of the earth vary as a result?

8.5. In the article "The Climatic Effects of Nuclear War" (*Scientific American*, Aug. 1984), the authors calculate a global energy balance corresponding to the first few months following a 5000-megaton nuclear exchange. The resulting smoke and dust in the atmosphere absorb 75 percent of the incoming sunlight (257 W/m²) while the albedo is reduced to 20 percent. Convective and evaporative heating of the atmosphere from the earth's surface is negligible as is the energy reflected from the earth's surface. The earth's surface radiates 240 W/m², all of which is absorbed by the atmosphere. Assuming the earth can be modeled as a blackbody emitter as shown in Figure P8.5, find the following (equilibrium) quantities:

  a. The temperature of the surface of the earth (this is the infamous "nuclear winter").

  b. X, the rate at which radiation is emitted from the atmosphere to space.

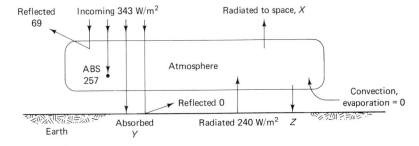

Figure P8.5

    c. *Y,* the rate of absorption of short-wavelength solar radiation at the earth's surface.

    d. *Z,* the rate at which the atmosphere radiates energy to the earth's surface.

8.6. In Figure 8.8, the average rate at which energy is used to evaporate water is given as 80 W/m². Knowing that the latent heat of vaporization of water is about 2.61×10^6 J/kg and that the surface area of the earth is about 5.1×10^{14} m² , estimate the total world annual precipitation in m³/year (which is equal to the total water evaporated).

8.7. Suppose a greenhouse effect enhancement reduces the energy radiated directly from the earth's surface to space from 20 to 15 W/m². Suppose the earth's surface increases to 294 K as a result, but that the albedo and latent and sensible heat transfer to the atmosphere do not change. Also, assume solar energy absorbed by the atmosphere is not changed. Redraw Figure 8.8 to show these changes, keeping energy balanced in each of the three regions (space, atmosphere, surface).

8.8. The concentration of CO_2 in 1974 was about 330 ppm and in 1984 it was 345 ppm. Using a simple exponential growth model, if the rate of growth that prevailed during that 10 year period remains constant, in what year would CO_2 concentration reach 560 ppm, double its preindustrialization value?

8.9. The United States derives about 25 percent of its energy from coal, 45 percent from oil, 20 percent from gas, and 10 percent from non-carbon-emitting sources.

    a. Estimate the ratio of tons of carbon emitted per quad of energy.

    b. Suppose all of our coal consumption is replaced with nuclear and renewable energy systems, but everything else remains constant (total energy and distribution of sources). Estimate the new ratio of carbon to energy.

    c. Suppose that transition takes 100 years. Modeled as an exponential change, what rate of growth in the ratio of emissions to energy, $r_{em/en}$, would have prevailed?

8.10. Suppose in 100 years the world is virtually out of natural deposits of oil and gas and energy demand is double the current rate of 300 quads/year. At that time, imagine 28 percent of world energy comes from coal, 20 percent is synthetic gas from coal, 40 percent is synthetic oil from shale, and the remainder is derived from non-carbon-emitting sources (that is, our demands for different fuels is, percentage wise, the same as now).

    a. What would be the carbon emission rate then (Gt/year)?

    b. If carbon emissions grow exponentially from the current rate of 5.5 Gt/year to the rate found in (a), what rate of growth *r* would have prevailed during that 100 years?

    c. If the airborne fraction is 50 percent, how much of the carbon emitted in the next 100 years would remain in the atmosphere?

    d. What would be the atmospheric concentration of CO_2 in 100 years if there is no net contribution from biomass (and there are 730 Gt of carbon in the atmosphere now)?

    e. If the equilibrium temperature increase for a doubling of CO_2 is 4 °C, what would be the equilibrium temperature increase in 100 years if the initial concentration is 345 ppm?

8.11. Suppose in 100 years energy consumption is still 300 quads, but coal supplies only 20 percent of total demand, natural gas supplies 15 percent, and oil 10 percent. The remaining 55 percent of demand is met by nuclear and solar energy. Using this conservation scenario, repeat parts a–e of Problem 8.10.

8.12. For the two scenarios shown in the table:

Population r_p (%)	GNP/person $r_{GNP/cap}$ (%)	Energy/GNP $r_{en/GNP}$ (%)	C/Energy $r_{em/en}$ (%)	Airborne fraction (%)	Equilibrium ΔT_d
(A) 1.0	0.3	−2	−0.7	40	3 °C
(B) 1.5	1.5	−0.2	0.4	50	3 °C

    a. Predict the global equilibrium temperature increase in 70 years, if the initial CO_2 concentration is 345 ppm, the initial emission rate is 5.5 Gt C/year, and the initial atmospheric carbon content is 730 Gt C.

    b. If those growth rates continue, in what years would CO_2 concentrations be equal to double the initial 345 ppm?

8.13. Consider the following four ways to heat a house. The first is a new high-efficiency pulse-combustion gas furnace; the second is a conventional gas furnace; the third is an electric heat pump that delivers 2.5 units of energy to the house for each unit of electrical energy that it consumes (the other 1.5 units are heat taken from the ambient air); and the fourth is conventional electric heating. Using the new pulse-combustion gas furnace as a "standard," rate the other options in terms of carbon emissions, that is, by the ratio (C emissions option i)/(C emissions pulse furnace). Assume the electrical power plant burns coal.

Option	Description	Furnace efficiency (%)	Power plant efficiency (%)	$\dfrac{\text{C emissions } i}{\text{C emissions } 1}$
1	Pulse-gas	95	—	1
2	Conventional gas	70	—	?
3	Heat pump	250	33	?
4	Electric	100	33	?

8.14. Based on a 1980 world carbon emission rate of 5.2 Gt/year, a maximum emission rate equal to 22 Gt/year, and total fossil fuel carbon ever emitted equivalent to the consumption of 200 000 quads of coal, draw a graph of carbon emissions versus time using a Gaussian emission function. In what year would maximum emissions occur? Compare your answers to the curve labelled "D" in Figure 8.20.

8.15. For total carbon emissions of fossil fuels equivalent to the consumption of 200 000 quads of coal, and using an airborne fraction of 73 percent (this may seem high, but it is the result of the diminishing effectiveness of the oceans to absorb excess carbon), estimate the ultimate concentration of CO_2 added to the "natural" 280 ppm. Except for our assumption that the 73 percent of the carbon emitted that stays in the atmosphere is there forever (we neglect the approximately 1000-year time constant associated with the gradual assimilation of carbon in the oceans), this problem is similar to the assumptions built into Figure 8.20*b*. Find the resulting equilibrium global temperature increase assuming $\Delta T_{2x} = 3$ °C compared to preindustrial times when there were 280 ppm CO_2.

8.16. Write the chemical formula for CFC-134 and CFC-114.

8.17. What would be the CFC number for $C_2H_4F_2$?

8.18. Assume the following data for CFC-11: atmospheric residence time = 75 years; atmospheric concentration in 1985 = 0.24 ppb; 1985 emission rate = 342 000 t/year.
  a. If emission rates remain constant, what would be the ultimate, steady-state concentration?
  b. What drop in emission rate would be required to maintain existing concentrations?
  c. If emissions immediately drop to 50 percent of their initial (1985) level, sketch the concentration versus time and indicate the ultimate, steady-state level that would be achieved.

8.19. Repeat Problem 8.19 for CFC-113, which has an atmospheric residence time of about 90 years, a 1985 concentration of 0.032 ppb, and a 1985 emission rate of about 165 000 t/year.

8.20. If CFC-11 and CFC-12 emissions remain at their current rate, their ultimate concentrations in the atmosphere would be about 1 ppb for CFC-11 and 3.1 ppb for CFC-12 (See Example 8.10, and Problem 8.18). What equilibrium global temperature increase would be expected due to each of these greenhouse gases?

8.21. Given the 1985 concentrations and growth rates for CO_2, CH_4, N_2O, CFC-11, and CFC-12 given in Example 8.11, estimate the incremental temperature increase between 2075 and 2085 associated with each component. What fraction of the decade's increase is CO_2 and what fraction is other trace greenhouse gases?

8.22. In Example 8.10, calculations were made for CFC-12 with various levels of constant production assumed. Suppose production of CFC-12 were to have been stopped completely in 1985. Plot the atmospheric concentration versus time and identify the year in which its concentration would be half of its 1985 value.

REFERENCES

Arrhenius, S., 1896, On the influence of carbonic acid in the air upon the temperature of the ground, *Philosophical Magazine and Journal of Science*, S.5, 41(251):237–276.

Bach, W., 1984, *Our Threatened Climate*, Reidel, Dordrecht, Holland.

Bach, W., A. J. Crane, A. L. Berger, and A. Longhetto, 1983, *Carbon dioxide, current views and developments in energy/climate research*, Reidel, Dordrecht, Holland.

Barnola, J. M., D. Raynaud, Y. S. Korotkevich, and C. Lorius, 1987, Vostok ice core provides 160,000-year record of atmospheric CO_2, *Nature* 329, 1 October.

Bolin, B., 1970, The carbon cycle, *Scientific American* 223, 130.

Bolin, B., 1986, How much carbon will remain in the atmosphere?, in *The Greenhouse Effect, Climatic Change, and Ecosystems*, B. Bolin et al. (eds.), Wiley, New York.

BOLLE, H. J., et al., 1986, Other greenhouse gases and aerosols, in B. Bolin et al. (eds.), *The Greenhouse Effect, Climatic Change, and Ecosystems,* Wiley, New York.

BROECKER, W. S., and G. H. DENTON, 1990, What drives glacial cycles?, *Scientific American,* January.

California Energy Commission, 1989, *The Impacts of Global Warming on California, Interim Report,* P500-89-004, August.

CLARKE, W. C. (ed.), 1982, *Carbon Dioxide Review 1982,* Oxford University Press, New York.

DICKINSON, R. E., 1982, Modeling climate changes due to carbon dioxide increases, in *Carbon Dioxide Review, 1982,* W. C. Clark (ed.), Oxford University Press, New York.

DICKINSON, R. E., 1986, How will climate change?, in B. Bolin et al. (eds.), *The Greenhouse Effect, Climate Change, and Ecosystems,* Wiley, New York.

EIA, 1986, *Energy Facts 1986,* Energy Information Administration, Department of Energy, Washington, DC.

EMMETT, E. A., 1986, Health effects of ultraviolet radiation, *Effects of Changes in Stratospheric Ozone and Global Climate,* Vol. 1, U.S. Environmental Protection Agency and United Nations Environment Programme, Washington, DC.

FLAVIN, C., 1989, Slowing global warming: A worldwide strategy, *Worldwatch Paper 91,* Worldwatch Institute, Washington, DC.

FLAVIN, C., 1990, Slowing global warming, *State of the World 1990,* Worldwatch Institute, Norton, New York.

FREDERICK, J. E., 1986, The ultraviolet radiation environment of the biosphere, *Effects of Changes in Stratospheric Ozone and Global Climate,* Vol 1., USEPA and UNEP, Washington, DC.

GOLDEMBERG, J., T. B. JOHANSSON, A. K. N. REDDY, and R. H. WILLIAMS, 1985, Basic needs and much more with one kilowatt per capita, *Ambio,* Vol. 14, No. 4-5, pp. 190–200.

HAMMITT, J. K., K. A. WOLF, F. CAMM, W. E. MOOZ, T. H. QUINN, and A. BAMEZAI, 1986, Product uses and market trends for potential ozone-depleting substances, 1985–2000, in *Assessing the Risks of Trace Gases that Can Modify the Stratosphere,* Vol. 6, U.S. Environmental Protection Agency, Washington, DC.

HANSEN, J., and LEBEDEFF, S., 1988, Global surface air temperatures: Update through 1987, *Geophysical Research Letters,* 15:323–326.

HANSEN, J., A. LACIS, D. RIND, G. RUSSELL, I. FUNG, P. ASHCRAFT, S. LEBEDEFF, R. RUEDY, and P. STONE, 1986, The greenhouse effect: Projections of global climate change, in *Effects of Changes in Stratospheric Ozone and Global Climate,* USEPA, UNEP, Washington, DC.

HARTE, J., 1985, *Consider a spherical cow, A course in environmental problem solving,* Kaufmann, Los Altos, CA.

HOFFMAN, J. S., 1987, *Assessing the Risks of Trace Gases That Can Modify the Stratosphere,* Vol. 1, USEPA, Washington, DC.

HOFFMAN, J. S., D. KEYES, and J. G. TITUS, 1983, *Projecting Future Sea Level Rise, Methodology Estimates to the Year 2100, & Research Needs,* Environmental Protection Agency, Washington, DC.

IIASA, 1981, *Energy in a Finite World,* W. Häfele (ed.), Institute for Applied Systems Analysis, Ballinger, Cambridge, MA.

JOUZEL, J., C. LORIUS, J. R. PETIT, C. GENTHON, N. I. BARKOV, V. M. KOTLYAKOV, and V. M. PETROV, 1987, Vostok ice core: A continuous isotope temperature record over the last climatic cycle (160,000 years), *Nature* 329, 1 October.

KEELING, C. D., and R. B. BACASTOW, 1977, Impact of industrial gases on climate, *Energy and Climate,* National Research Council, Washington, DC.

KEELING, C. D., R. B. BACASTOW, A. F. CARTER, S. C. PIPER, T. P. WHORF, M. HEIMANN, W. G. MOOK, AND H. ROELOFFZEN, 1989, A three dimensional model of atmospheric CO_2 transport based on observed winds: Observational data and preliminary analysis, *Aspects of Climate Variability in the Pacific and the Western Americas,* Geophysical Monograph, American Geophysical Union, Vol. 55.

KEEPIN, W., I. MINTZER, and L. KRISTOFERSON, 1986, Emission of CO_2 into the atmosphere, The rate of release of CO_2 as a function of future energy developments, in *The Greenhouse Effect, Climate Change, and Ecosystems,* B. Bolin et al. (eds.), Wiley, New York.

LACIS, A., J. HANSEN, P. LEE, T. MITCHELL, and S. LEBEDEFF, 1981, Greenhouse effect of trace gases, 1970–1980, *Geophysical Research Letters* 8(10):1035–1038.

LOVINS, A. B., L. H. LOVINS, F. KRAUSE, and W. BACH, 1981, *Least Cost Energy: Solving the CO_2 Problem,* Brick House, Andover, MA.

MANABE, S., and R. T. WETHERALD, 1986, Reduction in summer soil wetness induced by an increase in atmospheric carbon dioxide, *Effects of Changes in Stratospheric Ozone and Global Climate,* Vol. 1, USEPA and UNEP, Washington, DC.

MINTZER, I. M., 1987, *A Matter of Degrees: The Potential for Controlling the Greenhouse Effect,* World Resources Institute, Research Report 5, Washington, DC.

MITCHELL, J. M., JR., 1977, Records of the past, lessons for the future, *Proceedings of the Symp. on Living With Climate Change,* The Mitre Corp. McLean, VA, pp. 15–25.

MOLINA, M. J., and F. S. ROWLAND, 1974, Stratospheric sink for chlorofluoromethanes: Chlorine atom catalysed destruction of ozone, *Nature* 249:810–812.

National Research Council, 1983, *Changing Climate: Report of the Carbon Dioxide Assessment Committee,* National Academy Press, Washington, DC.

NEFTEL, A., H. OESCHGER, J. SCHWANDER, B. STAUFFER, and R. ZUMBRUNN, 1982, Ice core sample measurements give atmospheric CO_2 content during the past 40,000 Years, *Nature* 295:220–222.

NEFTEL, A., E. MOOR, H. OESCHGER, and B. STAUFFER, 1985, Evidence from polar ice cores for the increase in atmospheric CO_2 in the last two centuries, *Nature* 315, May 2.

PERRY, A. M., Carbon dioxide production scenarios, 1982, *Carbon Dioxide Review: 1982,* W. C. Clark (ed.), Oxford Press, New York.

RAMANATHAN, V., 1985, Trace gas effects on climate, *Atmospheric Ozone 1985,* World Meteorological Organization Global Ozone Research and Monitoring Project Report No. 16, Vol. 3.

RAMANATHAN, V., R. J. CICERONE, H. B. SINGH, and J. T. KIEHL, 1985, Trace gas trends and their potential role in climate change, *Journal of Geophysical Research,* 90:(D3):5547–5565.

REVELLE, R. R., and P. E. WAGGONER, 1983, Effects of a carbon dioxide-induced climatic change on water supplies in the western United States, *Changing Climate,* National Research Council, Washington, DC.

ROTTY, R. M., and C. MASTERS, 1984, *Past and Future Releases of CO_2 from Fossil Fuel Combustion,* Institute for Energy Analysis, Oak Ridge, TN.

ROWLAND, F. S., 1989, Chlorofluorocarbons, stratospheric ozone and the Antarctic "ozone hole," in *Global Climate Change,* S. F. Singer (ed.), Paragon House, New York.

SCHNEIDER, S. H., 1989, *Global Warming,* Sierra Club Books, San Francisco.

SEIDEL, S., and D. KEYES, 1983, *Can We Delay a Greenhouse Warming?* Environmental Protection Agency, Washington, DC.

SHEA, C. P., 1989, Protecting the ozone layer, *State of the World 1989,* Worldwatch Institute, Norton, New York.

SIEGENTHALER, U., 1986, Carbon dioxide: Its natural cycle and anthropogenic perturbation, *The Role of Air-Sea Exchange in Geochemical Cycling,* P. Baut-Menard (ed.), Reidel, Dordrecht, Holland.

SIEGENTHALER, U., and H. OESCHGER, 1987, Biospheric CO_2 emissions during the past 200 years reconstructed by deconvolution of ice core data, *Tellus,* 39B:140–154.

SINGER, S. F., 1989, Stratospheric ozone: Science and policy, *Global Climate Change,* S. F. Singer (ed.), Paragon House, New York.

STOCKTON, C. W., and W. R. BOGGESS, 1979, Geohydrological implications of climate change on water resource development, U.S. Army Coastal Engineering Research Center, Fort Belvoir, VA.

STORDAL, F., and I. S. A. ISAKSEN, 1986, Ozone perturbations due to increases in N_2O, CH_4, and chlorocarbons: Two-dimensional time-dependent calculations, *Effects of Changes in Stratospheric Ozone and Global Climate,* Vol. 1, USEPA and UNEP, Washington, DC.

SWISHER, J. N., 1989, Defining environmental security, *Techne,* Vol. 3, Spring, Stanford University Program in Values, Technology, Science, and Society.

SWISHER, J. N., and G. M. MASTERS, 1989, International carbon emission offsets: A tradeable currency for climate protection services, Stanford University Civil Engineering Department Technical Paper No. 309.

TITUS, J. G., 1986, The causes and effects of sea level rise, *Effects of Changes in Stratospheric Ozone and Global Climate,* Vol. 1, USEPA and UNEP, Washington, DC.

USEPA, 1983, *Projecting Future Sea Level Rise,* Office of Policy & Resource Management, EPA 230-09-007, October.

USEPA, 1986, *Effects of Changes in Stratospheric Ozone and Global Climate,* Vol. 1, U.S. Environmental Protection Agency and United Nations Environment Programme, Washington, DC.

USEPA, 1987, *Assessing the Risks of Trace Gases That Can Modify the Stratosphere,* Vol. 2, Washington, DC.

USEPA, 1989, *Potential Effects of Global Climate Change in the U.S.,* J. Smith and D. Tirpak, Washington, DC.

WALLACE, J. M., and P. V. HOBBS, 1977, *Atmospheric Science, An Introductory Survey.* Academic Press, Orlando, FL.

WASHINGTON, W. M., and G. A. MEEHL, 1984, Seasonal cycle experiment on the climate sensitivity due to a doubling of CO_2 with an atmospheric general circulation model coupled to a simple mixed layer ocean model, *Journal of Geophysical Research* 89:9475–9503.

WORREST, R. C., 1986, The effect of solar UV-B radiation on aquatic systems: An overview, *Effects of Changes in Stratospheric Ozone and Global Climate,* Vol. 1, USEPA and UNEP, Washington, DC.

APPENDIX

Useful Conversion Factors

Length

1 inch	= 2.540 cm
1 foot	= 0.3048 m
1 yard	= 0.9144 m
1 mile	= 1.6093 km
1 meter	= 3.2808 ft
	= 39.37 in.
1 kilometer	= 0.6214 mi

Area

1 square inch	= 6.452 cm^2
	= 0.0006452 m^2
1 square foot	= 0.0929 m^2
1 acre	= 43,560 ft^2
	= 0.0015625 sq mi
	= 4046.85 m^2
	= 0.404685 ha
1 square mile	= 640 acre
	= 2.604 km^2
	= 259 ha
1 square meter	= 10.764 ft^2
1 hectare	= 2.471 acre
	= 0.00386 sq mi
	= 10,000 m^2

Volume

1 cubic foot	= 0.03704 cu yd
	= 7.4805 gal (U.S.)
	= 0.02832 m^3
	= 28.32 L
1 acre foot	= 43,560 ft^3
	= 1233.49 m^3
	= 325,851 gal (U.S.)
1 gallon (U.S.)	= 0.134 ft^3
	= 0.003785 m^3
	= 3.785 L
1 cubic meter	= 8.11×10^{-4} Ac ft
	= 35.3147 ft^3
	= 264.172 gal (U.S.)
	= 1000 L
	= 10^6 cm^3

Linear Velocity

1 foot per second	= 0.6818 mph
	= 0.3048 m/s
1 mile per hour	= 1.467 ft/s
	= 0.4470 m/s
	= 1.609 km/hr
1 meter per second	= 3.280 ft/s
	= 2.237 mph

Mass

1 pound (avdp)	= 0.453592 kg
1 kilogram	= 2.205 lb (avdp)
	= 35.27396 oz (avdp)
1 ton (short)	= 2000 lb (avdp)
	= 907.2 kg
	= 0.9072 ton (metric)
1 ton (metric)	= 1000 kg
	= 2204.622 lb (avdp)
	= 1.1023 ton (short)

Flowrate

1 cubic foot per second	$= 0.028316$ m^3/s
	$= 448.8$ gal (U.S.)/min (gpm)
1 cubic foot per minute	$= 4.72 \times 10^{-4}$ m^3/s
	$= 7.4805$ gpm
1 gallon (U.S.) per minute	$= 6.31 \times 10^{-5}$ m^3/s
1 million gallons per day	$= 0.0438$ m^3/s
1 million acre feet per year	$= 39.107$ m^3/s
1 cubic meter per second	$= 35.315$ ft^3/s (cfs)
	$= 2118.9$ ft^3/min (cfm)
	$= 22.83 \times 10^6$ gal/d
	$= 70.07$ Ac-ft/d

Density

1 pound per cubic foot	$= 16.018$ kg/m^3
1 pound per gallon	$= 1.2 \times 10^5$ mg/L
1 kilogram per cubic meter	$= 0.062428$ lb/ft^3
1 gram per cubic centimeter	$= 62.427961$ lb/ft^3

Concentration

1 milligram per liter in water $= 1.0$ ppm
(specific gravity $= 1.0$)

$= 1000$ ppb
$= 1.0$ g/m^3
$= 8.34$ lb per million gal

Pressure

1 atmosphere	$= 76.0$ cm of Hg
	$= 14.696$ lb/in.^{2} (psia)
	$= 29.921$ in. of Hg (32 °F)
	$= 33.8995$ ft of H$_2$O (32 °F)
	$= 101.325$ kPa
1 pound per square inch	$= 2.307$ ft of H$_2$O
	$= 2.036$ in. of Hg
	$= 0.06805$ atm
1 Pascal (Pa)	$= 1$ N/m^2
	$= 1.45 \times 10^{-4}$ psia
1 inch of mercury (32 °F)	$= 3386.4$ Pa
(60 °F)	$= 3376.9$ Pa

Energy

1 British Thermal Unit	$= 778$ ft-lb
	$= 252$ cal
	$= 1055$ J
	$= 0.2930$ Whr
1 quadrillion Btu	$= 10^{15}$ Btu
	$= 1055 \times 10^{15}$ J
	$= 2.93 \times 10^{11}$ kWhr
	$= 172 \times 10^{6}$ barrels (42-gal) of oil equivalent
	$= 36.0 \times 10^{6}$ metric tons of coal equivalent
	$= 0.93 \times 10^{12}$ cubic feet of natural gas equivalent
1 Joule	$= 1$ N-m
	$= 9.48 \times 10^{-4}$ Btu
	$= 0.73756$ ft-lb
1 kilowatt-hour	$= 3600$ kJ
	$= 3412$ Btu
	$= 860$ kcal
1 kilocalorie	$= 4.185$ kJ

Power

1 kilowatt	$= 1000$ J/s
	$= 3412$ Btu/hr
	$= 1.340$ hp
1 horsepower	$= 746$ W
	$= 550$ ft-lb/s
1 quadrillion Btu per year	$= 0.471$ million barrels of oil per day
	$= 0.03345$ TW

Index